Future Mobile Communication: From Cooperative Cells to the Post-Cellular Relay Carpet

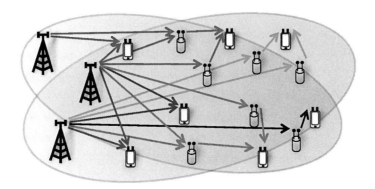

Raphael Rolny

λογος

Series in Wireless Communications
edited by:
Prof. Dr. Armin Wittneben
Eidgenössische Technische Hochschule
Institut für Kommunikationstechnik
Sternwartstr. 7
CH-8092 Zürich

E-Mail: wittneben@nari.ee.ethz.ch
Url: http://www.nari.ee.ethz.ch/

Bibliographic information published by the Deutsche Nationalbibliothek

The Deutsche Nationalbibliothek lists this publication in the
Deutsche Nationalbibliografie; detailed bibliographic data are
available in the Internet at http://dnb.d-nb.de .

ISBN 978-3-8325-3332-8
ISSN 1611-2970

Logos Verlag Berlin GmbH
Comeniushof, Gubener Str. 47,
10243 Berlin
Tel.: +49 030 42 85 10 90
Fax: +49 030 42 85 10 92
INTERNET: http://www.logos-verlag.de

Diss. ETH No. 23243

Future Mobile Communication: From Cooperative Cells to the Post-Cellular Relay Carpet

A thesis submitted to attain the degree of

DOCTOR OF SCIENCES of ETH ZURICH

(Dr. sc. ETH Zurich)

presented by

RAPHAEL THOMAS LIVIUS ROLNY

Master of Science (MSc), ETH Zurich

born on September 13, 1982

citizen of Regensdorf (ZH), Switzerland

accepted on the recommendation of

Prof. Dr. Armin Wittneben, examiner

Prof. Dr. Gerhard Bauch, co-examiner

2016

Day of Doctoral Examination: January 18, 2016

You see, wire telegraph is a kind of a very, very long cat. You pull his tail in New York and his head is meowing in Los Angeles. Do you understand this? And radio operates exactly the same way: you send signals here, they receive them there. The only difference is that there is no cat.

- Attributed to Albert Einstein when asked to describe radio.

Abstract

The increasing demand for ubiquitous data service sets high expectations on future cellular networks. They should not only provide data rates that are higher by orders of magnitude than today's systems, but also have to guarantee high coverage and reliability. Thereby, sophisticated interference management is inevitable. The focus of this work is to develop cooperative transmission schemes that can be applied to cellular networks of the next generation and beyond. For this, conventional network architectures and communication protocols have to be challenged and new concepts need to be developed. Starting from cellular networks with base station (BS) cooperation, this thesis investigates how classical network architectures can evolve to future networks in which the mobile stations (MSs) are no longer served by BSs in their close vicinity, but by a dynamic and flexible heterogeneity of different nodes.

Based on recent information theoretic results, we develop a practical and robust linear BS cooperation scheme based on block zero-forcing in a limited area (BS cluster) and an approximation of the residual interference with which a convex optimization problem can be formulated and efficiently be solved. The transmission scheme is then extended to heterogeneous networks that can also include remote radio heads and/or decode-and-forward (DF) relays. While such relays can improve coverage range as well as the data rates in interference limited areas, they need to be involved in the cooperation process and have to exchange signals with their cooperation partners (e.g. BSs), what makes the relays rather complex.

Amplify-and-forward (AF) relays of low complexity that cooperate in a distributed fashion are therefore an attractive alternative. With properly selected amplification gains, such AF relays can cancel interference and assist the communication between BSs and MSs. In order to find appropriate amplification factors, we develop a distributed gradient-based optimization algorithm that allows each node to calculate its factors with local channel state information (CSI) only, even when applied to multi-hop setups. We show that the overhead of the scheme does not scale with the number of involved nodes and that it can be further improved by subcarrier cooperation, i.e. when signals are combined over multiple subcarriers.

In order to increase the capacity of cellular systems by the required factors, we combine large antenna arrays (massive MIMO) at few BS locations with massively deployed small relay cells. In this "relay carpet" concept, we can benefit from the advantages of both approaches and can simplify channel estimation at the BSs, which would limit the performance gains in conventional networks. The relays, that are of very low complexity and low cost, turn the network into a two-hop network where the BSs as well as the MSs see the relays as the nodes they communicate with. This enables sophisticated multi-user MIMO beamforming at the BSs and the performance of such a network scales beneficially with the number of involved nodes. Especially very simple AF relays without the requirement of any CSI can thus lead to large rates when the network is dense. These relays are also beneficial for coverage and power savings.

In a further evolution, we abandon the cellular network layout completely and let backhaul access points operate in places where they can most easily be installed. For such a "post-cellular" network architecture, we apply dynamic cooperation clusters and many distributed low-complexity relays. By optimizing the BS clusters as well as the relay routing under practical conditions and power control, we show that coverage and high performance can be achieved with aggressive spatial multiplexing and cooperative leakage-based precoding.

With a transition from classical cell-based networks to relay enabled post-cellular networks, we trade off node complexity with density. Aggressive spatial multiplexing can thereby deliver high data rates to large areas in a very efficient way, even when the backhaul capacity is limited or when in certain areas no backhaul access is available at all. The beneficial performance scaling shows that such post-cellular networks can offer a flexible and dynamic solution for mobile communication of future generations.

Kurzfassung

Die immer grösser werdende Nachfrage an mobile Datendienste stellt hohe Anforderungen an zukünftige Mobilfunknetze. Diese sollen nicht nur Datenraten liefern, die um Grössenordnungen höher liegen als in gegenwärtigen Systemen, sondern auch hohe Verfügbarkeit und Zuverlässigkeit garantieren. Dabei ist insbesondere ein effektiver Umgang mit Interferenz von hoher Bedeutung. In dieser Arbeit entwickeln wir deshalb effiziente kooperative Übertragungstechniken, die im Mobilfunk der nächsten Generation und darüber hinaus angewendet werden können. Dafür müssen konventionelle Netzwerkarchitekturen und Kommunikationsprotokolle überdacht und neue Konzepte erarbeitet werden. Ausgehend von Netzwerken mit kooperierenden Basisstationen untersucht diese Arbeit, wie klassische zelluläre Netzwerkarchitekturen zu zukünftigen Netzwerken weiterentwickelt werden können, in denen die mobilen Nutzer nicht mehr von nahe gelegenen Basisstationen bedient werden, sondern durch eine dynamische und flexible Vielzahl von verschiedenen Infrastrukturknoten.

Basierend auf aktuellen informationstheoretischen Forschungsresultaten entwickeln wir praktische und robuste lineare Kooperationsverfahren. Dabei wird die Interferenz in einem beschränkten Gebiet durch eine Gruppe von mehreren Basisstationen komplett ausgelöscht. Durch eine Approximation der restlichen Interferenz von ausserhalb des Kooperationsgebietes kann ein konvexes Optimierungsproblem formuliert werden, das effizient gelöst werden kann. Das Übertragungsverfahren wird dann so erweitert, dass es auch in heterogenen Netzwerken angewendet werden kann, in denen zusätzliche Decode-and-Forward (DF) Relais, Femto-Zellen-Basisstationen oder andere Hilfsknoten die Basisstationen unterstützen. Damit die Hilfsknoten nicht nur die Netzabdeckung vergrössern, sondern auch die Datenraten in interferenzlimitierten Umgebungen verbessern können, müssen diese in den Kooperationsprozess der Basisstationen eingebunden werden und mit diesen Signale sowie Kanalinformation austauschen, was zu einer relativ komplexen Implementierung führt.

Amplify-and-Forward (AF) Relais, die durch verteilte Verfahren miteinander kooperieren, sind deshalb eine attraktive Alternative, da diese als einfache Umsetzer mit geringer Komplexität realisiert werden können. Mit geeignet gewählten Verstärkungsfaktoren können solche AF Relais Interferenz abschwächen oder auslöschen und somit die Kommunikation zwischen Basisstationen und Mobilgeräten unterstützen. Um die Verstärkungsfaktoren zu optimieren, entwickeln wir einen verteilten gradientenbasierten Algorithmus, der es jedem Knoten erlaubt, seine Faktoren mit lokalem Wissen zu berechnen. In Übertragungen über zwei oder mehr Stufen skaliert der Overhead dieses Verfahrens nicht mit der Anzahl Relais pro Stufe. In Breitbandsystemen kann darüber hinaus eine weitere Verbesserung der erzielbaren Datenraten durch optimierte Linearkombinationen über mehrere Unterträger erzielt werden.

Um die Kapazität der Systeme um die benötigten Grössenordnungen zu erhöhen, kombinieren wir grosse Antennenarrays an Basisstationen mit kleinen Relaiszellen, die flächendeckend verteilt sind. In diesem "Relaisteppich" kann man von den Vorteilen ausgeklügelter Mehrfachnutzerübertragung an den Basisstationen und der verteilten Signalverarbeitung der Relais profitieren. Die vielen Relais verwandeln das Netzwerk in ein Zwei-Hop-System, was grosse Antennenarrays an den Basisstationen ohne den sonst damit verbundenen hohen Overhead ermöglicht. Mit einfachen AF Relais, die keinerlei Kanalwissen benötigen, kann eine äussert günstige Skalierung der Leistungsfähigkeit des Netzwerkes erreicht werden, besonders wenn viele Knoten dicht beieinander liegen. Neben deutlich erhöhten Datenraten können die Relais auch benötigte Sendeleistung einsparen, was die Systeme auch noch energieeffizienter macht.

In einer weiteren Evolution verzichten wir schliesslich ganz auf die zelluläre Struktur und lassen die mobilen Nutzer nur durch spärlich platzierte Basisstationen bedienen, die auch weit weg von den Nutzern liegen können. In solchen "post-zellulären" Netzwerken müssen die Basisstationen nicht mehr überall verteilt sein, sondern können da platziert werden, wo es möglichst (kosten-) günstig möglich ist. Durch eine optimierte Zuteilung der kooperierenden Basisstationsgruppen und Relais unter praktischen Bedingungen und Leistungsregulierung können sehr hohe Netzabdeckungen und Datenraten erreicht werden.

Mit dem Übergang von klassischen zellbasierten Netzwerken zur post-zellulären Netzwerkarchitektur können wir die Komplexität der Knoten mit deren Anzahl aufwiegen. Aggressives Multiplexing in der räumlichen Dimension kann hohe Datenraten in grossen Gebieten erzielen, auch wenn die kabelgebundene Infrastruktur limitiert oder in Teilen des Netzwerkes gar nicht vorhanden ist. Die vorteilhafte Skalierung mit der Anzahl der Knoten zeigt deshalb, dass post-zelluläre Netzwerke eine äusserst dynamische und flexible Lösung für die Anforderungen zukünftiger Mobilfunknetze bieten.

Acknowledgements

First and foremost, I would like to thank my advisor Prof. Dr. Armin Wittneben for guiding me through this thesis. Your continuous support, your motivation and inspiration helped me in all the time of research and writing of this thesis. I could not have imagined a better research environment as the one provided by the Communication Technology Laboratory at ETH Zurich. Thank you for allowing me to grow as an engineer and as a scientist.

Besides my advisor, I would also like to thank Prof. Dr. Gerhard Bauch for being the co-referee of this thesis and for serving as committee member at my defense. I also want to thank you for the exam questions (even at hardship), your comments and suggestions. Thank you for letting my defense be an enjoyable moment.

A sincere thank you also goes to Dr. Marc Kuhn, who always provided me with valuable suggestions, comments, and assistance. The discussions with you, whether at work or during lunch, were always fruitful, interesting, and fun. I thank my fellow labmates Zemene, Eric, Yehia, Tim, Gregor, Christoph, Heinrich, and all the other current and former members of the Communication Technology Lab for the stimulating discussions, the vivid talks during or after work, the exciting conference trips, ski weekends, summer events, and for all the fun we have had in the last years. You were more than just colleagues. Special thanks also go to Jutta, Priska, Lara, Barbara, and Claudia. Your active support in all administrative things contributed greatly to an enjoyable working atmosphere.

Also I thank my friends Doris, Harri, Sandro, David, and Beni. The weekly Signalöl always provided very exhilarating discussions (of course only about serious scientific topics). Furthermore, I also thank Noppa, Kevin, Markus, Lea, and Alessia for being my bandmates of Walter Calls Ambulance and the patience when I was stressed during the writing of my thesis. The rehearsals, concerts, and all the time we have spent together were always a good balance for the time at work. You guys rock!

I want to give special thanks to my family: my parents and to my sister for the unconditional support throughout the writing of this thesis and my whole life. I would not have completed this thesis if it was not for you.

Last but not the least, I would also like to thank you, dear reader, for putting this book into your hands and anyone else who I have forgotten to mention here. Thank you very much! I try to consider you in the next book ;-)

Zurich, February 2016

Contents

Contents

1

Introduction

1.1 Background & Motivation

It lies in our nature as human beings that we want to communicate with each other and wish to do so in an ubiquitous way, fast, and with low cost. Starting with the invention of fires, smoke signals or the like as early forms of communication systems, a tremendous effort has been undertaken to develop and improve new technologies, to make them more efficient, and to increase their coverage range. With the introduction of the first practical telegraphs in the 1830s, electrical telecommunication systems started to appear and evolved to a huge business sector that has led us to the digital revolution and the Information Age [92].

An important example of this dynamic industry that influences our daily life and constantly undergoes far-reaching changes are certainly the mobile or cellular networks. Such networks are wireless communication systems that are distributed over large areas that are divided into cells. Each of these cells is served by a fixed infrastructure-based transceiver node, called base station (BS), that serves the mobile users within the confined region of a cell. When taken together, all these cells provide radio range over wide geographic areas, such as whole countries, and enable a large number of portable devices (mobile phones, tablets, laptops or others) to communicate with each other as well as with fixed-line telephones or computers. To this end, the BSs are connected to the core network of the telecommunication providers and the internet and act as central coordination units of their respective cell. This network architecture has the advantage that it achieves a higher capacity than a single large transmitter and that it can serve larger areas by adding additional BSs. Moreover, mobile stations (MSs) need less power for transmission if an infrastructure unit is in their close vicinity.

With modern smartphones, we nowadays have almost ubiquitous access to almost every place on the planet and beyond. We are not only able to make phone calls to almost any place on the earth, but we can also share and receive text, pictures, music, videos, or other multimedia contents with almost no delay. The enabler of this development was the invention of transistors and the possibility to integrate very complex circuits into single chips (very large scale integration –VLSI) and the never ending hunger for higher data rates and the corresponding need for a more efficient use of the limitedly available radio spectrum. Accordingly, academic research has led to an enormous amount of new insights, fundamental limits, and innovative ideas for further enhancements, which in turn have enabled the development of new technologies and inventions.

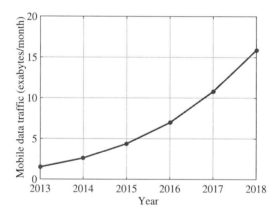

Figure 1.1.: Forecast of global mobile data traffic per month (*source:* [24]).

With the ever growing data rates that can be delivered and the possibilities for new applications that come along with that, also the demand for even faster connections grows further (see Fig. 1.1). The reason for this is that higher data rates encourage a more excessive usage of more high-bandwidth applications. Besides this, also the number of devices that communicate over cellular networks is literally exploding. These are not only the smartphones and tablets that become increasingly popular all over the world (Fig. 1.2), but the trend of massive growth is also due to the upcoming importance of machine-to-machine (M2M) communication [40, 85]. Thereby, smart devices also use the mobile networks to exchange data with each other. Such M2M connections can include e.g. sensor networks of all sorts, smart environments e.g.

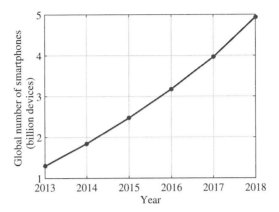

Figure 1.2.: Forecast of global number of IPv6-capable (latest internet protocol) devices such as smartphones, tablets, etc. (*source:* [24]).

washing machines that communicate with the local power supplier to schedule their operation according to the available power, or intelligent transportation systems, e-health, identification and authentication, or a variety of other applications. As shown in Fig. 1.3, it is expected that this form of communication will play a more and more important role in the future. The trend of exponential growth in the number of devices that concurrently communicate over the mobile networks as well as the data volume they generate and the required data rates will therefore persist. With the evolution of the mobile telecommunication standards from the first (analog) and second (digital) generation (2G) to the third and fourth generation (3G and 4G), this development has already been considered. Future networks of the fifth generation (5G) and beyond will however have to deal with the ongoing humongous growth by new concepts and new ideas and not just incremental improvements of existing technologies. Accordingly, networks of future generations should be able to provide data rates that are higher by orders of magnitude than today's systems and also have to guarantee high reliability and pervasive coverage.

The two major problems to achieve good performance in cellular networks are fading and interference. The former, including pathloss and shadowing, limits the coverage of a BS with certain maximal transmit power in point-to-point transmissions, which leads to poor signal-to-noise ratios (SNRs) when mobile users are far away. This problem can be overcome by letting BSs transmit with higher power or by installing additional

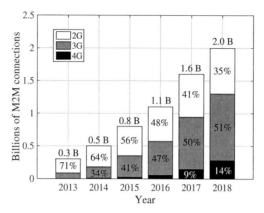

Figure 1.3.: Forecast of number of global machine-to-machine connections and migration from 2G to 3G and 4G (*source:* [24]).

infrastructure nodes to make the network denser. Interference, on the other hand, is the main limiting factor of the performance in dense networks. These networks are mostly interference-limited, as they usually serve many users which are in coverage range of each other at the same time. In such scenarios, increasing the transmit power does not help to get a better signal-to-interference-and-noise ratio (SINR) as higher signal levels also cause higher interference at other users. In order to cope with that, sophisticated interference management is inevitable to achieve a satisfactory performance. The classical approach thereby is to install BSs that serve mobile (or also static) users within a confined region by allocating certain resources such as frequency bands (frequency division multiple access – FDMA), time slots (time division multiple access – TDMA), different codes (code division multiple access – CDMA) etc. for the transmission/reception to/from them. In order to serve many users in a larger area, a spatial reuse is introduced that ensures a certain separation between transmissions that use the same physical channel, i.e. neighboring BSs use different resource blocks with which they avoid to disturb each others' signals. By exploiting the pathloss of the wireless medium, distant BSs can reuse the same resource blocks of another BS when they are sufficiently far apart and the signals decay to an extent that they do not harm other transmissions anymore. Each BS has thus to be closer to the users it serves than to other users it interferes with, which leads to the usual *cellular* network topology that gave these networks its name. A sketch of such a tra-

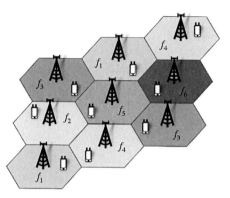

Figure 1.4.: Example of a cellular network in which different cells use different frequency bands around the center frequencies $f_1, \ldots, f_6$.

ditional cellular network is shown in Fig. 1.4 where the transmissions of different cells are separated to different frequency bands around the center frequencies $f_1, \ldots, f_6$. As a result, each cell can only use a fraction of the available spectrum, in this case described by the reuse factor 1/6, which might be further divided among multiple users within the same cell.

As the resources of the wireless medium over which mobile communication networks operate are limited, it is not sufficient to simply increase the bandwidth allocated for cellular networks or to increase the number of BSs to build denser networks. It is necessary to exploit the available resources and to make a drastically more efficient use of them, not least to also achieve a more energy efficient environment. To this end, the spatial dimension can be used to overcome the problems of interference, limited spectrum, and restricted transmit power. A major break through thereby was the invention of multiple-input multiple-output (MIMO) techniques, which allow to transmit several data streams over the same physical channel, concurrently in the same frequency band, without interfering each other and thus to increase the data rates by a factor that is in principle unbounded (see e.g. [63, 102, 143]). By expanding the networks in the spatial dimension (space division multiple access – SDMA), by increasing the number of antennas not only at the BS side but also at the user devices, the performance can thus significantly be boosted.

A further extension of MIMO to form large or even very large MIMO systems is nowadays seen as the key to success in the evolution of future networks [8, 120]. With this, the transmission to/from different users can be separated by accurate beamforming instead of allocating orthogonal resource blocks (time and frequency slots) to each of them. This allows to exploit the spatial degrees of freedom of the wireless channel and to serve many users concurrently without allocating different frequency bands. The reuse factor of the network can thus be improved such that every user can get a larger fraction of the available spectrum or, ideally, the whole licensed spectrum can be allocated to all of the active users simultaneously, leading to a reuse 1 network and thus a significantly higher data rate for all involved devices.

One approach to realize such large antenna systems is to equip the BSs with many antennas, tens or even hundreds of them, to form massive antenna arrays [94]. Such "massive MIMO" systems can preserve the basic organization of cellular networks but are extended in the spatial domain by many additional degrees of freedom. An alternative is to allow multiple nodes (BSs) to cooperate with each other such that a group of BSs with moderately sized antenna arrays can jointly act as a large "virtual" array. This has the additional advantage that mobile users that are located on a cell border can still receive strong signals as they are impinging from different directions and are ideally constrictively added at the receiving antenna(s). In this way, unwanted interference (signals from other BSs) can be turned into desired signals (see e.g. [35,39,57,155] and references therein for an overview of BS cooperation). To this end, however, multiple BSs have not only to exchange the user data or even the whole transmit signals, but also accurate channel state information (CSI) has to be acquired and shared and sophisticated signal processing is needed. Even though such schemes can achieve large performance gains in theory, the overhead that comes with the necessary signal processing and the exchange of the required information for the cooperation may limit the effective gains [89,90]. On top of that, also the backhaul connection of the BSs, the links that connect the BSs to the core network, need to support the necessary amount of data with sufficiently small delay.

It is thus necessary to develop algorithms that can efficiently deal with such practical conditions and can overcome the difficulties that come along with them. The overhead should be small while the expected performance gains should be achieved to a large extent. With the introduction of the Long-Term Evolution (LTE) Release 9 [1] and LTE-Advanced Release 11 [3] standard and the planning for 5G networks (see [5,45] for an overview of current activities), several of the schemes developed in theory have found its initiation in the specifications of practical systems. With the orthogonal frequency division multiplexing (OFDM) based modulation in these systems and carrier aggregation, a much more flexible and dynamic resource allocation is possible and higher bandwidths can be assigned to the users. By allocating different time-frequency resource blocks to different users, no static reuse partitioning is needed anymore and the level of interference can better be controlled, e.g. by a partial frequency reuse (PFR) schemes as proposed in [73, 160]. In these approaches, the reuse patterns can dynamically be adapted to the current user distribution and traffic demand. Moreover, by scheduling different users to different resource blocks, also a diversity gain can be achieved which further increases the performance. Additionally, MIMO is also included in the current standards. While LTE supports four spatial streams with 4x4 MIMO systems, eight streams can be sent in parallel in LTE-Advanced systems. In future releases of 4G systems, the multi-user concept is already taken one step further to forming larger virtual arrays through cooperation between different BSs, also known as coordinated multipoint (CoMP) transmission/reception in the jargon of LTE-Advanced [2,11]. This cooperation can range from coordinated scheduling and resource allocation to make the dynamic flexibility of interference aware reuse partitioning more efficient up to joint transmission and reception with cooperative signal processing. A further increasingly important role in these networks is beamforming, which allows steering the signals precisely to the intended users while interference to other users is reduced. In LTE-Advanced, also range extension with relays and the deployment of small cells is included. With this, the network densification is carried one step forward and mobile users can be served in a more flexible way.

In 5G networks, a further paradigm shift will take place towards the use of very high frequencies (ultra dense and small cells, e.g. for indoor hot spots), massive bandwidth (millimeter wave in 60 GHz transmissions), massive MIMO (with hundreds of antennas), extreme BS and device densities, highly integrative interfaces (tying air interface of 5G together with LTE and WiFi) [5]. CoMP, beamforming, relaying, and user cooperation (MSs forward signals as relays or communicate directly with other MSs) will also have higher importance. The details of the technologies that will be applied are however still under discussion. Nevertheless, future networks will become highly heterogeneous and dynamic with respect to the different types of nodes that have to coexist and it remains unclear how cooperative communication schemes will be realized in practical networks and many challenges are still unsolved.

The goal of this work is thus to develop schemes and concepts to increase the efficiency and performance of practical cellular networks. Thereby, we investigate practical scenarios and develop schemes and concepts that can be applied to realistic systems, possibly not directly to existing standards as LTE-Advanced, but potentially for future releases or generations. As the BS-to-mobile link (downlink) customarily require higher data rates than the mobile-to-BS link (uplink), the focus is primarily set on the downlink, but concepts for the uplink are also discussed. Even though the schemes and algorithms developed in this work are mainly designed for cellular networks, the concepts could as well be applied to different types of networks such as wireless local area networks (WLAN), sensor networks, or others. In the following, we summarize the main contributions of this thesis and give an outline of the remainder of this work.

1.2   Main Contributions

The focus of this work is to develop practical cooperative transmission schemes that can be applied to cellular networks of the next generation and beyond. For this, conventional network architectures and communication protocols have to be challenged and new concepts need to be developed. Starting from cellular networks with BS cooperation, this thesis investigates how classical network architectures can evolve to future networks in which the MSs are no longer served by BSs in their close vicinity as in classical networks, but by a dynamic and flexible heterogeneity of different nodes that might have different complexities or functionalities in which each node can contribute to achieve the efficiency, reliability, and quality of service (QoS) that is desired for future mobile communication networks. Thereby, we are mostly interested in physical layer (PHY) transmission schemes. Schemes for coordination on higher layers such as scheduling, resource management, etc. are mostly not considered.

1.2.1   Multi-Cell Cooperation

In theory, the capacity of interference limited cellular networks can be achieved by cooperative transmission schemes such as dirty paper coding (DPC) in the downlink (DL) or joint decoding in the uplink (UL) [39]. In practice, however, such schemes are too complex to be implemented and introduce overheads that diminish the theoretical performance gains [38, 93]. Hence, practical schemes are required that are of low complexity, with small overhead, and robust to imperfections. In order to solve the difficult problem of precoding in interference limited cellular networks, we develop a linear cooperation scheme that is based on block zero-forcing (ZF) in a limited area (a cluster of BSs) and an approximation of the residual interference. With this, we can formulate a convex optimization problem, which we can solve efficiently. The proposed scheme increases the spectral efficiency to a significant extent, especially on the cell edges. The transmission is also very robust with respect to imperfect CSI and can be extended to heterogeneous networks that also contain femto-cell BSs, remote radio heads (RRHs), or decode-and-forward (DF) relays. The schemes are extensively studied in practical scenarios for which we developed a realistic simulation framework and we assess the potential of the performance that BS cooperation can achieve in practice. The main results are presented in Chapters 3 and 4 and have partly been published in

- M. Kuhn, R. Rolny, A. Wittneben, M. Kuhn, and T. Zasowski, "The Potential of Restricted PHY Cooperation for the Downlink of LTE-Advanced," *IEEE Vehicular Technology Conference (VTC) Fall*, San Francisco, CA, USA, September 2011.

- R. Rolny, M. Kuhn, A. Wittneben, and T. Zasowski, "Relaying and Base Station Cooperation: a Comparative Survey for Future Cellular Networks," *Asilomar Conference on Signals, Systems, and Computers*, Pacific Grove, CA, USA, November 2012.

1.2.2 Distributed Cooperation with Relays

DF relays cannot only be applied for range extension, but also to increase data rates in interference-limited areas. To this end, DF relays need to be involved in the cooperation process with other nodes (e.g. BSs) and have to exchange signals with their cooperation partners. This makes the relays rather complex and expensive. Amplify-and-forward (AF) relays may therefore be an attractive alternative. With properly selected amplification gains, AF relays can cancel interference and assist the communication between BSs and MSs. In order to find appropriate amplification factors, we develop a gradient-based optimization algorithm. In its distributed version, each node calculates its factors with local CSI only and the overhead of the scheme does not scale with the number of involved nodes. We show that only few iterations are required to achieve close to optimal results, which makes the algorithm especially attractive for channel tracking in slow fading environments. The scheme can also be applied to multi-hop networks, which enable long-range communication. Further improvements can be achieved with subcarrier cooperation, i.e. signal combinations over multiple subcarriers. This is discussed in Chapter 5 and the main results have been published in

9

- R. Rolny, J. Wagner, C. Eşli, and A. Wittneben, "Distributed Gain Matrix Optimization in Non-Regenerative MIMO Relay Networks," *Asilomar Conference on Signals, Systems, and Computers*, Pacific Grove, CA, USA, November 2009.

- R. Rolny, J. Wagner, and A. Wittneben, "Distributed Gain Allocation in Non-Regenerative Multiuser Multihop MIMO Networks," *Asilomar Conference on Signals, Systems, and Computers*, Pacific Grove, CA, USA, November 2010, (invited paper).

- R. Rolny, M. Kuhn, A. U. T. Amah, and A. Wittneben, "Multi-Cell Cooperation Using Subcarrier-Cooperative Two-Way Amplify-and-Forward Relaying," *International Symposium on Wireless Communication Systems (ISWCS)*, Ilmenau, Germany, August 2013, (invited paper).

The foundation of the gradient based gain allocation has already been developed in

- R. Rolny, "Distributed Multiuser Multihop Networks: Performance Limits and Decentralized Algorithm Design," Master Thesis, ETH Zürich, August 2009,

but the setup here is extended to a more general case that also includes distributed source precoding. The approach with block ZF and the extension to subcarrier cooperation has not been considered in this previous work.

1.2.3   The Relay Carpet

In order to increase the capacity of cellular systems by orders of magnitude, approaches are required that go further than BS cooperation and/or simple relaying for range extension. Among others, we can

1. equip the BSs with very large antenna arrays (massive MIMO) or

2. install more BSs where they are needed most, until the cells are scaled down to pico- or femto-cells.

In the concept of ubiquitous relaying introduced and discussed in Chapters 6 and 7, we combine large BS antenna arrays and small relay cells. With this, we can benefit from the advantages of both approaches and can simplify channel estimation at the BSs, which limits the performance gains in conventional networks. As a result, few sophisticated BSs with large arrays are supported by many relays spread over the entire area of the network, similar to a carpet. To this end, the relays need to be of low complexity and low cost. This "relay carpet" turns the entire network into a two-hop network where the BSs as well as the MSs see the relays as the nodes they communicate with. Stationary relays enable sophisticated multi-user MIMO beamforming with large arrays as the BS-relay channels can be assumed to be slowly changing or even quasi static, which allows the necessary acquisition of accurate CSI.

In order to achieve high data rates, we compare different relay architectures, including AF, DF, one-way and two-way relaying, and combine them with appropriate BS beamforming schemes. In combination, the BSs transmit to the relays within their cells, while the relays apply simple filters to reduce the residual interference and forward the signals to the MSs. We observe that the performance scales beneficially with the number of involved nodes and that especially very simple relays (AF relays without the requirement of any CSI), achieve large rates when the network is dense. We further conclude that these simple relays are also beneficial for coverage extension, QoS, and power savings. Parts of these results have been published in

- R. Rolny, M. Kuhn, and A. Wittneben, "The Relay Carpet: Ubiquitous Two-Way Relaying in Cooperative Cellular Networks," *IEEE International Symposium on Personal, Indoor, and Mobile Radio Communications (PIMRC)*, London, UK, September 2013 (best paper award).

- R. Rolny, T. Rüegg, M. Kuhn, and A. Wittneben, "The Cellular Relay Carpet: Distributed Cooperation with Ubiquitous Relaying," *Springer International Journal of Wireless Information Networks*, June 2014.

- R. Rolny, C. Dünner, and A. Wittneben, "Power Control for Cellular Networks with Large Antenna Arrays and Ubiquitous Relaying," *IEEE Workshop on Signal Processing Advances for Wireless Communications (SPAWC)*, Toronto, Canada, June 2014.

1.2.4 Post-Cellular Networks

In a further evolution of mobile communication networks described in Chapter 8, we abandon the cellular network layout completely and let backhaul access points operate in places where they can most easily be installed. Thereby, we focus on practical scenarios and assume that the backhaul connections of the BSs as well as their computational capabilities are limited. Accordingly, we apply a flexible post-cellular network architecture in which the MSs are served by dynamic BS cooperation clusters and many distributed low-complexity relays. We show that coverage can be achieved with aggressive spatial multiplexing and cooperative leakage-based precoding at the sparsely located BSs. When few BSs are located in a confined region, they can still serve a large area with the help of relays. The BS cooperation clusters as well as the relay routing is optimized under practical conditions such as backhaul rate limitations and power control. The main results are published in

- R. Rolny, M. Kuhn, and A. Wittneben, "Constrained Base Station Clustering for Cooperative Post-Cellular Relay Networks," *IEEE Wireless Communications and Networking Conference (WCNC)*, New Orleans, USA, March 2015.

With the transition from classical cell-based networks to relay enabled post-cellular networks, we trade off node complexity with density. Aggressive spatial multiplexing can thereby deliver high data rates to large areas in a very efficient way, even when the backhaul capacity is limited or when no infrastructure is provided at all in large areas. The large number of static relays thereby enables sophisticated multiuser beamforming with massive MIMO arrays at the BSs. The beneficial performance scaling shows that such post-cellular networks can offer a flexible and dynamic solution for future mobile communication networks.

1.3 Outline

This thesis is organized as follows. In Chapter 2, we give an overview over the state of the art and discuss the most promising approaches to enhance the performance of cellular networks to meet the demands for future generations. Information theoretic concepts that form the foundation of PHY cooperation are also summarized. Most schemes discussed here are not really practical, but should provide a theoretical background for the remainder of this work. Additionally, this chapter also describes the underlying assumptions and models as well as main figures of merit that we apply in this thesis.

Practical cooperation schemes that can be applied to cellular networks are discussed in Chapter 3, where we develop a practical, locally restricted block ZF approach for joint transmission of multiple BSs. We study the performance under realistic conditions and study the influence of practical aspects such as network layout, cell and frequency planning, and orientation of sectors.

In Chapter 4, we extend the BS cooperation schemes to heterogeneous networks where additional supporting nodes (femto-cell BSs, RRHs, or DF relays) assist the transmission. To this end, we formulate a unified framework in which relays (or other types of supporting nodes) can either be used for range extension or are part of the block ZF together with the BSs. We study the performance of such heterogeneous networks with different types of nodes and compare different schemes with respect to achievable rates, coverage, and robustness.

As the DF relays are rather complex for implementation, we discuss in Chapter 5 AF relays as an alternative of low(er) complexity. This chapter develops how such AF relays can be used for distributed cooperation. We design a gain allocation scheme that can be distributed and does (almost) not scale with the number of involved nodes. This scheme is particularly interesting for channel tracking in slow fading environments and quite robust to channel changes. As an alternative, we also discuss how block ZF can be applied to AF relays and additionally show that subcarrier cooperation can give further performance gains.

In Chapter 6, we combine BS cooperation with a large number of relays. With this combination, we introduce and discuss ubiquitous relaying as a concept for the evolution of cellular networks that enable large BS antenna arrays. We compare different relay architectures and study their applicability to the relay carpet concept and under imperfections. Thereby, it turns out that especially AF relays of very low complexity can offer high performance gains when they are deployed in large numbers.

The concept of the relay carpet and its advantages is introduced and described in Chapter 7. There, we also provide a discussion about the behavior of the relay carpet in the case of growing number of nodes. We conclude that many relays can only beneficially be used in parallel when appropriate power control is applied. To this end, we extend the transmission schemes by power control to optimize the data rates, to minimize the outage probability, or required transmit power.

A further step of the evolution of cellular networks is presented in Chapter 8, where we abandon the cellular network layout completely but let BSs operate from sparse locations. A wide area that should be served is covered by many low-complexity relays. We combine BS cooperation with simple relay forwarding and develop and discuss flexible and dynamic cooperation schemes with relay routing and power control. The post-cellular network can thus be seen as an architecture that combines all the properties and advantages of the previously studied and developed components.

The achievements and contributions of this thesis are summarized and concluded in Chapter 9. A discussion about challenges for practical implementation and further work is also provided.

Parameters and specific settings for the simulations discussed in this work can be found in Appendix A.

2

Cooperative Cellular Networks

The performance of cellular networks is strongly affected by fading and interference. While fading, including pathloss and shadowing, limits the coverage of single point-to-point transmissions, interference is the main limiting factor in dense networks with multiple users operating concurrently over the same physical channel. Therefore, large efforts have been undertaken to overcome these impairments or to develop schemes that can even exploit these effects. In order to increase the capacity of mobile communication systems, a variety of approaches is possible. Thereby, also other factors than high data rates need to be considered. So are the cost of implementation, QoS, energy efficiency, robustness, or other factors also important in the development of new technologies and systems. A vast amount of research results is thus available in the literature, that all attempt to improve the performance and quality of mobile communication. This can be achieved on different levels and by a multiplicity of means.

In the following, we will give an overview over the most important approaches from current literature that seem promising for the enhancement of mobile communication for future generations. Together with a system model and a definition of the assumptions that we also introduce, this chapter forms the basis for the schemes and concepts that we develop and study in this work.

2.1 Enhance Future Cellular Networks

The conventional, non-cooperative, approach to face the problem of interference as well as fading is spatial reuse partitioning. Thereby, many BSs are placed in the entire area of service, more or less evenly distributed, which leads to the cellular structure which gave these networks its name. Accordingly, the mobile users receive signals from a BS close by, which leads to a good SNR in most places. In order to reduce the interference, especially for MSs that are in coverage range of multiple BSs, adjacent cells are assigned to orthogonal resources such as frequency bands (FDMA), time slots (TDMA), or different codes (CDMA) [143]. Typically, the reuse factor is chosen to be much smaller than unity (common values are e.g. 1/3, 1/4, 1/7, 1/9 or 1/12), so that the level of interference is low. Thus, the interference is controlled by fixing the reuse pattern and the maximal power with which the BSs can transmit. Following this approach, the interference seen by the different users is indeed reduced, but at the cost of reduced resources that can be used for a single transmission. Therefore, the spectral efficiency, expressed in bit/sec/Hz (bps/Hz) in conventional cellular networks is strongly limited, either by interference or by the reuse factor.

In modern OFDM based systems [86], this static reuse assignment can be improved by flexible reuse patterns that are dynamically changed according to the current user distribution and traffic demand. In [73], a scheme is introduced that divides the cells into different zones. The center zone close to the BS can then reuse the same full frequency band as all other cells, while cell edge zones use only a fraction of the frequency band that is orthogonal to the one of the neighboring cells. By optimizing the transmit powers in the different frequency bands (and hence the radii of the different zones) according to the user locations, a significant performance gain can be achieved as compared to static approaches. By a coordinated (between all or a subset of BSs) resource allocation, even higher gains can be achieved when each MS is assigned to an individual set of subcarriers that does not have to be contiguous but can be interleaved [160]. With carrier aggregation, even subcarriers from frequency bands far apart can be allocated to the same user. When optimized among multiple cells, the performance can benefit from significant diversity. Moreover, also interference seen by the different users can be reduced when the individual frequency bands are allocated such that the signals experience less fading when they are desired and are more attenuated where they are undesired.

While such optimized resource allocation schemes can indeed improve the instantaneous performance and shadowing can be avoided by macro diversity, the problem of interference limitedness remains. The available resources have to be shared among the different users, which reduces the individual data rates when the number of users increases. In order to improve the capacity of cellular systems by the required factors (e.g. a factor 1000 to 5G networks), other more drastic measures have to be taken. To this end, better channel codes and higher modulation alphabets can be developed that achieve close to optimal data rates with less redundancy, the traffic management can be improved by distributing the data load better between the different cells or BSs, the data traffic can be offloaded to secondary systems as WLAN, or big data driven intelligence can be applied to optimize routing or resource allocation, which are all under discussion in the planning of 5G networks as summarized in [5,45]. In this work, however, we limit ourselves to concepts that affect and improve the PHY layer and do not consider approaches on higher levels of the protocol stack. Thereby, the current research activities and promising enablers for the required performance gains can be summarized into the following categories.

2.1.1 Densification

A straightforward and efficient way to improve the overall capacity of cellular networks is to install many additional BSs and to make the cells smaller. By such a densification, individual BSs have to serve less users, which allows to allocate more bandwidth to each of them. To this end, however, the transmit power of the BSs has to be adapted such that the signal-to-interference ratio (SIR) is maintained. In principle, the cells can thereby shrink almost indefinitely until as many (or even more) BSs are deployed than MSs [5]. With this, each user could be served by its own BS. When the BSs can coordinate themselves to schedule the users and optimize handovers, very high macro diversity gains can be achieved. Small cells, such as pico- or femto-cells, can also be combined with larger macro-cells or other nodes such as relays or RRHs [41]. In such heterogeneous networks, all users can dynamically connect to the infrastructure node that offers the best conditions, e.g. to a WLAN-like femto-cell hot spot if the user is static and located inside a building or to a macro-cell BS if the user is outside and moving with high velocity. Flexible handovers between the different tiers of cells allows to allocate the best resources to each MS. In indoor environments, this can e.g. be a frequency band around 60 GHz where high bandwidths are available or to low frequencies that experience less attenuation and thus wider coverage range for outdoor users.

While small cells can provide high gains when the network densification becomes extreme [98], several challenges need to be overcome: In order to support mobility in such heterogeneous networks, intelligent handover mechanisms are required and sophisticated interference avoidance schemes are needed such that high data rates can also be preserved for cell edge users. If different radio access technologies are used in parallel (e.g. millimeter wave transmission in indoor environments and wide range transmission from outdoor BSs), the association between users and BSs needs to be determined appropriately and the user equipment must be able to flexibly switch between these different technologies. This might require multiple transceiver chains or the usage of software defined radio which allows the same transceiver chain to adapt to different technologies [91]. Another problem are the costs of the deployment of many BSs, the installation of the backhaul connection as well as the maintenance that is required. These costs can be reduced when the additional nodes in the network are wireless relays that do not need a wired backhaul connection but just forward the signals they receive to their destination.

2.1.2 Increased Bandwidth

A second immediate measure to higher capacity in cellular networks is to allocate more bandwidth. With this, the capacity can almost linearly be increased. As most of the spectrum that is used for mobile communication services (microwave frequencies between several hundred MHz to a few GHz) is already occupied by the different services in use, additional bandwidth has to be found. To this end, two approaches are possible. Frequency blocks that are licensed and reserved for a primary service but are currently idle in certain areas for a certain time can be allocated as additional bandwidth for a secondary unlicensed system. Thereby, it has to be ensured that the primary service is not disturbed and the frequency block is made free as soon as a user in the primary service becomes active. With cognitive radio, i.e. when systems intelligently sense the wireless medium to find unused frequencies, this can be realized [135]. With more sophisticated signal processing, this idea can even be carried one step further. If the secondary system can transmit signals that do not disturb the licensed users, e.g. by accurate beamforming that nulls the interference the secondary system causes to the primary, both systems can operate simultaneously in the same frequency band [137]. With carrier aggregation as included already in LTE-Advanced, any currently available frequency block can be allocated to the mobile users, which then can benefit from much higher bandwidths and thus higher data rates.

An alternative that has the potential to drastically increase the available bandwidth is to expand the systems to millimeter wave transmissions in the range of $30 - 300\,\mathrm{GHz}$. Particularly large unlicensed frequency bands that are idle and can be used to enhance data rates in mobile communication systems are around $60\,\mathrm{GHz}$. The reason these frequencies are idle is that the propagation conditions are rather hostile due to strong pathloss, atmospheric and rain absorption, low diffraction around obstacles and very low penetration through objects or walls, phase noise, and high equipment costs [5]. An inherent advantage on the other hand is that the antennas can be of very small dimension. Consequently, large antenna arrays can be realized with a small form factor. This allows to transmit signals with very narrow and focused beams that can compensate the high pathloss and reduce interference between different users [136]. In order to achieve good results, however, precise beamforming is required for which accurate CSI and precoding coefficients are necessary. Moreover, the beamforming schemes have to be quite robust as even small changes, even small movements of the equipment due to wind, can cause misalignments and thus severe performance drops [58]. An additional difficulty is that a very large amount of infrastructure nodes is required for 60 GHz transmission as line-of-sight (LOS) propagation is required due to the high losses signals at these frequencies experience when they penetrate through walls or are reflected. Hence, ubiquitous deployment is necessary such that every room or office in indoor environments is equipped with at least an own infrastructure node or every street corner or house entry is covered in outdoor environments. Such a densification however leads to exorbitant costs. This costs could e.g. be reduced when the millimeter wave nodes are fed wirelessly and are implemented as relays. In this case, no wired backhaul access is required and relays can potentially be implemented with low complexity as we will see later.

2.1.3 Beamforming

Besides allocating more bandwidth and densifying the networks, higher data rates can also be achieved when the available spectrum is exploited more efficiently. To this end, sophisticated multi-user beamforming can be applied that attempts to serve multiple users in the same physical channel without (significantly) disturbing each other. By the use of multiple antennas, signals intended for different users can be precoded in a way such that they are constructively combined where they are desired while they are reduced or even cancelled where they are undesired. In this way, the physical cannel can be used for multiple users at the same time without sharing resources; all users can thus be served simultaneously in the entire frequency band that is available. This increases spectral efficiency and thus data rates for all. As the problem of calculating beamforming filters is quite difficult in general, a vast amount of theoretical and practical work has been conducted which ranges from capacity achieving non-linear schemes as dirty paper coding (DPC) to linear techniques as zero-forcing (ZF), matched filter

(MF), minimum mean squared error (MMSE), leakage reducing precoding or others that are optimal or suboptimal in certain scenarios or aspects (see [38, 63, 102] for an overview).

Traditional beamforming techniques applied to cellular networks steer the transmitted energy in the horizontal plane and each BS calculates its own precoding filters. More recent approaches [47] intend to change that and extend it to full three dimensions (3D beamforming). With a more dynamic antenna pattern adaptation also in the vertical domain, more MSs can be separated in space, especially as many users are usually located inside buildings with multiple floors. If 3D antenna arrays are built, more antennas can be placed in a confined space, which in turn enables to exploit more degrees of freedom as with only 2D arrays with fewer antennas. Other approaches discussed in research are distributed beamforming schemes through cooperation among multiple BSs or, carried to the extreme, a single super BS that can precode the signals for a large area (up to an entire country) that are transmitted by RRHs distributed in the area of service [107]. In the other extreme, many simple nodes can exchange signals with each other such that these nodes can perform distributed beamforming. In [100], such an approach has been proposed in a hierarchical way in which few nodes start to exchange signals in a cluster and precode their signals to achieve cooperation between different clusters. These clusters can then in turn again apply distributed beamforming to transmit signals to larger clusters until all nodes are involved in the cooperation process. With this, a linear scaling of the capacity can be achieved, even if the different nodes are not connected together by a fixed infrastructure based backhaul.

2.1.4 Cooperation

When multiple nodes, e.g. BSs, are connected together via the backhaul, they can cooperate with each other to improve the overall performance of the network. Different levels of cooperation (also called CoMP) are discussed in literature as summarized in [79, 80]. They can broadly be classified into BS or transmit point selection, coordinated scheduling and coordinated beamforming, and joint transmission. In the simplest form, BS selection, the signal to a specific MS is transmitted by a single BS in a resource block that is chosen such that the performance to this MS is as high as possible while the interference caused to other users is small. For this selection, only receive signal strength feedback from the MSs is required. In coordinated scheduling and beamforming, multiple BSs exchange CSI in order to find an optimal BS-MS assignment and optimized beamforming weights that improve the desired signals but limit the interference to other users. The data intended for the users is however not exchanged between BSs and each BS transmits its signal to its associated MSs alone. If not only CSI but also user data can be exchanged, multiple BSs can act as a single large virtual antenna array that jointly precodes all signals to all involved users. Depending on the capacity of the backhaul connection between the BSs, this can then be similar to multiuser beamforming in a broadcast channel with the only exception that multiple individual power constraints have to be fulfilled, as the transmit power of each BS is limited. If the backhaul capacity is limited or if the CSI is imperfect, simplified forms of joint transmission have to be applied that can cope with quantized data symbols or are robust with respect to the imperfections. In more sophisticated scenarios, transmit filters at the BSs and receive combiners at the MSs can jointly be optimized, which can lead to particularly good results [42].

2.1.5   Massive MIMO

MIMO and beamforming can enhance the spectral efficiency significantly. When combined with CoMP, many users across multiple cells can be served simultaneously in the full frequency band without disturbing each other. In current 4G networks, the BSs are however equipped with no more than 8 antennas. When many more antennas are installed, much higher spectral efficiency can be achieved. In the massive MIMO approach, the proposal is to equip the BSs with a number of antennas that is much larger than the number of active MSs per resource block, up to several hundreds of them. With this, enormous enhancements of spectral efficiency are possible and a vast spatial diversity is available to benefit from. Moreover, when the number of antennas grows large, the channels between different users become more and more orthogonal to each other, which can simplify the transmit and receiver structure drastically [94,120].

Besides its promising advantages, there are however also several challenges that need to be overcome to make massive MIMO feasible for practice. One problem is channel estimation. Due to the large number of antennas, also many channel coefficients need to be estimated to achieve good performance. This introduces a large overhead that might diminish or even destroy the performance gains. Moreover, when many training sequences are used in parallel, as it is necessary with many antennas and users in multiple cells, the quality of the CSI is limited by the so called pilot contamination [5]. With many users and antennas involved, the same pilots have to be reused. As a consequence, the quality of the available CSI is limited by the interference these pilots cause to each other and the beamforming suffers from the resulting imperfections. Other issues to solve are enabling the coexistence with underlying tiers of small cells, issues with antenna coupling when many antennas are located close to each other, or the implementation of power amplifiers that support large numbers of antennas [5].

2.1.6 Relaying

An alternative to serving the mobile users directly by BSs, with or without cooperation or large antenna arrays, is to install intermediate nodes that do not transmit own information but forward processed versions of signals they have received from a source node. The signal thereby arrives at the destination via one or multiple relays. The direct path between the terminals can thereby be assisted by these relays or the signal can traverse the network in a two- or multi-hop fashion when links over more than two consecutive nodes are blocked or ignored. With such relays, the coverage range of a transmitting node can be extended and, depending on its implementation, the effective channel between the terminals can be shaped in a beneficial way to improve reliability and spectral efficiency by boosting the SNR or the SINR.

An overview of different relaying techniques and their applications can be found e.g. in [87]. The relays can be classified as full-duplex or half-duplex. While full-duplex relays can simultaneously transmit and receive, half-duplex relays cannot. For instance, half-duplex nodes may operate in *time-division duplex* (TDD) mode, i.e. each node transmits and receives in different time slots; in *frequency-division duplex* (FDD) systems, nodes can transmit and receive at the same time but use different frequency channels. The transmission of one symbol from source to destination thus requires the use of two resource blocks which amounts in a 50% loss in spectral efficiency.

Furthermore, different signal processing strategies are possible. The most widely used are the *decode-and-forward* (DF) strategy, which involves decoding of the source transmission at the relays before the re-encoded signals (possibly with a different code-book) are forwarded, and *amplify-and-forward* (AF) relaying, where the relays forward a linear combination of signals at their receive antennas. The former has the advantage that relay noise that affects the receive signal of the relay is removed by the decoding process. However, DF relays can suffer from error propagation, i.e. if a relay decodes the received signals wrongly, it will also forward a wrong codeword to the destination. The achievable end-to-end data rate is thus limited by the weakest link. AF relays on the other hand also amplify and forward their own noise which affects the overall SNR. However, the signals can in turn be processed by appropriate amplification factors and phase shifts such that multiple signal links are combined coherently. With this, the effective channel between sources and destinations can be brought into specific forms that boost the strength of the desired signals or cancel or mitigate interference between different users [10, 29]. Besides these two strategies, there are also other possibilities for signal processing in relays. Quantize-and-forwards (QF) or compress-and-forward relays for instance do not decode their receive signals but quantize them and forward quantization indices or a compressed version of the receive signal to the destination. Compute-and-forward (CF) relays are also possible, which forward generalized functions of their receive signals. In cases where source-relay and direct source-destination channels are of similar quality, the compressed relay signals can act as independent signal observations that can be used to assist the decoding at the destination. The forwarded signals can thereby comprise hard decision indices of quantization regions or soft information [152].

When considering bidirectional communication between the terminals (UL and DL), the relaying protocols can be further classified into *one-way* (conventional) and *two-way* relaying [112]. When applied to cellular networks where the terminals are associated with BSs and MSs, one-way relays separate the UL and DL and either forward the BS signals to the MSs or vice versa. In two-way relaying, both directions of communication are combined such that the relays receive the superposition of all BS and MS signals and broadcast a processed version of these signals back to all terminals. This can double the spectral efficiency as compared to one-way relaying. An inherent drawback of two-way relaying is that the signal received by a terminal (BS or MS) also contains the signal that this terminal node has previously transmitted and is backscattered by the relays [158]. This so-called *self-interference* needs to be subtracted at the terminal before the signal can be decoded.

In 4G standards such as LTE-Advanced, relaying is already foreseen to assist the communication between BSs and MSs [3]. In contrast to current systems where relays can be deployed as simple repeaters to extend the coverage to areas with poor reception, future networks allow relays to cooperate with other nodes e.g. to enhance diversity. To this end, relays can be coordinated with other nodes such that the signals are sent over different paths (direct links or via relays). A diversity gain can then be achieved when either one (the best) of the different paths is chosen (relay selection) or by (coherently) combining multiple of them [87]. As the performance of DF relays is limited by the weakest link, this type of relaying is particularly suitable for link selection. With AF relaying on the other hand, multiple links can be used at the same time. To this end, such relays can be used as "active scatterers" that improve the quality of the effective channel between the terminals and enforce well conditioned MIMO channel matrices [111]. When more sophisticated signal processing is possible and CSI is available at the relays, the signal transformation can be used to optimize the end-to-end performance. To this end, the AF relay gains can e.g. be used to zero-force interference terms or to maximize the spatial multiplexing gain by optimizing an MMSE criterion [9]. The relays can thereby be dedicated infrastructure nodes installed at fixed positions with the sole purpose to enhance the QoS in a certain region. In future networks such as in 5G designs, also user cooperation plays a role of increasing importance [5]. MSs that are idle at the moment can thereby jump in as relays in a dynamic way to assist the communication of other users.

Another approach to incorporate relaying techniques into mobile communication networks is to use additional nodes that are directly connected to the wired backbone. Especially as future 5G networks are expected to combine the proprietary cellular systems with WLAN-like femto-cell hot spots, small internet access points can be used to assist or complement the BS-to-MS links. To this end, internet access points can apply compress-and-forward strategies to provide users with additional signal observations even when the additional access points only have a very limited backhaul connection. With techniques as proposed in [75, 124], the receivers can benefit from compressed information that is forwarded by multiple access points. With such a decentralized information processing, different systems can elegantly be combined and assist each other.

2.2   System Model & Assumptions

In this thesis, we will evaluate and compare different PHY layer techniques that seem promising for the enhancement of cellular networks for future generations. To this end, we attempt to combine the different approaches introduced above to be able to benefit from the advantages that these techniques offer. In particular, we will combine ultra dense networks that consist of different types of nodes with sophisticated cooperative multi-user beamforming from large BS antenna arrays. As the resulting communication networks are rather complex, it is required to simplify them to a manageable and sufficiently simple model.

The main focus of this work is set on the DL. As the UL is generally used to upload data with the intention that this data is downloaded again, in many cases desirably even by many users (e.g. posted movies, audio files, or other multimedia content), we regard the DL as the more important direction of communication. Moreover, due to the asymmetric network topology with BSs as more sophisticated nodes than the mobile user equipment, we also regard the DL as the direction which poses more challenges, especially for cell edge users which are usually affected by strong interference. When we assume that the MSs cannot cooperate with each other, the enhancement of the data rates for the different users has to be enabled by the signaling of the infrastructure nodes (BSs and/or possible additional nodes such as relays). The UL on the other hand can be seen as a multi-user detection problem [144] that is conceptually easier to extend to the case of multiple BSs that cooperate with each other. To this end, we mostly consider transmission schemes that are designed and applied for the DL. The UL is only treated in certain example scenarios, particularly when we consider two-way relaying. In this case, both directions of communication are inherently included in the transmission protocol.

Modern communication systems communicate with multiple antennas over wide bandwidths (up to 20 MHz in LTE and up to 100 MHz in LTE-Advanced and possibly even more in 5G systems [1,3,60]). The wireless channels are thus generally frequency selective. In order to avoid intersymbol interference (ISI) coming with such channels, different techniques are possible. One of the most popular among them is OFDM that has gained the highest acceptance as the modulation technique for modern high-speed wireless networks and 4G mobile broadband standards [87]. When combined with MIMO, high spectral efficiencies can be achieved, which leads to high capacities and data rates for the users that are served with this technology. The key concept thereby is that the symbols transmitted are modulated in the frequency domain, are mapped to different *subcarriers*, and then transformed to the time domain by an inverse discrete Fourier transformation. A cyclic prefix is inserted before each block, which assures that the transmission leads to a circularly symmetric convolution of the transmitted signal with the channel. If this cyclic prefix is sufficiently long, ISI can be avoided even when the channels have a large delay spread. A major disadvantage of OFDM is that it comes with a large peak-to-average power ratio, which implies an increased energy consumption [93]. As this is particularly undesirable in the mobile user equipment, LTE-Advanced and other standards decided to apply single carrier FDMA in the UL which reduces high peak powers [1]. For the DL, however, OFDM is the technique of choice. As this work is focused on the DL, we thus assume that all systems are based on this form of modulation.

When OFDM is extended to a multiuser media access technique (orthogonal frequency division multiple access – OFDMA), such systems have the advantage, that different users can be allocated to orthogonal resource (time-frequency) blocks according to their needs. Each receiver can then consider its intended subcarriers, discard the cyclic prefix, apply a discrete Fourier transformation, and access the (scaled and noisy) transmitted symbols in the frequency domain [93]. Furthermore, the different subcarriers can be considered separately as they are orthogonal. The channels can then be assumed to be frequency flat over each subcarrier, which simplifies the mathematical notation and analysis.

In the remainder of this work, we thus usually consider a single subcarrier. An exception is Chapter 5, where we explicitly look at broadband channels and show that a certain gain can be achieved when linear transformations of symbols across different subcarriers are allowed. Otherwise, we study the performance that can be achieved with specific precoding and transmission schemes on narrowband channels of a single subcarrier. The extension to the wideband case can then be made by applying the schemes on the different subcarriers in parallel and to perform power loading across these subcarriers. This extension comes however with a significant increase in complexity, especially when multi-antenna systems are considered. The precoding has to be calculated on each subcarrier. With the consideration of a single subcarrier only, we can limit the computational complexity for the simulations to a manageable level. The diversity that comes with wideband channels and can be exploited with power loading over the whole bandwidth is however lost in the narrowband consideration.

By considering the transmission on a single frequency flat OFDMA subcarrier, we can typically focus on a subset of the cellular network in which B BSs communicate with M MSs. We thereby assume that a scheduler has assigned the MSs to the resource blocks that are the same for all involved nodes. In order to represent the signals, we apply the equivalent baseband representation [106]. Due to the sampling theorem, a symbol discrete system model provides a sufficient description of the systems [77]. Thereby, the properties of the modulators and demodulators, specifically their imperfections, are not considered. By assuming perfect matched filters in each signal branch and perfect sampling, all relevant information is contained in one sample per symbol duration T at the receivers (T-spaced sampling) [106].

By modeling the communication systems in the symbol discrete model with frequency flat narrowband channels, we particularly assume throughout this thesis:

- Perfect synchronization in time and frequency between all nodes that are involved in a transmission.

- Different propagation delays that arise from different propagation paths are compensated and do not have to be included in the signal descriptions.

- Perfect CSI is available at all receiving nodes (channel state information at the receiver – CSIR) to decode their intended messages. The corresponding channel coefficients can perfectly be estimated e.g. by pilot symbols that are included in the signals.

The channel knowledge at the transmitters (channel state information at the transmitter – CSIT) is usually more difficult to acquire. To this end, the CSIT varies for the different transmission schemes that we consider in the sequel and discuss cases in which perfect CSIT can be assumed, no CSIT at all is available, or the CSIT is noisy or quantized.

Regarding the different nodes, we further assume that BSs have strong computational capabilities that are sufficient for the applied signal processing schemes. If not stated otherwise (e.g. in Chapter 8 where we explicitly look a limited backhaul links), the backhaul connections are idealized such that required information can readily be exchanged without delays (infinite capacity). The MSs on the other hand, cannot cooperate with each other but have sufficient capabilities to perfectly decode their intended data symbols. Relays and other nodes that assist the communication between BSs and MSs are assumed to be dedicated infrastructure nodes. Their computational power and available CSI varies depending on the scenarios considered in the respective chapters.

2.3 Fundamentals

Based on the assumptions above, we can now formulate symbol discrete input-output (IO) relations of the communication systems and the wireless channels in between. Communication through a wireless channel is generally difficult because signals are affected by different impairments. These include noise, attenuation due to pathloss, fading, distortion, and, if multiple users are considered, interference. The simplest way to describe a wireless channel between a transmitter that is equipped with N_t antennas and a receiver with N_r antennas is by the single-tap linear IO relation at discrete time m as

$$\mathbf{y}[m] = \mathbf{H} \cdot \mathbf{x}[m] + \mathbf{n}[m], \tag{2.1}$$

which corresponds to a frequency flat, slow fading channel $\mathbf{H} \in \mathbb{C}^{N_r \times N_t}$, with transmit signal $\mathbf{x}[m] \in \mathbb{C}^{N_t}$, receive signal $\mathbf{y}[m] \in \mathbb{C}^{N_r}$, and additive noise $\mathbf{n}[m] \in \mathbb{C}^{N_r}$. Therein, the symbols in the symbol vectors $\mathbf{x}[m]$ and $\mathbf{y}[m]$ correspond to the sampled baseband representations of real passband signals transmitted by each antenna of the transmitter and received by each antenna of the receiver, respectively. Due to the sampling theorem for passband signals, this symbol discrete baseband representation provides a sufficient description of the system [77]. The noise $\mathbf{n}[m]$ is usually modeled to contain independent and identically distributed (i.i.d.) components that are circularly symmetric Gaussian (CSCG): $(\mathbf{n}[m])_i \sim \mathcal{CN}(0, \sigma_n^2)$, $\forall i$, also denoted as additive white Gaussian noise (AWGN).

Throughout this Thesis, we are mainly interested in information theoretic aspects, particularly in the channel capacity. In this case and with perfect CSIR but absence of CSIT, the capacity can be given by [139]

$$C = \log_2 \det \left(\mathbf{I}_{N_r} + \frac{P}{N_t \cdot \sigma_n^2} \mathbf{H} \mathbf{H}^{\mathsf{H}} \right) \text{ [bps/Hz]}, \tag{2.2}$$

for some sum transmit power P. With CSIT, the capacity can be increased to [139,143]

$$C = \max_{\mathbf{Q}:\mathrm{Tr}\{\mathbf{Q}\mathbf{Q}^{\mathsf{H}}\} \leq P} \log_2 \det \left(\mathbf{I}_{N_r} + \frac{1}{\sigma_n^2} \mathbf{H} \mathbf{Q} \mathbf{Q}^{\mathsf{H}} \mathbf{H}^{\mathsf{H}} \right) \text{ [bps/Hz]}, \tag{2.3}$$

where the precoding or beamforming matrix $\mathbf{Q} \in \mathbb{C}^{N_t \times N_t}$ can be optimized under the power constraint $\mathrm{Tr}\{\mathbf{Q}\mathbf{Q}^{\mathsf{H}}\} \leq P$, e.g. by a singular value decomposition (SVD) of $\mathbf{H}$ and power loading according to the waterfilling solution [143].

Another quantity that measures the quality of the channel is the outage probability

$$p_{\text{out}} = \Pr\{C < R_{\text{out}}\}, \tag{2.4}$$

i.e., the probability that a certain target rate R_{out} cannot be supported by the actual realization of the channel. The spatial degrees of freedom of such a MIMO channel can be used to improve the data rate or the outage probability. Measures for this are the spatial *multiplexing gain* or the spatial degrees of freedom of the channel defined as

$$d = \lim_{P \to \infty} \frac{\log_2 \det\left(\mathbf{I}_{N_r} + \frac{P}{N_t \cdot \sigma_n^2} \mathbf{H}\mathbf{H}^{\mathsf{H}}\right)}{\log_2\left(\frac{P}{\sigma_n^2}\right)}, \tag{2.5}$$

which measures the number of spatial streams that can reliably be transmitted in parallel and the *diversity gain* defined as

$$r = \lim_{P \to \infty} -\frac{\log\left(\Pr\left\{\log_2 \det\left(\mathbf{I}_{N_r} + \frac{P}{N_t \cdot \sigma_n^2} \mathbf{H}\mathbf{H}^{\mathsf{H}}\right) < R_{\text{out}}\right\}\right)}{\log\left(\frac{P}{\sigma_n^2}\right)}. \tag{2.6}$$

Both quantities have a maximum value that depends on the number of involved antennas:

- Maximum multiplexing gain: $d_{\max} = \min\{N_r, N_t\}$
- Maximum diversity gain: $r_{\max} = N_r \cdot N_t$.

Due to the fundamental diversity-multiplexing tradeoff, both gains can in general not be achieved simultaneously, as degrees of freedom that are used for diversity are not available anymore for multiplexing and vice versa [159].

2.3.1 Broadcast & Multiple Access Channel

When considering cellular networks, the channel models need to be extended such that the communication between multiple nodes can be taken into account. The simplest models that still contain the essential properties, at least for the communication in a single cell in which multiple users are served, are the broadcast channel (BC) for the DL and the multiple access channel (MAC) for the UL, which share a duality [65,145].

In a simple discrete time BC model of a single BS that is equipped with N_{B} antennas that communicates with K MSs, each with N_{M} antennas, the DL signal received by user k can be modeled as

$$\mathbf{y}_k = \mathbf{H}_k \cdot \mathbf{x} + \mathbf{n}_k,, \qquad k = 1, \dots, K, \tag{2.7}$$

where we drop the discrete time index and $\mathbf{x} \in \mathbb{C}^{N_{\mathrm{B}}}$ is the transmit signal vector of dimension N_{B}, $\mathbf{H}_k \in \mathbb{C}^{N_{\mathrm{M}} \times N_{\mathrm{B}}}$ the frequency flat channel from the BS to MS k, and $\mathbf{n}_k \in \mathbb{C}^{N_{\mathrm{M}}}$ the noise induced in the MS. As the transmitted signal $\mathbf{x}$ contains the data intended for multiple users, it can assume the superposition

$$\mathbf{x} = \sum_{k=1}^{K} \mathbf{x}_k, \tag{2.8}$$

with $\mathbf{x}_k$ being the signal carrying the message for user k, possibly linearly or non-linearly encoded with covariance matrix $\mathbf{K}_k = \mathsf{E}\left[\mathbf{x}_k \cdot \mathbf{x}_k^{\mathsf{H}}\right]$. The power allocated to MS k follows thus as $P_k = \mathrm{Tr}\left\{\mathbf{K}_k\right\}$. The other messages contained in $\mathbf{x}_j$, for $j \neq k$, pose undesired interference to user k.

In the UL, each MS transmits its signal $\mathbf{x}_k$ and the BS receives the superposition

$$\mathbf{y} = \sum_{k=1}^{K} \mathbf{H}_k^{\dagger} \cdot \mathbf{x}_k + \mathbf{n}, \tag{2.9}$$

with $\mathbf{H}_k^{\dagger} \in \mathbb{C}^{N_\mathrm{B} \times N_\mathrm{M}}$ being the uplink channel from MS k to the BS, which may or may not be reciprocal to $\mathbf{H}_k$, and $\mathbf{n}$ the noise induced in the BS. This channel model is also referred to as the MAC.

In contrast to point-to-point communication systems where a single source transmits to a single destination, the capacity of a multiuser setup cannot be described by a single number as in (2.3), but by a multidimensional rate region in which each point therein is a rate-tuple with an achievable rate for each individual user. In multiuser setups as described by the BC or MAC, each of the K users can thus communicate simultaneously with a rate that lies within that rate region.

In the case of the BC, this rate region can be written as [93]

$$\mathcal{C} = \mathrm{conv}\left\{\bigcup_{\mathbf{Q},\pi} \{R_1(\mathbf{Q},\pi), \ldots, R_K(\mathbf{Q},\pi)\}\right\}, \tag{2.10}$$

where $\mathrm{conv}(\cdot)$ is the convex hull operation and $\bigcup$ states the union of multiple rate regions. Thereby, all choices of possible precoding matrices $\mathbf{Q}$ and all encoding orders π need to be considered for the evaluation. For a fixed choice of these parameters, the individual rates are bounded by

$$R_k(\mathbf{Q},\pi) \leq \log_2 \det \left(\mathbf{I}_{N_r} + \left(\sigma_n^2 \mathbf{I}_{N_r} + \sum_{\pi(j) > \pi(k)} \mathbf{H}_k \mathbf{Q}_j \mathbf{Q}_j^{\mathsf{H}} \mathbf{H}_k^{\mathsf{H}}\right)^{-1} \mathbf{H}_k \mathbf{Q}_k \mathbf{Q}_k^{\mathsf{H}} \mathbf{H}_k^{\mathsf{H}}\right), \tag{2.11}$$

where the interference terms that come, with a specific encoding ordering π, *before* the signal of user k, i.e. all interference terms for $\pi(j) < \pi(k)$, are pre-cancelled at the transmitter.

Even though difficult to achieve in practice, this rate region can be characterized by the *duality* that the BC shares with the corresponding (or dual) MAC in which the sum of the transmit powers of all users is equal to the total transmit power of the BC. The capacity region of the MAC is easy to achieve (at least in theory), by applying successive interference cancellation (SIC), which decodes each stream sequentially after subtracting the reconstructed signal contributions of the previously decoded data streams [143]. But also in this case, all possible user orderings have to be considered, which makes a full characterization of the achievable rate region difficult when many users are involved.

The corresponding dual scheme with which the capacity of the BC can be achieved is dirty paper coding (DPC) [18, 26]. Thereby, the interference is pre-cancelled at the BS and user scheduling and power loading is also implicitly included in the orderings π and the precoders $\mathbf{Q}$. While this form of precoding is the optimal strategy from an information theoretic perspective, it is difficult to implement in practice [38, 93].

Accordingly, focus has been set on suboptimal but less complex schemes such as vector perturbation methods [55, 103], sphere decoding [46], or Tomlinson-Harashima precoding [50, 142]. Thereby, the interference is precanceled at the transmitter and, since a simple subtraction might lead to highly increased transmit powers, applied to a modulo operation until the transmit signal strength is again in the range of the one without interference subtraction and does satisfy the power constraints. While such precoding schemes achieve close to optimal results, they require highly precise CSIT which is difficult to obtain in practice. The least inaccuracies in the CSI lead to a failure of the interference cancellation. Moreover, these schemes require complicated encoding and decoding processes which are regarded to be too complex for practical implementation in real-time systems (see e.g. [81, 153, 157]).

An alternative is *linear* precoding in which linear precoding matrices $\mathbf{Q}$ are applied to transform the data symbols. By formulating appropriate objective functions, linear precoding matrices can be optimized with relatively low complexity as compared to the better non-linear schemes. The downside is however that residual interference between data streams have to be accepted which are in the receivers usually treated as noise. Many different schemes have been proposed and discussed that, despite being suboptimal, achieve close to optimal results under certain conditions. Prominent

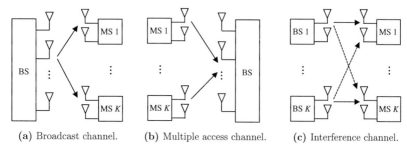

(a) Broadcast channel.   (b) Multiple access channel.   (c) Interference channel.

Figure 2.1.: Basic information theoretic channels.

examples that are promising due to their relative ease for practical implementation are zero-forcing (ZF) or block ZF in which the interference terms between data streams or between different users are cancelled by a precoder that forces the signals to lie in the null space of the respective channels [25, 132], minimum mean squared error (MMSE) precoding that chooses the precoding matrices such that the MMSE of the received data symbols is minimized [66, 134, 140], or leakage based precoding designs which attempt to increase the desired signal strength and reduce the interference caused to others users at the same time, i.e. to maximize the signal-to-leakage-and-noise ratio (SLNR) [121, 123]. Further details are provided in the chapters where we apply these schemes.

Even though such linear schemes come with a loss in the achievable rates compared to DPC, the scaling law of the achievable rates for large SNRs behaves the same, i.e. the maximal multiplexing gain can be characterized by

$$d_{\mathrm{max}} = \min \{N_{\mathrm{B}}, K \cdot N_{\mathrm{M}}\} \qquad (2.12)$$

and the maximal diversity gain that can be achieved is

$$r_{\mathrm{max}} = N_{\mathrm{B}} \cdot K \cdot N_{\mathrm{M}}, \qquad (2.13)$$

when non-degenerate channels that offer full rank are assumed.

2.3.2   Interference Channel

The BC and MAC are models that describe the communication within a single cell where a single BS serves multiple MSs. In order to describe situations in which multiple sources and destinations wish to communicate concurrently with each other, these models are no longer sufficient but need to be combined to the interference channel (IC), which in its simplest form consists of K source-destination pairs that wish to communicate simultaneously with each other. In order to model such a scenario, the receive signal of destination k, which can be attributed to a BS or an MS, can be described by

$$\mathbf{y}_k = \mathbf{H}_{k,k} \cdot \mathbf{x}_k + \sum_{\substack{j=1 \\ j \neq k}}^{K} \mathbf{H}_{k,j} \cdot \mathbf{x}_j + \mathbf{n}_k, \tag{2.14}$$

where the first term reflects the intended signal from its corresponding source node k and the second and third term are the interference (from and intended to other source-destination pairs) and noise. This channel model can be seen as a combination of the BC and MAC where $\mathbf{H}_{j,k} \in \mathbb{C}^{N_r \times N_t}$ describes the channel between source j and destination k, which are equipped with N_t and N_r antennas, respectively.

Even though extensively studied (see e.g. [17] and references therein), the exact capacity of the general IC, and thus an optimal transmission strategy, remains unknown until now. Several methods to cope with the interference in such a network have been considered [17]:

Orthogonalize different users: This is the conventional (cellular) approach in which each user pair is assigned to different resources, e.g. with FMDA, TDMA, or CDMA. With this, each user sees its intended signal without interference but at the cost of sharing resources. The data rates can thus only scale with $\frac{1}{K} \cdot \min\{N_r, N_t\} \cdot \log_2(P) + o(\log_2(P))$.

Decode the interference: If the interference received by a destination is strong, the nodes can try to decode the interfering signals along with the desired one in order to improve the rate of the desired signal. Thereby, the strength of the interference still limits the data rates of the other signals and such an approach comes with a extensive receiver complexity.

Treat as noise: If the interference seen by the receiving nodes is weak, it can be treated as noise. In case of linearly precoded, i.i.d. Gaussian signals $\mathbf{x}_k = \mathbf{Q}_k \cdot \mathbf{s}_k$, with $\mathbf{s}_k \sim \mathcal{CN}(\mathbf{0}, \mathbf{I})$, the achievable rate of user k follows as

$$R_k = \log_2 \det \left(\mathbf{I}_{N_r} + \left(\sum_{\substack{j=1 \\ j \neq k}}^{K} \mathbf{H}_{k,j} \mathbf{Q}_j \mathbf{Q}_j^{\mathsf{H}} \mathbf{H}_{k,j}^{\mathsf{H}} + \sigma_n^2 \cdot \mathbf{I}_{N_r} \right)^{-1} \cdot \mathbf{H}_{k,k} \mathbf{Q}_k \mathbf{Q}_k^{\mathsf{H}} \mathbf{H}_{k,k}^{\mathsf{H}} \right). \tag{2.15}$$

Thereby, the precoding matrices $\mathbf{Q}_k$ can be chosen in a selfish fashion, i.e. only with single user precoding and ignoring the other users, or in a more altruistic fashion in which the precoding is designed to keep the interference caused to other users low.

A particularly interesting result thereby is reported in [17]. Even though such an IC is considered to be interference limited, a rate scaling of $\frac{K}{2} \cdot \min\{N_r, N_t\} \cdot \log_2(P) + o(\log_2(P))$ can be achieved, i.e. each user can get one half of the degrees of freedom, irrespective of how many users share the wireless medium. A way to achieve this is *interference alignment*. Thereby, the desired signals and the interference as seen by each receiver are precoded such that they fall into orthogonal signal dimensions. As a consequence, each transmitter-receiver pair can use half of the system resources for interference free communication [17]. In order to achieve that, symbols have to be coded over long blocks (e.g. time or frequency blocks) which grow faster than exponentially with the number of users [32]. Alternatively, spatial multi-antenna transmit and receive filters can be used to obtain aligned solutions. In this case, the number of antennas required to achieve interference alignment grows only linearly with the number of users, similar to approaches as ZF.

The concept of interference alignment can be seen as a coordination technique in which multiple source-destination pairs coordinate their transmission such that the desired signals are not affected by interference. Thereby, accurate CSI is required but different nodes do not need to exchange user data. Even though sum rate increases linearly with the number of involved users, more sophisticated cooperation schemes where nodes can also exchange user data and apply joint precoding can achieve higher performance. With full cooperation, a rate scaling of $K \cdot \min\{N_r, N_t\} \cdot \log_2(P) + o(\log_2(P))$ can be achieved.

2.3.3 BS Cooperation with Joint Linear Precoding

While the basic channel models discussed above provide insightful results for multiuser communication, practical cellular networks are too complex to be captured in these simple scenarios and consist of a combination of all previously discussed channel models. A multi-cell setup consists of multiple BSs that each serve multiple MSs. Depending on wether the DL or the UL is considered, such a network can be seen as multiple interfering BCs or MACs that are distributed in space and are thus affected by pathloss, fading, and interference. The first two challenges can be mitigated by installing more BSs within a region with which the cells become smaller. The users can accordingly

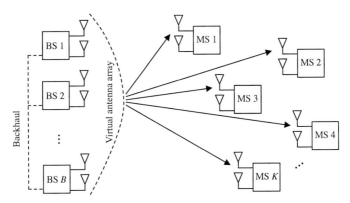

Figure 2.2.: Virtual antenna array consisting of multiple cooperating BSs: when multiple BSs cooperate together over a backhaul of infinite capacity, the resulting network can be seen as a BC.

benefit from higher signal powers due to the smaller pathloss as well as from macro diversity if they are in reception range of multiple BSs. In order to mitigate the impairments of interference, users that are in coverage range of each other can be assigned to orthogonal resources. With a spatial reuse partitioning, the same resources can be used for other users that are sufficiently far apart. This however limits the achievable rates by the reuse factor which linearly lowers the data rates for each user when the number of MSs increases. A more sophisticated interference management which allows all users to profit from high bandwidths is to apply joint beamforming across multiple BSs. With this, the interference between different users can be mitigated by precoding.

If multiple BSs can cooperate with each other via a backhaul of infinite capacity, i.e. any information exchange can be achieved without any delay, a multi-cell network acts similar to a BC with only a single BS (cf. Fig. 2.2). In this case, all information that is available at one BS can immediately be transferred to all others and all BSs can then jointly precode all signals for all MSs. The communication of the cooperating BSs is then very similar to a conventional BC, with the difference that not a single antenna array serves the MSs but a large virtual one that is distributed in space. If multiple BSs are involved, however, a fundamental difference arises from the power constraint that is usually imposed. While a certain maximal transmit power of a single node must not be exceeded by regulatory restrictions that limit the emissions, multiple nodes cannot share their available or allowed transmit power. As a consequence, each node must fulfill its own power constraint individually, which makes the optimization of the transmission strategy mathematically more difficult. As in the original BC, DPC is the capacity achieving transmission scheme. As discussed before, non-linear schemes are of high complexity and have very stringent requirements on CSI accuracy which are hard to meet in practical systems, especially when multiple BSs cooperate with each other.

In order to develop practical transmission schemes, we thus restrict ourselves in this work to linear precoding. With this, a tradeoff between near-capacity performance and reasonable complexity can be achieved. With linear precoding, the transmit signals are restricted to form linear combinations of the individual symbol vectors for the different users, i.e. the signal transmitted by a BS b with N_B antennas is of the form

$$\mathbf{x}_b = \sum_{k=1}^{K} \mathbf{Q}_{k,b} \cdot \mathbf{s}_k, \tag{2.16}$$

where $\mathbf{s}_k \in \mathbb{C}^{N_\mathrm{s}}$ is the signal vector with $N_\mathrm{s} \leq N_\mathrm{B}$ data streams intended for user k and $\mathbf{Q}_{k,b} \in \mathbb{C}^{N_\mathrm{B} \times N_\mathrm{s}}$ the corresponding precoding or beamforming matrix from BS b. When the network consists of B BSs, the receive signal of user k can be given as

$$\mathbf{y}_k = \sum_{b=1}^{B} \mathbf{H}_{k,b}\mathbf{Q}_{k,b}\mathbf{s}_k + \sum_{b=1}^{B} \sum_{\substack{j=1 \\ j \neq k}}^{K} \mathbf{H}_{k,b}\mathbf{Q}_{j,b}\mathbf{s}_j + \mathbf{n}_k, \tag{2.17}$$

with $\mathbf{H}_{k,b}$ being the frequency flat block fading channel between BS b and MS k and $\mathbf{n}_k$ the noise induced in MS k. The resulting achievable rate of MS k follows as

$$R_k = \log_2 \det \left(\mathbf{I}_{N_\mathrm{r}} + \left(\mathbf{K}_k^{(\mathrm{i})} + \mathbf{K}_k^{(\mathrm{n})} \right)^{-1} \cdot \mathbf{K}_k^{(\mathrm{s})} \right), \tag{2.18}$$

with the covariance matrices of the desired signal, interference, and noise given by

$$
\begin{aligned}
\mathbf{K}_k^{(\mathrm{s})} &= \mathsf{E}_\mathrm{s} \left[\sum_{b=1}^{B} \sum_{b'=1}^{B} \mathbf{H}_{k,b}\mathbf{Q}_{k,b}\mathbf{s}_k\mathbf{s}_k^\mathsf{H}\mathbf{Q}_{k,b'}^\mathsf{H}\mathbf{H}_{k,b'}^\mathsf{H} \right] \\
&= \sum_{b=1}^{B} \sum_{b'=1}^{B} \mathbf{H}_{k,b}\mathbf{Q}_{k,b}\mathsf{E}_\mathrm{s} \left[\mathbf{s}_k\mathbf{s}_k^\mathsf{H} \right] \mathbf{Q}_{k,b'}^\mathsf{H}\mathbf{H}_{k,b'}^\mathsf{H} \\
&= \sum_{b=1}^{B} \sum_{b'=1}^{B} \mathbf{H}_{k,b}\mathbf{Q}_{k,b}\mathbf{Q}_{k,b'}^\mathsf{H}\mathbf{H}_{k,b'}^\mathsf{H}, \tag{2.19}
\end{aligned}
$$

$$\mathbf{K}_k^{(i)} = \mathsf{E}_s \left[\sum_{\substack{j=1 \\ j \neq k}}^{K} \sum_{b=1}^{B} \sum_{\substack{j'=1 \\ j' \neq k}}^{K} \sum_{b'=1}^{B} \mathbf{H}_{k,b} \mathbf{Q}_{j,b} \mathbf{s}_j \mathbf{s}_{j'}^{\mathsf{H}} \mathbf{Q}_{j',b'}^{\mathsf{H}} \mathbf{H}_{k,b'}^{\mathsf{H}} \right]$$

$$= \sum_{\substack{j=1 \\ j \neq k}}^{K} \sum_{b=1}^{B} \sum_{b'=1}^{B} \mathbf{H}_{k,b} \mathbf{Q}_{j,b} \mathbf{Q}_{j,b'}^{\mathsf{H}} \mathbf{H}_{k,b'}^{\mathsf{H}}, \tag{2.20}$$

and

$$\mathbf{K}_k^{(n)} = \mathsf{E}_n \left[\mathbf{n}_k \mathbf{n}_k^{\mathsf{H}} \right]$$

$$= \sigma_n^2 \cdot \mathbf{I}_{N_M}. \tag{2.21}$$

By applying per BS power constraints of the form

$$\mathrm{Tr} \left\{ \mathsf{E}_s \left[\sum_{k=1}^{K} \sum_{j=1}^{K} \mathbf{Q}_{k,b} \mathbf{s}_k \mathbf{s}_j^{\mathsf{H}} \mathbf{Q}_{j,b}^{\mathsf{H}} \right] \right\} = \mathrm{Tr} \left\{ \sum_{k=1}^{K} \mathbf{Q}_{k,b} \mathbf{Q}_{k,b}^{\mathsf{H}} \right\}$$

$$\leq P_{\mathrm{B}}, \qquad \forall b \in \{1, \ldots, B\}, \tag{2.22}$$

with P_{B} the maximal allowed transmit power of each BS, the performance of the entire network can thus in principle be optimized. For the example of the sum rate, this leads to the following optimization problem:

$$\max_{\{\mathbf{Q}_{j,b}\}_{j,b}} \sum_{k=1}^{K} R_k \tag{2.23}$$

$$\text{s.t.} \quad \mathrm{Tr} \left\{ \sum_{j=1}^{K} \mathbf{Q}_{j,b} \mathbf{Q}_{j,b}^{\mathsf{H}} \right\} \leq P_{\mathrm{B}}, \qquad \forall b \in \{1, \ldots, B\}.$$

Solving this problem is however very difficult, as it is highly non convex as all user rates inter-depend on each other in the different covariance matrices of the desired signals and interference. Many local optima can be found with narrow regions of attraction [127]. Moreover, many of these local optima, if one is found by an optimization tool, do not lead to the desired performance as not all locally optimal points exploit all degrees of freedom; the performance can thus be rather poor if not a "good" optimum is reached. Hence, attempting to solve (2.23) directly is mostly infeasible in practice. Research has therefore focused on sub-optimal but simpler approaches in which the performance criterion (e.g. sum rate) is usually not directly optimized, but a simpler optimization problem is formulated that can be solved more efficiently [38]. In this thesis, we thus focus on schemes that are of comparably low complexity. Thereby, we put a special focus on block ZF and SLNR precoding that still yield spatial multiplexing, array, and diversity gains, which can enhance the performance of cellular networks.

2.4   Channel Model

In order to evaluate the performance of different transmission schemes, the underlying channel model is of particular importance. Especially in cellular networks, propagation phenomena as pathloss and shadowing severely affect the performance and can even be exploited by the network topology, e.g. by applying specific reuse patterns to minimize interference between locally separated users. A channel model that accurately models these aspects is thus vital in order to study transmission techniques and network architectures of cellular networks. Several different channel models have thus been developed that all attempt to reflect the most important physical propagation properties in a realistic way. Prominent channel models that are widely used in literature include the 3GPP's spatial channel model (SCM) and its extended version (SCME) [4], the channel model IEEE 802.16j developed for WiMAX [59], and the models developed by the WINNER phase II project (WINNER II) [76]. These models have initially been developed for studies on 4G networks and are supported by extensive measurement campaigns. According to [149] and others, particularly the WINNER II model is thereby widely accepted to be well suited for the evaluation of cooperative transmission with multiuser MIMO schemes for LTE-Advanced systems and thus for cellular networks of the current and next generation. More recent projects such as the METIS project [99] that explore and study technologies for 5G networks also adopted the WINNER II channel model. For the evaluations in this thesis, we thus apply this model.

The WINNER II channel model is a generic, geometry based stochastic channel model that generates time varying and frequency selective channel impulse responses in the equivalent baseband representation and supports bandwidths up to 100 MHz and carrier frequencies between 2 and 6 GHz [76]. The model can describe an arbitrary number of propagation environment realizations for multiple radio links with single or multi-antenna nodes. The channel impulse response realizations are generated with geometrical principles by taking into account contributions of different rays (plane waves) with specific parameters like delay, attenuation, shadowing, angle of arrival, angle of departure, and others. As the system models used in this thesis are limited to a single frequency flat subcarrier, we apply the WINNER II model with single tap channels only. As a consequence, the channel model reduces to contain Rayleigh or Rice fading with specific parameters such as pathloss and shadowing that depend on the chosen propagation environment.

Among the different propagation scenarios that are provided in the WINNER II channel model, the following are of particular importance for the networks that we consider in this thesis:

- urban micro-cells,

- residential macro-cells, and

- rural macro-cells.

The radio signal is thereby affected by different pathloss and shadow fading (shadowing) parameters as well as different scattering behaviors. The pathloss behavior for these scenarios is shown in Fig. 2.3. It can be seen that the urban case with non-line-of-sight (NLOS) condition and the rural environment with line-of-sight (LOS) condition form the most extreme cases. In the following, we are thus mainly interested in these two scenarios, with a special focus on the former, as dense urban environments pose the most demanding challenges for high data rate service in cellular networks.

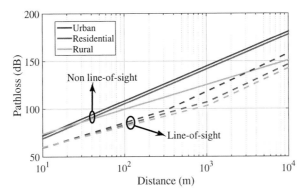

Figure 2.3.: Comparison of the pathloss in different environments [76].

Urban Micro-Cells

The environment of urban micro-cells corresponds to scenario C2 in the WINNER II channel model [76]. In such a micro-cell, mobiles are typically located outdoors at 1.5 m above street level, while BSs are fixed mostly on rooftop level. The environment has a high density of buildings. These can have irregular locations or can be placed in blocks between a rectangular grid of streets.

In the NLOS propagation condition, the channels are modeled as Rayleigh fading, i.e. the elements h_{ij} of a channel matrix $\mathbf{H}$ are i.i.d. $\mathcal{CN}(0, \sigma_h^2)$, where the variance is determined by the pathloss and shadowing. The pathloss is given by [76]

$$L = \left(44.9 - 6.55 \log_{10}\left(\frac{h_{\mathrm{BS}}}{\mathrm{m}}\right)\right) \log_{10}\left(\frac{d}{\mathrm{m}}\right) + 34.46 + 5.83 \log_{10}\left(\frac{h_{\mathrm{BS}}}{\mathrm{m}}\right) + 23 \log_{10}\left(\frac{f_c}{5\,\mathrm{GHz}}\right) \,[\mathrm{dB}],$$
(2.24)

with d being the distance between transmitter and receiver, h_{BS} the height of the BS antennas (in our simulations considered as 25 m) and f_c the carrier frequency which we usually assume to be 2.6 GHz. The shadowing has a log-normal distribution with mean $\mu = 0\,\mathrm{dB}$ and a standard deviation of $\sigma = 8\,\mathrm{dB}$. Antenna correlation is not applied.

In certain scenarios we also consider LOS links in the urban environment, particularly for links between BSs and dedicated relay nodes as discussed in Chapter 4. In this case, the fading is given by Rice fading with a log-normal Ricean K-factor with mean $\mu = 7\,\mathrm{dB}$ and standard deviation $\sigma = 3\,\mathrm{dB}$. The pathloss is given by [76]

$$L = \begin{cases} 26 \log_{10}\left(\dfrac{d}{\mathrm{m}}\right) + 39 + 20 \log_{10}\left(\dfrac{f_c}{5\,\mathrm{GHz}}\right), & 10\,\mathrm{m} < d < d_{\mathrm{BP}} \\[2ex] 40 \log_{10}\left(\dfrac{d}{\mathrm{m}}\right) + 13.47 - 14 \log_{10}\left(\dfrac{h'_{\mathrm{BS}}}{\mathrm{m}}\right) - \\[2ex] 14 \log_{10}\left(\dfrac{h'_{\mathrm{MS}}}{\mathrm{m}}\right) + 6 \log_{10}\left(\dfrac{f_c}{5\,\mathrm{GHz}}\right), & d_{\mathrm{BP}} < d < 5\,\mathrm{km}, \end{cases}$$
(2.25)

which depends on the break point distance $d_{\mathrm{BP}} = 4 \cdot h'_{\mathrm{BS}} \cdot h'_{\mathrm{MS}} \cdot f_c/c$, with the effective antenna heights $h'_{\mathrm{BS}} = h_{\mathrm{BS}} - 1\,\mathrm{m}$ and $h'_{\mathrm{MS}} = h_{\mathrm{MS}} - 1\,\mathrm{m}$ and c the speed of light. The shadowing is modeled as a log-normal random variable with zero mean and standard deviation

$$\sigma = \begin{cases} 4\,\mathrm{dB}, & 10\,\mathrm{m} < d < d_{\mathrm{BP}} \\ 6\,\mathrm{dB}, & d_{\mathrm{BP}} < d < 5\,\mathrm{km}. \end{cases}$$
(2.26)

Rural Macro-Cells

The propagation scenario of rural macro-cells corresponds to WINNER II scenario D1 [76]. It represents radio propagation in large areas with low building density. The environment is assumed to be mostly flat, consisting of sparsely located houses along roads that lead through fields and some forests or small villages. As a consequence, LOS conditions are expected to appear more frequently than in urban areas. To this end, we model the rural channels by Rice fading. The log-normal Ricean K-factor has a mean $\mu = 7\,\mathrm{dB}$ and a standard deviation $\sigma = 6\,\mathrm{dB}$. The pathloss is in this case [76]

$$
L = \begin{cases} 21.5\log_{10}\left(\frac{d}{\mathrm{m}}\right) + 44.2 + 20\log_{10}\left(\frac{f_c}{5\,\mathrm{GHz}}\right), & 10\,\mathrm{m} < d < d_{\mathrm{BP}} \\ 40\log_{10}\left(\frac{d}{\mathrm{m}}\right) + 10.5 - 18.5\log_{10}\left(\frac{h_{\mathrm{BS}}\cdot h_{\mathrm{MS}}}{\mathrm{m}^2}\right) + 1.5\log_{10}\left(\frac{f_c}{5\,\mathrm{GHz}}\right), & d_{\mathrm{BP}} < d < 10\,\mathrm{km}, \end{cases}
$$
$$(2.27)$$

with the break point distance $d_{\mathrm{BP}} = 4 \cdot h_{\mathrm{BS}} \cdot h_{\mathrm{MS}} \cdot f_c/c$. The log-normal shadowing has the same standard deviation as in (2.26).

Further Comments

In all simulations, we assume that the shadowing parameters as well as the K-factors are the same for all antennas of the same node but independently drawn for different nodes. For the specifications of the transmit powers and noise variances, we assume for all systems an underlying bandwidth of 100 MHz.

As the computer simulations are based on the described channel models, the obtained results depend on the chosen parameters. With other channel models or different parameters, different results might arise. Particularly when systems are considered that use radio frequencies which are not supported by the WINNER II model (e.g. 60 GHz transmissions), the conclusions drawn from this work might not be valid anymore. Nevertheless, the applied channel models can be considered to realistically reflect the most important propagation phenomena for cellular networks that operate in frequencies between 2 and 6 GHz.

2.5 Figures of Merit

Measuring the performance of cellular networks is a non-trivial task. The optimization of such networks can be seen as a multidimensional problem that consists of many different objectives that might even contradict each other. So are not only the achievable rates for each user and the total throughput of BSs of importance, but also coverage range, the network wide spectral efficiency, as well as the QoS that can be guaranteed. Moreover, also costs that come with them need to be considered. These include the required infrastructure, additional backhaul traffic, introduced delays, transmit power, complexity of hardware and algorithms, as well as economic costs.

In this thesis, we consider PHY layer cooperation schemes and do not consider functionalities of higher layers. To this end, the figures of merit that are used in this work are based on achievable rates. Achievable rate, as an information theoretic concept, provides an upper bound on the data rate in the limit of long code block lengths. Overhead due to channel estimation, node assignment or traffic management in higher layers, or cyclic prefixes when OFDM based systems are considered are not taken into account. Nevertheless, achievable rates provide a measure for the potential of PHY layer transmission. With the development of modern turbo codes and low-density parity check codes, research has found ways to come close to these theoretical rates [143]. Therefore, we use the achievable rates as the basic performance measure that form the basis of the figures of merit with which we will evaluate and compare the different schemes in this work.

In order to optimize the performance of the networks considered in the following chapters, we apply different objective functions that reflect the different criteria a cellular network should fulfill. Thereby we focus mainly on throughput, QoS, and power consumption. In order to allow feasible formulations of mathematical optimization problems, we focus on

- Sum rate maximization, which reflects the total throughput that can be achieved in a network or in a part of it. This objective usually leads to solutions that assign the most resources (such as power) to the strongest users while weaker ones are penalized.

- Maximizing the minimum rate (max-min optimization), which maximizes the data rate of the weakest user in a specific set. An optimization with respect to this objective leads to fair rate distributions among the different users. The performance of strong ones is however reduced such that the available resources can be used to improve the weaker ones. This objective usually leads to significantly smaller throughput than sum rate maximization.

- Outage minimization in which the probability that a user can achieve a certain target rate is minimized. In contrast to max-min optimization where very poor data rates for all users are possible when one of them is particularly weak, outage minimization does not penalize strong users. Resources of strong users are only reduced to an extent that still fulfills the targets. This approach can be seen as a certain tradeoff between sum rate maximization and max-min optimization.

- Transmit power minimization that leads to energy efficient solutions. Thereby, a target data rate is usually applied to each user and the lowest possible transmit power that is required to achieve this target rates is to be found. Optimization problems with this objective cannot always be solved. Due to the interference, certain target rates might be infeasible for any power allocation.

Depending on the specific networks and transmission protocols of the following chapters, we will attempt to optimize the network performance with respect to the objectives discussed above. In order to measure different aspects of the performance of cellular networks that can be achieved with these objectives, we use different figures of merit. Empirical cumulative distribution functions (CDFs) provide thereby insights into the statistics of instantaneous achievable user rates. In slow fading environments, the CDFs show which data rate the users can achieve with which probability. The CDFs however give no information about the locations where certain rates are achievable. Especially in cellular networks, the performance heavily depends on the position where the users are located relative to the infrastructure nodes and other users. Close to BSs, higher rates can be expected than in areas further away from them.

In order to reflect also the spatial distribution of the performance, we often show area plots with average rates, outage probabilities p_{out}, or outage rates R_{out} (a data rate that is achieved with a certain probability, e.g. with 95%). Such two dimensional representations give further insights into the achievable rates and where they can be obtained. Different objective functions will thereby lead to drastically different rate distributions. For sum rate maximization for example, it can be expected that high data rates are concentrated closely around BSs or other infrastructure nodes, while the max-min optimization leads to a more homogeneous distribution. When outage probabilities or outage rates are plotted in this fashion, important conclusions regarding coverage can be made as these measures reflect in which locations data rates can be guaranteed with high probability. Note however that these measures (p_{out} and R_{out}) might improve with additional diversity that would be present with wideband channels. As we limit the discussions and simulations to a single subcarrier, the diversity that comes with frequency selective channels is not taken into account. Other measures such as average rates which are averaged over many channel realizations are equivalent to the wideband case with a high frequency diversity when no power loading across different subcarriers is applied. To this end, average rates are also presented in CDFs as well as in area plots.

In order to reflect coverage and data rates that can be achieved in the entire network with high probability and to allow a comparison between different schemes with single numbers, we further define two key performance indicators: On one hand, we compare the *coverage* of the network, which we define as the percentage of the area in which the target rate of 1 bps/Hz can be achieved with a probability of 95 %:

$$\text{Coverage} = \frac{\underset{\text{Sector}}{\iint} \mathrm{I}\left\{R_{5\%}(x, y) \geq 1\right\} dx dy}{\underset{\text{Sector}}{\iint} 1 \cdot dx dy}, \qquad (2.28)$$

where $R_{5\%}(x, y)$ is the 5% outage rate in grid point (x, y) and $\mathrm{I}\{\cdot\}$ the indicator function which is one when the argument is true and zero otherwise. This reflects in which parts of the network a user can expect to have acceptable performance, but not how large the data rates are. Note that this measure heavily depends on the chosen target rate; a very low target rate can be achieved in 100 % of the area with all schemes, while a too large target might never be achieved in any of them. Nevertheless, 1 bps/Hz seems to us to be a meaningful target for a wide coverage with 2 spatial data streams that we usually assume for each MSs.

On the other hand, we use the *average 5% outage rate*, averaged over the area, i.e.

$$\text{Average } R_{5\%} = \frac{\underset{\text{Sector}}{\iint} R_{5\%}(x, y) dx dy}{\underset{\text{Sector}}{\iint} 1 \cdot dx dy}. \qquad (2.29)$$

In this measure, the relation to the location or the area is lost, but it reflects better how large the provided data rates are. A user can thus expect to be served by this rate with a probability of 95% when he is located in an arbitrary and unknown position.

The chosen figures of merit and their representations cannot provide a thorough characterization of the exact performance of cellular networks, particularly delays, overhead (e.g. due to channel estimation), or computational complexity or costs are not taken into account. Nevertheless, they offer a wide overview over the different performance gains that can be achieved with the transmission schemes presented in the following.

3

Locally Restricted BS Cooperation

From an information theoretic perspective, the capacity of the DL in cellular networks can be achieved by cooperative transmission schemes such as DPC [39, 87, 143]. When all BSs can exchange all their data, have full and perfect CSI of the entire network, and share a joint sum transmit power constraint, the network resembles a broadcast channel. When BSs have access to all this information and if they have sufficient computational capabilities, global cooperation combined with sufficiently large antenna arrays can provide all the gains that theoretic academic research suggests. However, when practical networks are considered, each BS has its own local maximal transmit power and the required CSI needs to be estimated and distributed to all involved nodes. While the former leads to a reduction in capacity and makes the mathematical optimization problem more difficult, the latter introduces additional overhead that might diminish or even destroy the potential performance gains with which practical cooperation has its limits [90, 109].

In order to achieve optimal or close to optimal results, unlimited cooperation of all BSs in the whole network is required. This means, that all cells/sectors of a cellular network have to be connected together, such that a global multi-cell MIMO communication can be realized. Only then, all interference terms can be turned into useful signal contributions and the full degrees of freedom of the network can be exploited. Even though such a global cooperation would be optimal, it is difficult to realize in practice if large networks are considered. Since all sectors of cooperating BSs are coupled with each other, a large number of cooperating BSs results in very stringent requirements on delay and very high computational complexity that would be hard to meet in practice. Hence, it is desirable if cooperation is limited to a subset of BSs. With this, the computational complexity for joint precoding and the overhead for channel estimation and dissemination can be kept limited.

Most academic research in this field has however focused either on strongly simplified networks such as Wyner type models (e.g. [39]) with which theoretic bounds can be derived, or only on individual aspects of cellular networks. Such aspects include the impact of sectorization on certain schemes [116], robustness of certain schemes against inaccuracies (e.g. imperfect CSI) [12], or optimizing the sum rate with linear precoding [156]. References [25] and [6] apply a block ZF approach and study its performance in a more realistic cellular setup where also the interference from other cooperation clusters is considered. Power allocation is either done uniformly or by maximizing the sum rate in a suboptimal way. Maximizing the sum rate is however not an appropriate criterion if good coverage and high QoS should be provided. If sum rate is maximized, users who are located close to a BS usually achieve much higher rates than users that are further away from them. In order to provide a high QoS to all users, max-min optimization leads to more fairness among the different user locations. This criterion is thus better suited to address and improve the performance of individual users. This comes with the consequence that high peak rates are reduced for the sake of the poor users. The authors in [69] and [68] compare approaches based on ZF combined with such max-min rate optimization and ZF-DPC. The goal in both cases is to increase the QoS of a cellular network by BS cooperation. Inter-cluster interference is in these works however not taken into account.

As available scientific work differs in its models and assumptions, it is difficult to compare several techniques with each other. Moreover, the performance heavily depends on various parameters, such as traffic load or effective SINR, which are hard to predict and model [11]. To this end, it is important to have a unified framework, which allows comparing different schemes in a fair and consistent fashion, and capture practical considerations in a realistic way. These include, among others, network topology, backhaul connection, channel models that attempt to accurately model fundamental properties of cellular networks such as distance dependent pathloss and shadowing, and imperfections. An attempt to provide a simulation framework that is close to reality can be found in [128], which describes an complete LTE link level simulator that includes very realistic models of the PHY and higher layers. In [129], different precoding designs are evaluated and compared in this environment, which shows the potential gain of cooperative transmission schemes such as ZF or SLNR over non-cooperative approaches. The simulations were conducted only for a 2×2 system (either a 2×2 point-to-point MIMO link, or two interfering 2×1 links), as the simulation environment is very complex and studying larger networks is hardly possible.

In this chapter, we develop a framework that allows to study the performance of larger networks that contain multiple cells. To this end, we focus on the PHY layer and apply a signal model for linear narrow band transmission. The framework comprises a general signal model for an arbitrary number of nodes and antennas, channel models that reflect realistic pathloss and shadowing, noise, and signal parameters recommended by the 3GPP as well as realistic network geometries. Higher layer considerations and the evaluation of broadband systems is omitted, such that an efficient evaluation of different schemes in different setups and network configurations is possible. This allows to study the potential of locally restricted, cluster based, cooperation schemes in realistic scenarios with practical considerations. Thereby we focus on suboptimal transmission schemes that are of high practical relevance. We are interested in the potential of CoMP in realistic cellular networks and what influence the organization and architecture of the network has on the performance.

In order to reduce the outage probability and maximize the coverage range, we propose a cooperation scheme that maximizes the minimal rate achievable in a cooperation set, based on block ZF. Since this cooperation scheme requires an adaptation of the transmission to each change in the channel realization, it is not applicable to users that move with high velocity. To this end, we also apply a second approach that exploits macro-diversity. This approach does not require CSIT at the BSs and can thus also be used for users that are affected by fast fading channels. With this two approaches, we reflect two cases of the different levels of cooperation that are foreseen for 4G networks which range from joint scheduling with BS selection up to joint multi-point transmission as discussed in Chapter 2.

3.1 Network Model

The considered network consists of multiple BSs that serve multiple mobile users. The entire area is divided into different *cells* that are assumed to be of the form of regular hexagons. Each cell is further divided into multiple *sectors*. If not stated otherwise, the number of sectors per cell is three, each with an angle of aperture of 120°. The BSs are located in the center of the cells and consist of a separate antenna array for each sector that belongs to the cell. For notational convenience, we refer to the antenna arrays corresponding to different sectors of the same cell as separate BSs, even though they are located on the same spot. The mobile users located in a specific sector are thus served by the BS that corresponds to this sector. Further, we consider only one

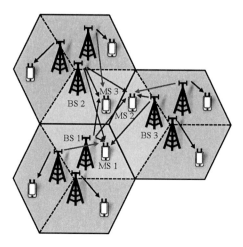

Figure 3.1.: Example of a cellular network with three cells that are divided into three sectors. Each sector comprises one BS and one MS. The cooperation set $\mathcal{M}$ comprises BSs 1, 2, and 3. The other BSs interfere with the communication within this cooperation area.

resource block (time and frequency slot) and assume that each sector of the network serves exactly one MS at a given time instant. Multiple MSs assigned to the same sector could share the resources by e.g. a TDMA or FDMA scheme. A sketch of the considered network can be seen in Fig. 3.1.

In order to limit the complexity and overhead for BS cooperation, we group the BSs into different clusters or cooperation sets $\mathcal{M}_1, \ldots, \mathcal{M}_S$ that each jointly serves a certain area, where S is the number of cooperation sets. The cooperation sets are assumed to be fixed during a transmission period and a specific resource (time/frequency) block. Different clusters can consist of a different number of nodes; some may contain only a single BS while others can contain a plurality of BSs and MSs. In the following, we refer to a specific cooperation set $\mathcal{M}_c$ that contains $|\mathcal{M}_c| = M$ BSs (or equivalently sectors) that are able to cooperate with each other. Note that the special cases $M = 1$ and $M = M_{\text{tot}}$, where M_{tot} is the total number of BSs in the entire network, mean no cooperation at all or global cooperation across all BSs, respectively. The set $\overline{\mathcal{M}_c}$ is then the set of all other BSs that belong to other cooperation sets and are not able to cooperate with the BSs in $\mathcal{M}_c$ but cause interference to the communication within the area of focus. The sets of MSs in the corresponding cooperation areas are denoted by $\mathcal{K}_c$. The cooperation area of focus consists of $|\mathcal{K}_c| = K$ MSs. Due to the aforementioned assumption of a single active MS per resource block in each sector, $K = M$. The BSs and MSs, respectively, are equipped with N_B and N_M antennas. The antennas of the BSs are assumed to be directional with antenna patters according to the 3GPP specification in [1], while those of the MSs are omnidirectional.

Regarding the signaling and signal processing, we make the following assumptions:

- Perfect CSIR with which the data rates that are calculated for the different schemes are achievable.

- If the joint transmission scheme with block ZF is applied, the BSs within a cluster have perfect CSIT of all users in this cluster.

- With joint transmission, the BSs within the same cluster can fully share CSI and user data. The BSs in different clusters do not share anything.

- BSs within the same cluster are perfectly synchronized in time and phase, and different propagation delays from the BSs are compensated.

- For the macro diversity schemes, not CSIT at all is assumed, but perfect rate knowledge of all MSs in the cluster is available to assign the MSs to the BSs.

With this assumptions, we can formulate the IO relation for the narrowband case as follows.

3.1.1 Narrowband Input-Output Relation

In the downlink, the b-th BS belonging to $\mathcal{M}_c$, for $b \in \{1, \ldots, M\}$, transmits a sum of different signals, one intended for each of the K MSs in the cooperation area:

$$\mathbf{x}_b = \sum_{j=1}^{K} \mathbf{x}_{j,b}, \tag{3.1}$$

where $\mathbf{x}_{j,b} \in \mathbb{C}^{N_B}$ is the signal from BS b intended for MS j. We assume linear precoding and factorize these signals to

$$\mathbf{x}_{j,b} = \mathbf{Q}_{j,b} \cdot \mathbf{s}_j, \tag{3.2}$$

with $\mathbf{s} \in \mathbb{C}^{d_s}$ being the symbol vector intended for MS j of d_s data streams, and $\mathbf{Q}_{j,b}$ the corresponding precoding matrix. The elements of $\mathbf{s}_j$ are assumed to be i.i.d. $\mathcal{CN}(0,1)$. The variance is normalized to unity such that the power allocation for each symbol takes place in the precoding $\mathbf{Q}_{j,b}$, additional to the beamforming.

The receive signal at MS k, for $k \in \{1, \ldots, K\}$, follows as

$$\mathbf{y}_k = \underbrace{\sum_{b \in \mathcal{M}_c} \mathbf{H}_{k,b} \mathbf{Q}_{k,b} \mathbf{s}_k}_{\text{desired signal}} + \underbrace{\sum_{b \in \mathcal{M}_c} \sum_{\substack{j=1 \\ j \neq k}}^{K} \mathbf{H}_{k,b} \mathbf{Q}_{j,b} \mathbf{s}_j}_{\text{intra-cluster interference}} + \underbrace{\sum_{i \notin \mathcal{M}_c} \mathbf{H}_{k,i} \mathbf{x}_i}_{\text{out-of-cluster interference}} + \underbrace{\mathbf{w}_k}_{\text{noise}}. \tag{3.3}$$

Therein, the first term captures all desired signals (transmitted by the cooperating BSs in $\mathcal{M}_c$). The second term contains the signals transmitted by nodes within $\mathcal{M}_c$ but intended for other MSs in $\mathcal{K}_c$ and the third sum describes the interference caused by the nodes outside the cooperation area of focus. The noise with elements i.i.d. $\mathcal{CN}(0, \sigma_w^2)$ induced in MS k is denoted by $\mathbf{w}_k$. The matrix $\mathbf{H}_{k,b} \in \mathbb{C}^{N_M \times N_B}$ describes the channel from BS b to MS k, where we assume that the frequency flat channel remains constant for one transmission period, i.e. we consider a slow fading channel of one subcarrier of an OFDM-based system. The interference in (3.3) contains two contributions that are distinguished as interference from all BSs belonging to $\mathcal{M}_c$, referred to as *intra-cluster interference* (ICI) and interference caused by the transmission in other sectors, referred to as *out-of-cluster interference* (OCI). The purpose of cooperation is to control or even exploit the ICI. The OCI, on the other hand, remains as the BSs in $\overline{\mathcal{M}}_c$ cannot cooperate with the BSs in $\mathcal{M}_c$ and the OCI can therefore not be cancelled.

The main performance measure that will be used in this work is the achievable rate which is for user k given by

$$R_k = \log_2 \det \left(\mathbf{I}_{N_M} + \left(\mathbf{K}_k^{(i)} + \mathbf{K}_k^{(n)} \right)^{-1} \cdot \mathbf{K}_k^{(s)} \right) \text{ [bps/Hz]}, \tag{3.4}$$

where $\mathbf{K}_k^{(s)}$, $\mathbf{K}_k^{(i)}$, and $\mathbf{K}_k^{(n)}$ are the covariance matrices of the desired signal, interference, and effective noise (including OCI) of MS k. These covariance matrices are given by

$$\mathbf{K}_k^{(s)} = \mathsf{E} \left[\sum_{b \in \mathcal{M}_c} \sum_{b' \in \mathcal{M}_c} \mathbf{H}_{k,b} \mathbf{Q}_{k,b} \mathbf{s}_k \mathbf{s}_k^{\mathsf{H}} \mathbf{Q}_{k,b'}^{\mathsf{H}} \mathbf{H}_{k,b'}^{\mathsf{H}} \right]$$
$$= \sum_{b \in \mathcal{M}_c} \sum_{b' \in \mathcal{M}_c} \mathbf{H}_{k,b} \mathbf{Q}_{k,b} \mathbf{Q}_{k,b'}^{\mathsf{H}} \mathbf{H}_{k,b'}^{\mathsf{H}} \tag{3.5}$$

and similarly

$$\mathbf{K}_k^{(i)} = \sum_{b \in \mathcal{M}_c} \sum_{b' \in \mathcal{M}_c} \sum_{\substack{j=1 \\ j \neq k}}^{K} \mathbf{H}_{k,b} \mathbf{Q}_{j,b} \mathbf{Q}_{j,b'}^{\mathsf{H}} \mathbf{H}_{k,b'}^{\mathsf{H}} \tag{3.6}$$

$$\mathbf{K}_k^{(n)} = \sum_{i \notin \mathcal{M}_c} \mathbf{H}_{k,i} \cdot \mathsf{E}\left[\mathbf{x}_i \mathbf{x}_i^{\mathsf{H}}\right] \cdot \mathbf{H}_{k,i}^{\mathsf{H}} + \sigma_w^2 \cdot \mathbf{I}_{N_{\mathrm{M}}}. \tag{3.7}$$

The achievable rate R_k corresponds to the spectral efficiency measured in bps/Hz.

In cooperative networks, the BSs that belong to the same cooperation set $\mathcal{M}_c$ can exchange their transmit symbols $\mathbf{s}_k$, as well as the channel coefficients. Based on this knowledge, the BSs can optimize their transmit beamforming matrices $\mathbf{Q}_{j,k}$ to maximize a certain performance criterion. To this end, different criteria can be considered such as the minimal rate among the users within a cooperation area, the sum rate, but also the required transmit power can be minimized subject to a certain target rate. Corresponding optimization schemes are derived in the following.

3.2 Block Zero-Forcing

While many beamforming strategies exist as already mentioned in Chapter 2, we focus here on a scheme that is based in block ZF. This approach allows to formulate convex optimization problems that can efficiently be solved by standard optimization tools. The conceptual limitations imposed by block ZF are discussed in Section 3.6, based on simulation results and available literature.

In the block ZF approach, the precoding matrices $\mathbf{Q}_{j,b}$ are chosen such that interference is completely eliminated. In order to cancel all the interference in the entire network, however, all BSs would have to cooperate with each other. To this end, we group the BSs into clusters and distinguish between ICI and OCI. For a practical ZF based transmission scheme, we focus on the ICI only to being cancelled.

To this end, we rewrite the receive signal of MS k in (3.3) by combining the OCI with the actual noise of MS k to the equivalent noise $\mathbf{n}_k = \sum_{i \notin \mathcal{M}_c} \mathbf{H}_{k,i} \mathbf{x}_i + \mathbf{w}_k$ and summarizing the channels from the BSs in $\mathcal{M}_c$ to MS k in the concatenated channel matrix $\overline{\mathbf{H}}_k = [\mathbf{H}_{k,1}, \ldots, \mathbf{H}_{k,M}]$. The more compact IO relation reads then

$$\mathbf{y}_k = \overline{\mathbf{H}}_k \cdot \overline{\mathbf{x}}_k + \sum_{\substack{j=1 \\ j \neq k}}^{K} \overline{\mathbf{H}}_k \cdot \overline{\mathbf{x}}_j + \mathbf{n}_k, \tag{3.8}$$

where $\overline{\mathbf{x}}_k = \left[\mathbf{x}_{k,1}^{\mathsf{T}}, \ldots, \mathbf{x}_{k,M}^{\mathsf{T}}\right]^{\mathsf{T}}$ is the vector that contains all transmit vectors from all BSs in $\mathcal{M}_c$ intended for MS k.

With this notation, we can formulate the block ZF conditions as

$$\sum_{j \neq k} \overline{\mathbf{H}}_k \cdot \overline{\mathbf{x}}_j = \mathbf{O}, \quad \forall k \in \{1, \ldots, K\}$$

$$\overline{\mathbf{H}}_k \cdot \overline{\mathbf{x}}_k \neq \mathbf{O}, \quad \forall k \in \{1, \ldots, K\}. \tag{3.9}$$

Assuming channel coefficients that are drawn i.i.d. from a non-degenerate continuous distribution, these conditions are fulfilled if all transmit signals that are unintended for a MS k lie in the null space of the corresponding $\overline{\mathbf{H}}_k$ [133]. If this is the case, the ICI is completely eliminated and the desired signals are non-zero with probability 1, since the channel matrices $\overline{\mathbf{H}}_i$ and $\overline{\mathbf{H}}_j$ are almost surely linearly independent if $i \neq j$.

In order to zero-force the unintended signals and optimize the power loading for each stream, we decompose the precoding matrices to the product $\mathbf{Q}_{k,b} = \mathbf{Z}_{k,b} \cdot \mathbf{G}_{k,b}$, i.e. the transmit signal over all involved BSs is

$$\overline{\mathbf{x}}_k = \overline{\mathbf{Z}}_k \cdot \mathbf{G}_k \cdot \mathbf{s}_k, \tag{3.10}$$

where $\overline{\mathbf{Z}}_k = \left[\left\{ \mathbf{Z}_{k,b}^\mathsf{T} \right\}_{b \in \mathcal{M}_c} \right]^\mathsf{T}$ is the concatenated ZF matrix of all involved BSs and $\mathbf{G}_k = \mathbf{G}_{k,b}, \forall b \in \mathcal{M}_c$ the power loading matrix that forms linear combinations of all transmit symbols intended for MS k.

As described in [132, 133, 154], the ZF matrices can be found by the SVD of

$$\tilde{\mathbf{H}}_j = \left[\overline{\mathbf{H}}_1^\mathsf{T}, \ldots, \overline{\mathbf{H}}_{j-1}^\mathsf{T}, \overline{\mathbf{H}}_{j+1}^\mathsf{T}, \ldots, \overline{\mathbf{H}}_K^\mathsf{T} \right]^\mathsf{T}. \tag{3.11}$$

The SVD results in

$$\tilde{\mathbf{H}}_j = \mathbf{U}_j \cdot \mathbf{D}_j \cdot \left[\mathbf{V}_j^{(1)} \; \mathbf{V}_j^{(0)} \right]^\mathsf{H}, \tag{3.12}$$

where $\mathbf{V}_j^{(0)}$ contains the singular vectors that correspond to the singular values that are zero and thus form an orthonormal basis of the null space of $\tilde{\mathbf{H}}_j$, i.e.

$$\overline{\mathbf{Z}}_j = \mathbf{V}_j^{(0)} = \text{null} \left\{ \tilde{\mathbf{H}}_j \right\}. \tag{3.13}$$

The IO relation (3.8) can then be written as

$$\mathbf{y}_k = \overline{\mathbf{H}}_k \cdot \overline{\mathbf{Z}}_k \cdot \mathbf{G}_k \cdot \mathbf{s}_k + \mathbf{n}_k, \tag{3.14}$$

where the ICI is completely eliminated. Note that by the rank-nullity-theorem [43], the matrix $\tilde{\mathbf{H}}_k$ must have a rank smaller than the number of columns, i.e. $\mathrm{rank}\left\{\tilde{\mathbf{H}}_k\right\} < N_\mathrm{B}$. Otherwise, only the trivial solution (all transmit symbols are zero) fulfills conditions (3.9). This implies that the total number of BS antennas in a cooperation area must be larger than the total number of antennas at the MSs that are served, i.e. $M \cdot N_\mathrm{B} \geq K \cdot N_\mathrm{M}$. More MSs could be served by sharing resources as time or frequency.

Due to the block ZF, the multiuser problem decomposes into K independent point-to-point MIMO links. Without loss of generality, each MIMO link can be diagonalized by SVD precoding and receive combining (possibly including prior noise and interference whitening at the MSs). Both in the case of individual transmit power constraints imposed on the signals to each MS and in the case of a sum power constraint across all cooperating BSs jointly, we can maximize the DL sum rate by waterfilling [143]. Other objective functions such as the maximization of the minimum rate to any of the K MSs can be handled with similar ease [68].

If multiple BSs cooperate together, the situation appears more difficult. Each BS is required to fulfill its own individual sum power constraint of the form

$$
\begin{aligned}
\mathsf{E}\left[\mathrm{Tr}\left\{\sum_{j\in\mathcal{K}_c}\sum_{i\in\mathcal{K}_c}\mathbf{Z}_{j,b}\mathbf{G}_{j,b}\mathbf{s}_j\mathbf{s}_i^\mathsf{H}\mathbf{G}_{i,b}^\mathsf{H}\mathbf{Z}_{i,b}^\mathsf{H}\right\}\right] &= \mathrm{Tr}\left\{\sum_{j\in\mathcal{K}_c}\sum_{i\in\mathcal{K}_c}\mathbf{Z}_{j,b}\mathbf{G}_{j,b}\mathsf{E}\left[\mathbf{s}_j\mathbf{s}_i^\mathsf{H}\right]\mathbf{G}_{i,b}^\mathsf{H}\mathbf{Z}_{i,b}^\mathsf{H}\right\} \\
&= \mathrm{Tr}\left\{\sum_{j\in\mathcal{K}_c}\mathbf{Z}_{j,b}\mathbf{G}_{j,b}\mathbf{G}_{j,b}^\mathsf{H}\mathbf{Z}_{j,b}^\mathsf{H}\right\} \\
&\leq P_\mathrm{B}, \qquad \forall b \in \mathcal{M}_c, \forall c, \quad\quad (3.15)
\end{aligned}
$$

when the elements of the data symbol vectors are assumed to be i.i.d. $\sim \mathcal{CN}(0,1)$. With these constraints, the precoding matrices are coupled again with each other, which prevents finding simple-closed form or quasi closed-form solutions such as waterfilling. An iterative procedure is therefore required to allocate the transmit power to the different streams optimally [25]. Nevertheless, convex optimization problems can be formulated for the power allocation which can be solved in a relatively efficient way.

3.3 Precoder Optimization with Block Zero-Forcing

In this section, we consider the optimization of the precoding matrices with respect to different objective functions discussed in Chapter 2:

- maximizing the minimum rate (max-min),

- maximizing the sum rate, and

- power minimization.

Thereby, we set a special focus on max-min optimization to improve coverage and QoS, while the other criteria are included to show what other benefits cooperation can have. In all cases, we model the OCI as $\mathcal{CN}(\mathbf{O}, \sigma_n^2 \mathbf{I})$ for the optimization process. In this way, the optimization of the precoding in different cooperation clusters is decoupled and we obtain convex optimization problems that can efficiently be solved by standard optimization tools such as *Yalmip* [88] or *SDPT3* [141]. For the evaluation of the achievable rates, the true OCI is taken into account.

3.3.1   Max-Min Optimization with Block Zero-Forcing

For given ZF precoding matrices $\mathbf{Z}_{j,b}$ and the per BS power constraint (3.15), the optimization problem of maximizing the minimum rate can be stated as

$$\mathbf{G}_{j,b}^{\star} = \arg \max_{\{\mathbf{G}_{j,b}\}_{j,b}} \min\{R_1, \ldots, R_K\}$$

$$\text{s.t. } \operatorname{Tr}\left\{\sum_{j \in \mathcal{K}_c} \mathbf{Z}_{j,b} \mathbf{G}_{j,b} \mathbf{G}_{j,b}^{\mathsf{H}} \mathbf{Z}_{j,b}^{\mathsf{H}}\right\} \leq P_{\mathrm{B}}, \qquad \forall b, \tag{3.16}$$

where the individual rates are given by

$$R_k = \log_2 \det\left(\mathbf{I}_{N_{\mathrm{M}}} + \left(\mathbf{K}_k^{(\mathrm{n})}\right)^{-1} \cdot \mathbf{K}_k^{(\mathrm{s})}\right) \tag{3.17}$$

with

$$\mathbf{K}_k^{(\mathrm{s})} = \sum_{b \in \mathcal{M}_c} \sum_{b' \in \mathcal{M}_c} \mathbf{H}_{k,b} \mathbf{Z}_{k,b} \mathbf{G}_{k,b} \mathbf{G}_{k,b'}^{\mathsf{H}} \mathbf{Z}_{k,b'}^{\mathsf{H}} \mathbf{H}_{k,b'}^{\mathsf{H}}$$
$$\mathbf{K}_k^{(\mathrm{n})} = \sum_{i \notin \mathcal{M}_c} \mathbf{H}_{k,i} \cdot \mathsf{E}\left[\mathbf{x}_i \mathbf{x}_i^{\mathsf{H}}\right] \cdot \mathbf{H}_{k,i}^{\mathsf{H}} + \sigma_w^2 \cdot \mathbf{I}_{N_{\mathrm{M}}}. \tag{3.18}$$

Note that $\mathbf{K}_k^{(\mathrm{i})} = \mathbf{O}$ due to zero forcing. The solution of this optimization problem assigns the available transmit power to linear combinations of the different transmit symbols such that $\min\{R_1, \ldots, R_K\}$ is maximized. As, however, the optimization over different cooperation areas are coupled through the interference covariance matrices $\mathbf{K}_k^{(\mathrm{n})}$ that depend on the beamforming matrices of other cooperation areas, the optimization is non-convex und thus difficult to solve. Moreover, as each optimization in a certain cooperation area depends on the previously optimized precoding and power loading in other BS clusters, each cluster would have to adapt its precoding once the transmission in another area is updated. This leads to a large overhead in order to find appropriate precoding matrices in all cooperation clusters and prevents an efficient usage of the transmission scheme. To this end, we decouple the different optimizations of adjacent cooperation clusters by approximating the OCI by AWGN. That is, for the optimization problem, we assume the noise covariance matrix to be

$$\mathbf{K}_k^{(\mathrm{n})} \approx \tilde{\mathbf{K}}_k^{(\mathrm{n})} = \sigma_n^2 \cdot \mathbf{I}_{N_\mathrm{M}}, \tag{3.19}$$

with σ_n^2 being the variance of the assumed noise. This leads to the approximated achievable rate

$$\tilde{R}_k = \log_2 \det \left(\mathbf{I}_{N_\mathrm{M}} + \left(\tilde{\mathbf{K}}_k^{(\mathrm{n})} \right)^{-1} \cdot \mathbf{K}_k^{(\mathrm{s})} \right)$$
$$= \log_2 \det \left(\mathbf{I}_{N_\mathrm{M}} + \frac{1}{\sigma_n^2} \cdot \overline{\mathbf{H}} \mathbf{Z}_k \mathbf{G}_k \mathbf{G}_k^\mathsf{H} \overline{\mathbf{Z}}_k^\mathsf{H} \overline{\mathbf{H}}_k^\mathsf{H} \right). \tag{3.20}$$

The resulting (simplified) optimization problem is thus

$$\mathbf{G}_{j,b}^\star = \arg \max_{\left\{ \mathbf{G}_{j,b} \right\}_{j,b}} \min \left\{ \log_2 \det \left(\mathbf{I}_{N_\mathrm{M}} + \frac{1}{\sigma_n^2} \cdot \overline{\mathbf{H}} \mathbf{Z}_k \mathbf{G}_k \mathbf{G}_k^\mathsf{H} \overline{\mathbf{Z}}_k^\mathsf{H} \overline{\mathbf{H}}_k^\mathsf{H} \right) \right\}_{k \in \mathcal{K}_c}$$
$$\text{s.t. } \mathrm{Tr} \left\{ \sum_{j \in \mathcal{M}_c} \mathbf{Z}_{j,b} \mathbf{G}_j \mathbf{G}_b^\mathsf{H} \mathbf{Z}_{j,b}^\mathsf{H} \right\} \leq P_\mathrm{B}, \quad \forall b \in \mathcal{M}_c. \tag{3.21}$$

The approximated (or relaxed) optimization problem (3.21) is a convex problem as it does not contain any inverted covariance matrices that depend on the optimization variables $\mathbf{G}_j$ anymore. The arguments are quadratic forms of the precoding matrices that preserve the concavity/convexity of log det and the trace, respectively, with which a convex optimization problem can be formulated [15]. Note that instead of the OCI modeled as AWGN, also the current interference covariance matrix can be used as long as it is positive definite, constant, and independent of the optimization variables $\mathbf{G}_j$. This covariance matrix would however change when a neighboring cooperation cluster adapts its precoding.

Solving this optimization problem results in equal rates for all users in the cooperation area. However, the OCI is ignored and the resulting rates $\left\{ \tilde{R}_1^\star, \ldots, \tilde{R}_K^\star \right\}$ are not the true rates that can be achieved by the users. These rates can be derived by applying (3.17) with (3.18), where also the true OCI is taken into account.

3.3.2   Sum rate Optimization with Block Zero-Forcing

In the previous section, block ZF with max-min optimization was considered that leads to a fair data rate distribution within a cooperation area. The same optimization framework can however also be used to optimize other objective functions. Once the block ZF matrices $\mathbf{Z}_k$ are obtained, the power loading matrices $\mathbf{G}_k$ can also be used to maximize the sum rate.

The optimization problem for the sum rate maximization can be stated as

$$\mathbf{G}_{j,b}^{\star} = \arg \max_{\left\{ \mathbf{G}_{jm,b} \right\}_{j,b}} \sum_{k \in \mathcal{K}_c} \tilde{R}_k$$

$$\text{s.t. } \operatorname{Tr} \left\{ \sum_{j \in \mathcal{K}_c} \mathbf{Z}_{j,b} \mathbf{G}_{j,b} \mathbf{G}_{j,b}^{\mathsf{H}} \mathbf{Z}_{j,b}^{\mathsf{H}} \right\} \leq P_{\mathrm{B}}, \qquad \forall b. \tag{3.22}$$

With the block ZF and the approximated data rates $\tilde{R}_k$ in which the OCI is assumed to be AWGN, this optimization problem is also a convex one and can efficiently be solved by the same framework as the max-min problem. In contrast to max-min where power for strong users is reduced and allocated to the weak ones, sum rate maximization favors users with strong receive signals and further increases their performance. Weak users however will suffer from bad or even no service. A generalization to the maximization of weighted sum rate is straight forward.

3.3.3   Power Minimization with Block Zero-Forcing

A third approach that can be solved with the block ZF optimization framework is to optimize the power allocation for energy efficiency, i.e. to reduce the transmit power to achieve a certain target rate R_{tar} for all involved users. This leads to another interesting aspect of cooperation as economically using transmit power has gained much importance in order to limit the ever growing energy consumption of communication networks. This aspect of "green" communication networks has also been studied e.g. in [34, 49]. Here, we apply the block ZF based cooperation scheme and formulate the optimization problem

$$\mathbf{G}_{j,b}^{\star} = \arg \min_{\left\{ \mathbf{G}_{j,b} \right\}_{j,b}} \operatorname{Tr} \left\{ \sum_{j \in \mathcal{K}_c} \mathbf{Z}_{j,b} \mathbf{G}_{j,b} \mathbf{G}_{j,b}^{\mathsf{H}} \mathbf{Z}_{j,b}^{\mathsf{H}} \right\}$$

$$\text{s.t. } \tilde{R}_k \geq R_{\mathrm{tar}}, \qquad \forall k. \tag{3.23}$$

Thereby, we apply again the approximated achievable rates $\tilde{R}_k$ without the correct OCI. As before, this optimization problem is convex and it can be solved by the same framework. Note however, that the solution of (3.23) does here not necessarily lead to true rates R_k that fulfill the constraint $R_k \geq R_{\mathrm{tar}}$ exactly due to the approximation. Nevertheless, the achieved rates are in most cases close to the desired target rate. QoS requirements could therefore be enforced with high probability by setting the target rate to a slightly higher value for the optimization.

3.4 Macro Diversity for High Mobility Users

The cooperation schemes described so far require the knowledge of the channel coefficients of all communication links within a cooperation area as well as the complete messages that have to be transmitted. This information can be exchanged between the involved BSs via the fixed backhaul. To this end, however, the backhaul needs to be of high capacity and small delay. Moreover, in order to benefit from the cooperation gains, slow or even quasi-static fading is required, since the channels need to be fixed over the duration of the CSI estimation, dissemination, and the transmission of the message. Otherwise, the CSI would be outdated and the precoding would not lead to optimal or close to optimal results. In scenarios where mobile users are affected by fast fading with a coherence time smaller than the duration of a transmission, e.g. if some users or certain scattering objects move with a high velocity, cooperation schemes that rely on accurate CSI are not appropriate. To this end, we also analyze the performance that can be achieved by a scheme that exploits the available macro diversity without requiring reliable channel knowledge.

The macro diversity scheme is based on fast handovers. Thereby, each MS dynamically connects to the BS with the strongest signal in each time slot. For each channel realization, the MSs are ideally connected to the BS that can offer the best rate under the instantaneous conditions of the channel and the interference situation. Under the assumption of exactly one MS in each sector, this leads to the problem of finding the BS-MS assignment that provides the best data rates among all possible permutations in the cooperation set. Following the max-min approach, this means that out of all possible permutations of MS-BS assignments within $\mathcal{M}_c$, the permutation that leads to the largest minimal rate is chosen. In the following, this schemes is referred to as the *macro diversity* scheme. Note that in this scheme $M!$ different permutations need to be evaluated. This scheme is thus only applicable for small cooperation sets, e.g. $M \leq 6$, as the number of permutations grows very large for increasing M. There is no further cooperation in $\mathcal{M}_c$ to control ICI and the transmit symbol vectors of the chosen BS are $\mathbf{s}_j \in \mathbb{C}^{N_B}$ where the precoding matrices are scaled unity matrices $\mathbf{Q}_{j,b} = \sqrt{\frac{P_B}{N_B}} \cdot \mathbf{I}_{N_B}$, i.e. no beamforming is applied and therefore no CSIT is required.

The macro diversity scheme does however not fully reflect the potential that BS selection can achieve. The assumption that only one MS is located in each cell limits the possibilities for the user assignments and the probability that the rate of a single user can be improved by a reassignment is to be expected quite low. To this end, we also consider an upper bound (UB) of this scheme in which a single MS, the test MS whose performance is measured, is privileged and can choose the BS that provides the best instantaneous rate without considering the performance of the other users. This means that user k located in cell c chooses the best BS according to

$$R_k = \max_j \log_2 \det \left(\mathbf{I}_{N_M} + \left(\sum_{i \neq j} \frac{P_B}{N_B} \mathbf{H}_{k,i} \mathbf{H}_{k,i}^H + \sigma_w^2 \cdot \mathbf{I}_{N_M} \right)^{-1} \cdot \frac{P_B}{N_B} \mathbf{H}_{k,j} \mathbf{H}_{k,j}^H \right). \quad (3.24)$$

Thereby, it is ignored that other MSs might also choose this same BS as the node that provides the best performance. The performance is thus an UB on the achievable rates. With this, we can account for users that move with high velocity and are passed through different cells, while other users are more static and served in a cooperative way. Besides this, the scheme could also be interesting for scenarios where certain premium users are privileged among others and are able to choose a BS with higher priority. This schemes is referred to as the *macro diversity UB*. For practical application, its achievability is unrealistic. The macro diversity scheme in which all active users are assigned to a BS is however achievable and serves therefore as a lower bound. The performance of practical networks lies accordingly in between these two schemes.

3.5 Performance Evaluation

In the following, we assess the performance of the described cooperation schemes by means of computer simulations. To this end, we consider realistic setups of practical relevance. The basic network configuration consists of 12 regular hexagonal cells, each with 3 sectors (thus 36 sectors in total): nine neighboring cells are located in a ring around three cells in the center (the center cells are shown in Fig. 3.3). Each sector is served by a BS that is equipped with $N_B = 4$ directive antennas with antenna patterns recommended by 3GPP [4]:

$$A(\theta) = G - \min\left\{12\left(\frac{\theta}{\theta_{3\,\mathrm{dB}}}\right)^2, A_\mathrm{m}\right\} [\mathrm{dB}], \quad -180 \leq \theta \leq 180, \qquad (3.25)$$

with

- θ being the angle between the direction of interest and the boresight of the antenna (angle between the main lobe of the antenna array and the MS of interest)

- $A_\mathrm{m} = 25\,\mathrm{dB}$ the maximal attenuation,

- $\theta_{3\,\mathrm{dB}} = 65°$ the 3 dB (half-power) beamwidth, and

- $G = 17, \mathrm{dBi}$ the antenna gain with respect to an isotropic antenna element.

The antenna pattern is depicted in Fig. 3.2. The MSs are equipped with $M_\mathrm{M} = 2$ antennas that are omnidirectional with a gain of 1 (0 dBi). The antennas are assumed to be half a wavelength apart from each other and we assume no correlation between antenna elements.

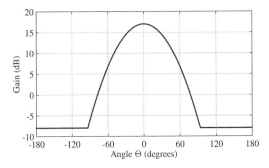

Figure 3.2.: Pattern of a directive antenna element at the BSs.

For the orientation of the BS sector antenna arrays, we distinguish two cases that are depicted in Fig. 3.3:

- 0°: a typical cell setup usually applied to 2G, 3G, and current networks (see (a) and (c) in Fig. 3.3).

- 30°: a setup in which three cooperating BSs point their antenna arrays towards each other (see (b) and (d) in Fig 3.3).

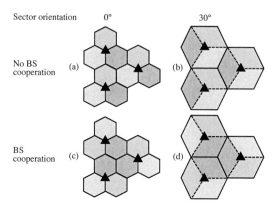

Figure 3.3.: Investigated cell orientations. For FDMA, the colors indicate sectors using the same frequency band. The triangles mark the positions of BSs.

Regarding frequency allocation, different reuse schemes are possible with OFDMA as it is used in the DL of LTE-Advanced systems (see [73] and references therein), for instance PFR. The goal is to allocate as much bandwidth to a user as reasonable. In most cases, users near the cell edge get only 1/3 of the overall bandwidth to control interference to other cells, whereas users near the BS can get up to the whole available bandwidth, i.e. frequency reuse 1. In [73], the idea of a load dependent frequency planning for OFDMA systems is introduced. The principle is coordination of frequency allocation for BSs. Inspired by these considerations, we investigate two different frequency allocations:

1. a static FDMA approach where each of the three sectors in a cell gets one third of the overall bandwidth

2. an approach where the whole bandwidth is available for a given user, i.e. frequency reuse 1.

Realistic schemes may operate between frequency reuse 1 and 3. Note that if the FDMA approach is used, the BSs belonging to the same cooperation set should be assigned to the same frequency band in order to allow for cooperative transmission ((c) and (d) in Fig. 3.3).

For the simulations, we assume that all BSs transmit with a maximal power $P_{\text{max}} = 49\,\text{dBm}$ (80 W) over an assumed bandwidth of 100 MHz that lies around a carrier frequency of 2.6 GHz, i.e. no power control is applied here. At the MSs, we assume a noise variance of $\sigma_w^2 = 5 \cdot 10^{-12}\,\text{W}$ which arises from the thermal noise over 100 MHz and a noise figure of 5 dB. These quantities are recommended for simulations of LTE-Advanced systems by the 3GPP [1].

For the computer simulations, we model frequency flat fading by Rayleigh-fading with a distance dependent pathloss and shadowing that corresponds to the WINNER II channel model [76] as described in Chapter 2. Among the different scenarios, we are mostly interested in urban environments with micro-cells as dense environments pose the most demanding challenges for high data rate services due to strong interference. Less dense rural environments are however also considered.

In order to estimate the spectral efficiency, the achievable rate of a test MS is evaluated in the cellular interference scenario for each point on a 25 m grid in a given sector of the cooperation area. The ICI is modeled by positioning $K-1$ MSs at random locations in the other sectors of the cooperation area and the corresponding BSs perform the same transmission scheme as the BS serving the test MS. The OCI, i.e. the signals transmitted by BSs that are not in the cooperation area of focus, is modeled as spatially white signals transmitted by the remaining BSs with maximum power $P_{max} = 49$ dBm. This arises from a worst case assumption as these BSs might serve MSs that are located on their respective sector edges. The spatially white transit signals are legitimated as these signals are independent of the channels to the sectors of interest. We thus assume that the beamforming weights do not change the statistics of the signals arriving at the test MS significantly. Moreover, when many BS antennas transmit to only few antennas of the MSs in the center area, these signals are expected to become more and more white when the number of BS antennas increases. Nevertheless, we apply independent fading with pathloss and shadowing from the WINNER II model also to these OCI signals. In order to get statistically significant results, we simulate 1000 random channel realizations and random positions of the $K-1$ other MSs for each grid point of the test MS. For each of the different settings, we are interested in the average achievable rate of the test MS as well as its outage probability $p_{out} = \Pr[R_k < R_{out}]$ for a target rate of $R_{out} = 1$ bps/Hz and the 5%-outage rate, i.e. the data rate in each grid point that can be guaranteed with 95% probability.

3.5.1 Urban Micro-Cells

In the urban environment, the distance between adjacent BSs is 700 m. The channels are modeled by the WINNER II based model as described in Chapter 2. We assume NLOS propagation condition and the pathloss is given by (2.24). Log-normal shadowing is also considered with a standard deviation of $\sigma = 8\,\text{dB}$. In the reuse 1 case, the transmission in all sectors is over the entire frequency band. In the FDMA case, the frequency assignment is such that the three sectors of a cell transmit in different frequency bands of one third of the overall bandwidth; a prelog factor of $\frac{1}{3}$ thus applies to all achievable rates in this case.

Non-Cooperative Reference

As a baseline with which we compare the different cooperative schemes, we define a reference setup that corresponds to a conventional cellular network. This reference scheme does not apply any cooperation or macro diversity handovers and transmits spatially white signals without beamforming, i.e. the precoding matrices at all BSs are $\mathbf{Q}_b = \sqrt{\frac{P_{\mathrm{B}}}{N_{\mathrm{B}}}} \cdot \mathbf{I}_{N_{\mathrm{B}}}$ with $P_{\mathrm{B}} = 80\,\text{W}$ for all BSs. The orientation is $0°$ and FDMA with orthogonal frequency bands in each sector as in Fig. 3.3a.

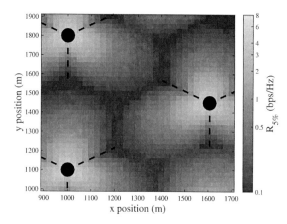

Figure 3.4.: Reference scenario: 5% outage rate, no cooperation, no handovers, FDMA frequency allocation, $0°$ orientation.

In Fig. 3.4, we show the 5% outage rates of this non-cooperative reference without any cooperation and no CSIT. The cell and sector borders can clearly be seen by the low rates in the areas that are mostly interference limited. Near the BSs, however, quite high data rates can be achieved as the signal power is high. The sectorization and the FDMA frequency allocation thus help in reducing the interference between the different sectors. We can further observe that all sectors in the figure look very similar. From this, we can conclude that the network under consideration is large enough to avoid edge effects. In the following, we thus focus on a single cell in the center of the network as this is enough to study the performance of larger networks.

Influence of Beamforming

Applying spatially white signaling at the BSs is pessimistic when BSs and MSs with multiple antennas are considered. With MIMO communication, the performance can be improved by beamforming. To this end, we extend the reference scheme by a precoding that maximizes the achievable rate within each sector individually without cooperation. This can be achieved by running the block ZF based max-min (or equivalently sum rate) maximization with a cooperation set that consists only of a single BS and a single MS, while the OCI again is assumed to be spatially white. In this way, the rate to the MS is maximized (under the assumption of spatially white OCI) and no cooperation is required.

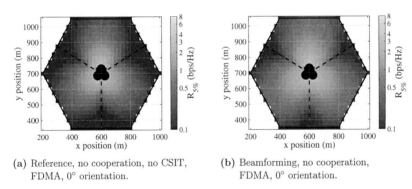

(a) Reference, no cooperation, no CSIT, FDMA, 0° orientation.

(b) Beamforming, no cooperation, FDMA, 0° orientation.

Figure 3.5.: Non-cooperative reference without CSIT and non-cooperative beamforming with max-min optimization: 5%-outage rates, FDMA, 0° orientation.

Fig. 3.5 shows the improvement that beamforming has on the reference setup. The FDMA frequency allocation and 0° orientation as well as all other parameters are unchanged. We can see that the coverage range is improved, i.e. the area with higher 5% outage rates in enlarged. The figure however looks similar as the one of the reference scenario and still has large areas with poor rates. The interference limitedness remains and beamforming alone does not help to improve the performance significantly.

63

This is also visible in Fig. 3.6, where the outage probabilities for a target rate of 1 bps/Hz as well as the average user rate in each grid point are plotted. All plots show the same behavior: The area with low outage probability (and thus coverage) is enlarged, but the improvements are small. Also the average user rate is not improved significantly with joint beamforming. In order to achieve significant improvements, the interference needs to be controlled. To this end, we apply the block ZF approach with cooperation of 3 BS sectors in the following.

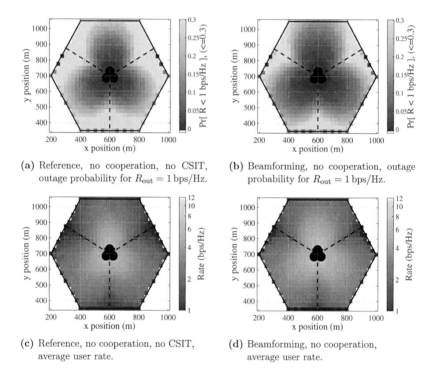

(a) Reference, no cooperation, no CSIT, outage probability for $R_{\mathrm{out}} = 1$ bps/Hz.

(b) Beamforming, no cooperation, outage probability for $R_{\mathrm{out}} = 1$ bps/Hz.

(c) Reference, no cooperation, no CSIT, average user rate.

(d) Beamforming, no cooperation, average user rate.

Figure 3.6.: Non-cooperative reference without CSIT and non-cooperative beamforming: outage probabilities and average user rates, FDMA, $0°$ orientation.

Sector Cooperation

The simplest configuration for joint transmission that arises from the network geometry is when the three sectors of the same cell cooperate with each other and jointly serve the three MSs therein. This sector cooperation is however sabotaged by the directive antennas at the BSs which prevent a meaningful contribution of the signals across the sector borders. Hence, the performance gain due to cooperative block ZF is marginal. To this end, we apply omnidirectional antennas for this case (antenna gain of 0 dBi in all directions).

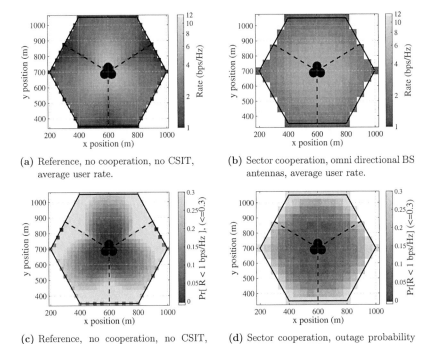

(a) Reference, no cooperation, no CSIT, average user rate.

(b) Sector cooperation, omni directional BS antennas, average user rate.

(c) Reference, no cooperation, no CSIT, outage probability for $R_{\text{out}} = 1\,\text{bps/Hz}$.

(d) Sector cooperation, outage probability for $R_{\text{out}} = 1\,\text{bps/Hz}$.

Figure 3.7.: Sector cooperation compared to the non-cooperative reference without CSIT, FDMA, 0° orientation.

In Fig. 3.7, we compare the reference without CSIT to this sector cooperation where the three sector antenna arrays of the same cell form a virtual array and perform max-min optimization. This case can be considered as a multiuser MIMO scenario where the BS of each cell uses all its 12 antennas to serve the three sectors jointly. Sector cooperation ((b) and (d) in the figure) leads to a performance gain regarding the data rates, especially on the sector borders. Due to the max-min optimization, the high rates close to the BSs are reduced but increased in the rest of the cell. This can be seen more clearly on the outage probability plots ((c) and (d)) where a much larger area has low outage probabilities. The performance gain, however, is not as good as with cooperation among three BS sectors from adjacent cells, as we will see next. Nevertheless, the co-located multiuser MIMO scenario does not need to exchange CSIT or data information between different BS sites.

3 BS Cooperation

A second cooperation setup that immediately arises from the network geometry is when a cooperation set is formed by three sectors from adjacent cells whose BSs are located around the cooperation area. In the 30° orientation (Fig. 3.3d), this seems particularly promising. In Fig. 3.8, the block ZF precoding with max-min optimization in this configuration is compared with non-cooperative beamforming. In the cooperative case, the spectral efficiency is much more homogeneously distributed and much lower outage probabilities are achieved in the entire cell. Also the 5% outage rates are evenly distributed and much higher in large areas as compared to the reference. Only the rates higher than 2 bps/Hz very close to the BSs in the reference are reduced for the sake of enlarged coverage in the max-min approach. A considerable improvement can also be observed with regard to sector cooperation. The max-min optimization with BSs that are distributed in space thus provides a homogeneous rate distribution and good performance for all involved mobile users. Due to the ZF, the sector borders vanish and as a result of the max-min optimization, the data rates of the cell edge users are significantly increased. This comes however with the price of reduced rates in the center of the cell, as less power is allocated to these users. The FDMA frequency allocation further reduces the OCI in the cooperation area.

The influence of the network layout and frequency allocation strategy is studied in Fig. 3.9. Therein, the 5% outage rates of all combinations that arise with the 0° and 30° sector orientation and the reuse 1 and FDMA frequency allocation are compared with each other. While all cases achieve a better coverage (larger area with higher rates) than the non-cooperative reference, the results are particularly good for the 30° orientation. The homogeneous rate distribution allows a balanced coverage over almost the entire area of the network. This effect is even more pronounced when the FDMA frequency allocation is applied. In this case, a 5% outage rate of about 2 bps/Hz can be achieved in large parts of the network. Only in the cell corners where no BSs are located, the rates are a bit lower. With the reuse 1 frequency allocation, the data

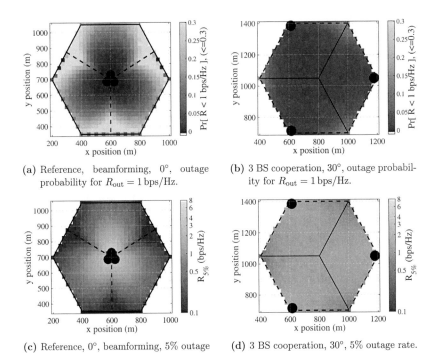

(a) Reference, beamforming, 0°, outage probability for $R_{out} = 1\,\text{bps/Hz}$.

(b) 3 BS cooperation, 30°, outage probability for $R_{out} = 1\,\text{bps/Hz}$.

(c) Reference, 0°, beamforming, 5% outage rate.

(d) 3 BS cooperation, 30°, 5% outage rate.

Figure 3.8.: 3 BS cooperation compared to non-cooperative beamforming: FDMA frequency allocation, 0° and 30° orientation.

rates drop close to the BSs. This has two reasons: On one hand, the optimization does not consider the true OCI which is particularly strong in these areas (especially the interference from the other sectors of the same BS site). On the other hand, the signals from the different cooperating BSs differ significantly in their strength such that the assisting BSs can only provide weak signal contributions to the improvement of the rate of this MS while the close BSs, due to the max-min optimization, has to sacrifice much of its transmit power to increase the rates of the other MSs that might be far away. A possible solution to avoid this problem could be to equip the BSs with excess antennas. With that, also MSs in neighbor cells that are close to the cell border

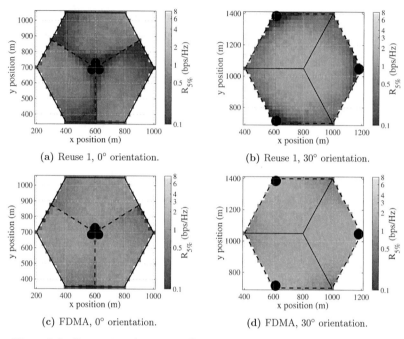

(a) Reuse 1, 0° orientation.

(b) Reuse 1, 30° orientation.

(c) FDMA, 0° orientation.

(d) FDMA, 30° orientation.

Figure 3.9.: Cooperation of 3 sectors: 5% outage rate, different network configurations.

can be zero-forced. Additionally, power control might also improve the performance. Nevertheless, the cooperation scheme in networks with the 30° orientation combined with FDMA shows large improvements regarding coverage compared to conventional cellular networks.

6 BS super-cells

Considering the 30° orientation with three cooperating BSs that are located in adjacent cells, the area between the BSs that belong to the same cooperation set forms a virtual cell that is served by three BSs located in corners. As seen in the simulation results before, this concept can offer large benefits as compared to the conventional network layout. With the exception that the sectors are rotated by 30°, the same structure as traditional cellular networks can be maintained and the same number of BSs is required to serve the entire area. However, the concept of serving a cell from its corners instead from the center can be carried one step further to "super-cells" that are served by six BSs, i.e., in each corner of such a super-cell there is a BS antenna array that cooperates with the others corresponding to this super-cell.

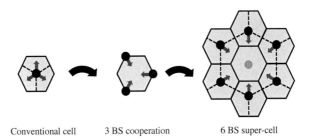

Conventional cell   3 BS cooperation   6 BS super-cell

Figure 3.10.: From conventional cells to super-cells.

If the BSs are arranged as depicted in Fig. 3.10, the entire area of the network is again divided into hexagonal cells (the super-cells), as in the conventional case. The super-cells are however larger as in the conventional network configuration and six BSs form a virtual array. In this setup, an even more homogeneous distribution of data rates can be expected than with 3 BS cooperation. As a side benefit from this configuration, a larger area can be served by fewer BSs as in the conventional setup, as a whole BS can be saved in the middle of the super-cell.

The performance of the cooperation scheme in such 6 BS super-cells is shown in Fig. 3.11, where the average rates as well as the 5% outage rates for the ZF based cooperation scheme are compared with the non-cooperative reference in the same setup. If no cooperation is applied, the performance is worse than in the traditional reference, because interference is increased due to the specific orientation of the BS arrays. With cooperation, however, a very homogeneous rate distribution can again be achieved with significantly higher rates than with 3 BS cooperation. On the cell corners, however, the max-min optimization leads again to poor performance, as power for these users is taken away to support weaker MSs that are further away from the BSs while the cell edge users are affected by more interference from neighbor BSs. In order to provide also good performance to users on the edge of the super-cell, the cooperation could be combined with non-cooperative signaling (e.g. with the macro diversity scheme) when users are close to a BS and to apply block ZF only in the area between the BSs where joint transmission is beneficial.

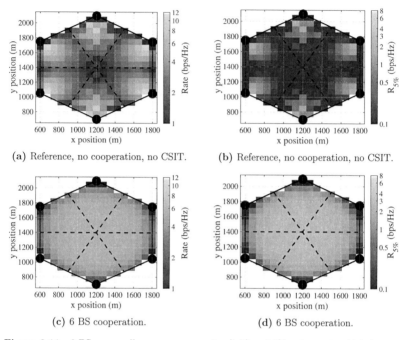

(a) Reference, no cooperation, no CSIT.   (b) Reference, no cooperation, no CSIT.

(c) 6 BS cooperation.   (d) 6 BS cooperation.

Figure 3.11.: 6 BS super-cells: average user rates (left) and 5% outage rates (right), reuse 1 frequency allocation.

Macro Diversity for High Mobility Users

For high mobility users, or if the acquisition of accurate CSIT is not feasible, we apply the macro diversity scheme with fast handovers within the cooperation area. In Fig. 3.12, we compare the performance of the non-cooperative reference without CSIT to the macro diversity scheme with $|\mathcal{M}_c| = 3$ cooperating (or rather coordinated) BSs. In the latter case, the three MSs in the cooperation area are assigned in the max-min sense to the three BSs. With this, almost no improvement can be seen. The reason is that the sectorization and the FDMA frequency allocation separate the different sectors. When now an MS with poor rate is assigned to another BS that offers a better rate, the MS that initially has been served by this BS has to connect to another BS which in most cases offers a data rate that is smaller than the initial rate of the first MS. In almost all cases, the conventional assignment, which connects each MS to the BS of

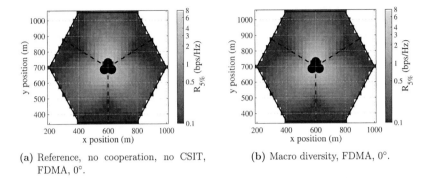

(a) Reference, no cooperation, no CSIT, FDMA, 0°.

(b) Macro diversity, FDMA, 0°.

Figure 3.12.: Macro diversity for high mobility users: 5% outage rate, FDMA, 0°.

the sector it is located in, offers the best performance. The macro diversity scheme in this small cooperation set does therefore not lead to benefits.

To this end, we enlarge the cooperation area to $|\mathcal{M}_c| = 36$ sectors and study the macro diversity UB to see the potential of coordinated scheduling. In Fig. 3.13, we show the performance of this bound for the different network setups. Note that in this UB, the test MS whose data rates are plotted in the figures is privileged among the others, as it can choose the BS that offers the best rate without caring about the other MSs that might be worse with this assignment. The resulting 5% outage rates show a performance that is very close to the one of 3 BS cooperation. The data rates are also very homogeneously distributed with some higher peak rates close to the BSs. Here, reuse 1 outperforms FDMA, while the differences between the different sector orientations are small. In contrast to the optimized block ZF approach, no CSIT is required for this scheme, as no beamforming is performed and only rate feedback has to be evaluated. Benefiting from macro diversity is thus much simpler than to apply joint transmission with block ZF.

Comparison of Schemes

The empirical CDFs in Fig. 3.14 give more insight into the results for both considered cooperation methods (macro diversity and BS cooperation with max-min block ZF) in the standard setup of urban micro-cells. Therein, the instantaneous rates that the test MS can achieve for each realization are considered. While the 30° sector orientation shows advantages for the joint beamforming cooperation regardless of reuse 1 or FDMA

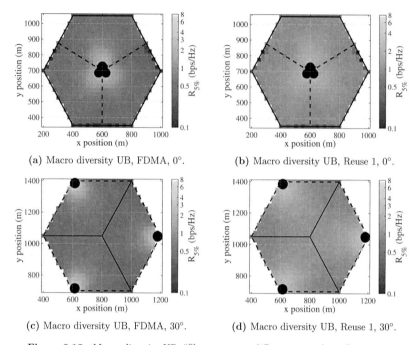

(a) Macro diversity UB, FDMA, 0°.

(b) Macro diversity UB, Reuse 1, 0°.

(c) Macro diversity UB, FDMA, 30°.

(d) Macro diversity UB, Reuse 1, 30°.

Figure 3.13.: Macro diversity UB: 5% outage rate, different network configurations.

frequency allocation, the 0° orientation performs better for the macro diversity UB and for the case of no cooperation. In case of FDMA without cooperation, the setup (a) in Fig. 3.3 achieves always higher spectral efficiencies than (c) and (d) - while joint beamforming in (a) is not feasible due to the different frequencies that are assigned to the sectors in the cooperation area. The reuse 1 frequency allocation results in higher mean and maximum spectral efficiencies for joint beamforming and the macro diversity approach; FDMA shows only advantages in two cases:

- for no cooperation in the reference scenario (a) of Fig. 3.3;

- for joint beamforming as far as spectral efficiencies below 3 bps/Hz are concerned.

However, as the International Telecommunication Union (ITU) requires higher rates for low mobility [61], FDMA may still be an interesting choice for LTE-Advanced due to the higher rates for low mobility MSs at the cell edges (Fig. 3.9). With the FDMA 30° orientation, cooperation with joint beamforming with 3 BSs shows the best

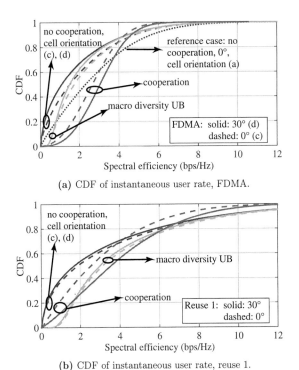

(a) CDF of instantaneous user rate, FDMA.

(b) CDF of instantaneous user rate, reuse 1.

Figure 3.14.: Empirical CDFs of different schemes for FDMA and reuse 1, 0° and 30° orientation.

performance regarding these cell edge users as it is reflected by the 5% outage rate. This performance is closely followed by the macro diversity UB in the reuse 1 case, which does not require any CSIT.

In Fig. 3.15, we compare the different schemes in their respective best configuration directly with each other. Here also the six BS super-cells as well as the non-cooperative beamforming are included. Non-cooperative beamforming and joint block ZF in the reuse 1 setup show certain improvements over the reference scenario in all regimes (in the low as well as in the high rates). These schemes are however clearly outperformed by the block ZF approach in the FDMA setting and the macro diversity UB. While the joint beamforming achieves higher rates above the 5% outage border, the much simpler macro diversity does hardly legitimate the additional complexity that comes with

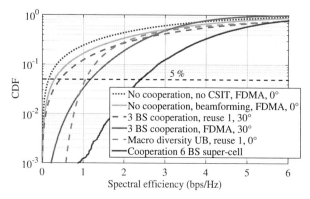

Figure 3.15.: Empirical CDFs of instantaneous user rates for urban micro-cells with 3 BS cooperation, 6 BS super-cells, and macro diversity UB.

the max-min block ZF. The macro diversity UB achieves a similar performance and in some places even higher rates, even though much less effort is involved to coordinate the scheduling. This changes however if the larger six BS super-cells are considered, where the performance (see the rates on the 5% outage border) of the simpler macro diversity UB scheme can be more than doubled by the joint beamforming. The cell setup in the super-cells seems thereby particularly beneficial. The overhead and additional complexity of coherent BS cooperation is thus only worth the effort when the cooperation clusters are large enough. For practical networks, the higher complexity has thus to be traded with the higher data rates that are achievable. Note however, that the macro diversity scheme requires coordination between 36 sectors, while for the joint beamforming only three or six sectors have to exchange signals.

3.5.2 Sum Rate Optimization

In contrast to max-min optimization which leads to a fair distribution of the data rates among the users, sum rate maximization favors users with strong receive signals and further increases their performance. Weak users on the other hand suffer from bad or even no service. This is reflected in the simulation results in Fig. 3.16 where the average user rates (sum rate in the cooperation area of interest divided by $K = 3$) for the urban micro-cell scenario are shown. While some users achieve only very small rates after the optimization, the improvement of the average rate is more pronounced

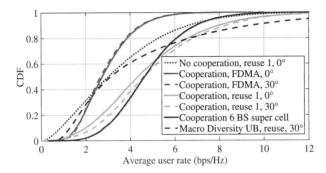

Figure 3.16.: Empirical CDFs of average user rate after sum rate maximization.

than with the max-min approach. Moreover, the reuse 1 frequency allocation leads here to much higher rates, as no prelog penalty of 1/3 due to the divided spectrum applies and the sum rate maximization can boost the rates for those users who experience only small interference. The joint beamforming outperforms here clearly the macro diversity UB (only the rates of the privileged test MS are shown). Consequently, sum rate maximization provides the best average user rates, but a certain outage probability, i.e. the probability that certain users that are too weak to achieve an useful performance, has to be accepted. As far as throughput is considered, the effort of joint beamforming is justified by the significantly higher rates with the optimized scheme.

The spatial distribution of the user rates is shown in Fig. 3.17 and compared to the non-cooperative reference and the max-min optimized cooperation scheme. By the more pronounced beams close to the BSs, it can be seen that the sum rate maximization leads to much higher rates for some users. For the MSs located in the corners of the virtual cell where no BS is present, very low rates or no service at all is provided. The data rates are not as homogeneously distributed as with the max-min approach. The data rates of the users that already achieve good performance in the non-cooperative reference are further increased, whereas no gain can be provided to the users with weak signals. Even though the sum rate (or the average user rate) is improved significantly, the sum rate maximization is therefore not suitable to improve coverage.

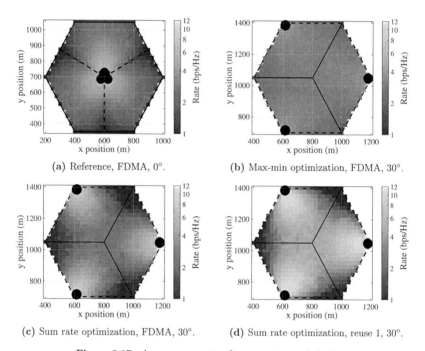

(a) Reference, FDMA, 0°.

(b) Max-min optimization, FDMA, 30°.

(c) Sum rate optimization, FDMA, 30°.

(d) Sum rate optimization, reuse 1, 30°.

Figure 3.17.: Average user rates after sum rate maximization.

3.5.3 Power Minimization

The optimization framework applied for the block ZF based max-min and sum rate optimization can also be used to minimize the transmit power required to achieve certain target rates within the cooperation area. In this section, we provide some results that indicate that cooperative cellular networks can also be made energy efficient. For the simulations here, we however only consider the power minimization within the cooperation set of interest $\mathcal{M}_c$, while all other BSs transmit with full power of $P_B = 49$ dBm, i.e. the OCI remains unchanged. This has the consequence that the target rate, here chosen to be $R_{tar} = 1$ bps/Hz, can alway be achieved in $\mathcal{M}_c$ (but not necessarily in the other sectors not belonging to $\mathcal{M}_c$). If power control would be applied to all cooperation sets, they would again interdepend on each other, as each transmit power in a certain set influences the interference in the other cooperation areas. The different cooperation clusters would then have to scale their transmit powers jointly

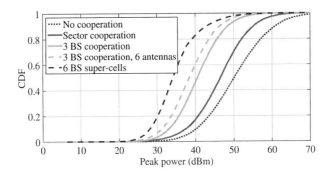

Figure 3.18.: CDF of maximal transmit power for a target rate of 1 bps/Hz, urban micro-cells, reuse 1.

or iteratively after each other, with which even lower power could be achieved than here. In certain interference limited scenarios, it could however happen that the target rate cannot be achieved at all MSs. Nevertheless, the results shown here indicate that power minimization can provide large savings regarding energy in such networks.

In Fig. 3.18, the maximum transmit power that is used by a BS in the cooperation set is plotted for the case of reuse 1 in urban micro-cells. These results have also been published in [74]. Determining the peak power (the highest BS transmit power in the cooperation cluster) required to achieve a target of $R_{\text{tar}} = 1$ bps/Hz in 80% of all simulations runs, this results in about 40 dBm for the 6 BS super-cells and about 45 dBm for three cooperating BSs. If the three cooperating BSs would be equipped with 6 antennas instead of the typical 4, the peak power would reduce to 43 dBm. In the case of sector cooperation or no cooperation at all, a peak power of 52 dBm or even 57 dBm would be required. BS cooperation shows therefore also here large improvements. Additionally, the results show that due to the cooperation also the OCI that the considered cooperation set generates to other sets is reduced. This, in turn, can further improve the overall performance due to lower interference, as the other cooperation sets can potentially also scale their transmit power down, which might again enable a further down scaling of the power in the first set.

3.5.4   Rural Macro-Cells

The propagation conditions of an urban NLOS and rural LOS environment are the most extreme conditions with respect to the pathloss as discussed in Chapter 2. After studying the former case quite extensively, we will now apply the same schemes also to the rural case. To this end, we apply the scenario D2 of the WINNER II channel model [76], which represents radio propagation in large areas with low building density. As a consequence, LOS conditions are expected to appear more frequently than in urban areas. The pathloss with LOS is given by (2.27) and the shadow fading is modeled as a log-normal random variable with the standard deviation

$$
\sigma = \begin{cases} 4\,\mathrm{dB}, & 10\,\mathrm{m} < d < d_{\mathrm{BP}} \\ 6\,\mathrm{dB}, & d_{\mathrm{BP}} < d < 10\,\mathrm{km}. \end{cases} \tag{3.26}
$$

The distance between adjacent BSs is in this environment assumed to be 1.5 km, the other assumptions and parameters are in line with the urban case described before.

Considering the CDFs of the rural setup in Fig. 3.19, it can be seen that cooperation among three BSs offers also here some benefits as compared to the conventional reference. The performance gains of the max-min scheme are however not as large as in urban micro-cells. This can be explained by the higher pathloss the signals suffer due to the larger distances between adjacent BSs. While this reduces the interference it also reduces the desired signal strength. In the case of cooperation, this leads to lower achievable rates as cooperation shows the best performance in situations where the signal strengths from all cooperating BSs have the same order of magnitude. In the rural case, some of these signals are much more attenuated and room for optimization is reduced, as the BSs have to sacrifice much of their power to completely cancel the signals to users that are far away and would not have much interference anyway. Also the super-cells with six cooperating BSs offer only small improvements. The macro diversity scheme on the other hand shows a larger potential in rural macro-cells. This is due to the lower interference that affects the users. The diversity gain offers especially in this case high peak rates as well as good outage performance. If the network is not dense, i.e. the performance is not dominated by interference but by weak signal strengths due to large distances between BSs and MSs, the macro diversity scheme is preferable as it is much simpler than the optimization procedure and achieves the best performance among the considered schemes.

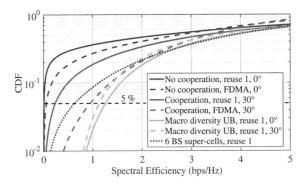

Figure 3.19.: Empirical CDFs of instantaneous user rate for rural macro-cells.

3.5.5 Comparison & Conclusions

In the previous sections, we have studied and compared different BS cooperation schemes in various network settings. The performance gains depend thereby on the cell layout and the frequency allocation strategy. In this section, we summarize the most important results and compare the best network configurations for the different schemes. It is however difficult to indicate which scheme is better or worse than another, as one scheme might be better in certain scenarios or regimes, while another has a better performance with respect to other considerations. As discussed in Chapter 2 where the different figures of merit and objectives are introduced, the most challenging aspect of cellular networks is to provide cell edge users with a good QoS. Thereby, it is important to provide reasonably high rates in the entire area of the network, such that the user experience is good.

In order to reflect these considerations, we compare the key performance indicators defined in Chapter 2. The coverage and the average 5% outage rates are summarized in Table 3.1 for all schemes in their respective best network configuration, for the urban as well as the rural environment. It can be seen that coverage as well as the average $R_{5\%}$ steadily improve when more complex transmission schemes are applied, at least for the urban case. The macro diversity scheme where all MSs in the cooperation area are considered does not lead to any improvement as compared to the reference. When beamforming is applied in the non-cooperative way, the coverage and the average $R_{5\%}$ are increased by approximately 50%. The 3 BS cooperation and the macro diversity UB lead to a significant further increase, where 3 BS max-min block ZF leads to a

Scheme	Configuration	Coverage	Av. $R_{5\%}$ (bps/Hz)
Urban micro-cells			
Reference	FDMA, 0°	15 %	0.53
Beamforming	FDMA, 0°	24 %	0.77
Macro diversity	FDMA, 0°	15 %	0.53
Macro diversity UB	reuse, 0°	61 %	1.19
3 BS cooperation	FDMA, 30°	86 %	1.26
6 BS super-cell	reuse, super-cell	77 %	1.88
Sum rate maximization	FDMA, 0°	5 %	0.18
Rural macro-cells			
Reference	FDMA, 0°	6 %	0.25
Macro diversity UB	reuse, 0°	97 %	1.34
3 BS cooperation	FDMA, 30°	63 %	1.01
6 BS super-cell	reuse, super-cell	66 %	1.34

Table 3.1.: Key performance indicators for the different schemes.

coverage of 86 %. This is a larger fraction of the area than with the macro diversity UB with 61 %, but the average $R_{5\%}$ is with about 1.2 bps/Hz very similar.

The performance of the macro diversity UB thus shows that joint scheduling can have a similar potential as 3 BS cooperation, which achieves its gains with a significantly higher complexity. For the six BS super-cells, the situation looks different. The coverage is not as good as with 3 BS cooperation (77 %), which arises from the poor regions on the edge of the super-cell, where the cooperation cannot unfold its potential due to the weak signal contributions of the BSs that are further away. The average $R_{5\%}$ is however significantly higher as with the other schemes. The high data rates in the center of the super-cells thus contribute to a high value in this indicator and more than compensate the lower rates on the cell edge. Users served by such super-cells have a somewhat smaller coverage but can expect a significantly higher QoS where service is available. Users close to the BSs should therefore be served in a non-cooperative way by only one BS. With sum rate maximization, coverage is very poor, even worse than the reference. As this form of optimization privileges the best users, it is not suitable for the enhancement of coverage and 5% outage rates, as it is reflected in the table.

In the case of the rural macro-cells, the macro diversity UB leads to a very large coverage of almost the entire area (97 %), while only 2/3 of the area can be covered by max-min optimization. Even with the six BS super-cells, not more than 66 % can be covered, but the average $R_{5\%}$ is the same as with the macro diversity UB. With the larger cells and the lower pathloss in these rural macro-cells, joint transmission is thus not very effective and the simpler coordinated scheduling seems better suited as indicated by the results of the macro diversity UB.

From the simulative study conducted in this chapter, we can conclude that block ZF allows the formulation of convex optimization problems and that it can efficiently cancel interference within a cooperation area when the cooperating BSs are dense. The max-min optimization leads to homogeneous rate distribution in the entire area and all users can be served in a fair way. The joint transmission can thus achieve large performance gains as compared to conventional networks, especially on the cell/sector edges. To profit from these gains, however, accurate CSIT within the cooperation area, backhaul connections of sufficient capacity between the cooperating BSs, adapted antenna/sector orientation, and sufficiently high SNRs (not too large distances between cooperating BSs) are required. Moreover, high data rates can only be achieved with this CoMP scheme when the cooperation areas are sufficiently dense. Cooperation areas with three BSs or less are not enough to exploit the potential of CoMP, as interference from neighboring clusters is still strong. Six BSs that cooperate with each other in super-cells offer much higher data rates. This observation is also in line with the literature. E.g. in [25], where also block ZF is considered, it is stated that an optimal cluster size consists of around 7 BS. The results can however not directly be compared, as only sum rate is considered and the network configuration differs in the pathloss and channel model and no sectorization is assumed therein. By increasing the cooperation cluster, however, also the complexity of the block ZF approach is increased as more CSIT has to be acquired and larger optimization problems have to the solved. If the complexity of signal processing and data as well as CSI exchange in BSs is not an issue, cooperation with joint beamforming should be applied to cooperation areas that are as large as possible. Combined with a dense deployment of BSs, high performance and good coverage can then be achieved.

If CSIT cannot be acquired or if the backhaul or the computational capability at BSs is not sufficient to perform joint block ZF in large enough clusters, macro diversity is a simpler alternative that can achieve similarly good performance but does not require CSIT and exchange of user data. The macro diversity UB is however not achievable in practice, the development of sophisticated scheduling algorithms is thus required to benefit from this potential gain. A drawback from the macro diversity scheme is that no explicit formulation or optimization of objectives such as sum rate maximization or power minimization is possible when no further optimization of the signals is conducted.

With the block ZF approach, the optimization problem can be reformulated and solved also for these cases. In practice, the different schemes can also be combined to benefit from all their advantages. A sophisticated scheduler can e.g. select three cell edge users with poor rates that are then served by the max-min scheme, while users that are close to a BS or are moving fast are served by a single BS that instantaneously offers the best performance. For users in between, also the sum rate maximization can be applied to offer higher peak rates.

3.6 Critical Discussion

In this chapter, we have applied different schemes to assess the performance of PHY layer cooperation in cellular networks. The specific schemes (block ZF with linear precoder optimization and the macro diversity scheme) are however only example schemes for which no optimality can be claimed. With these schemes, we were mainly interested in the potential that comes with them in realistic scenarios and what influence cooperation has on the cell planning and the architecture of future networks. The performance of the applied schemes demonstrates the gains that cooperation can have over conventional networks. The insights from this chapter form thereby the basis for the studies performed in the following chapters of this work. The obtained results, however, have to be appreciated under the limitations and assumptions imposed by the specific choice of transmission schemes and network setups. To this end, we summarize and discuss the limitations of the considered scenarios and transmission strategies.

3.6.1 Limitations of Block ZF

Block ZF is a suboptimal linear transmission scheme that completely cancels the interference between multiple users. This is per se suboptimal, as completely nulling the interference leads to less room for optimization of the desired signals, as the signal space is restricted due to the fulfillment of the stringent ZF conditions. Allowing some interference in the range of the noise power at the MSs would offer a higher potential to maximize the data rates achievable for the users. This would lead to improvements particularly in the rural case. Several results from recent research however indicate that block ZF can achieve data rates that are close to the optimum given by DPC when the SNR of the system is high and/or when the BSs have a larger number of antennas than the MSs [68, 69]. For the setups considered here, we can assume that this is fulfilled to a certain extent, as the BSs are equipped with $N_B = 4$ antennas as compared to the $N_M = 2$ antennas of the MSs. Regarding the high SNR regime, at least the denser urban micro-cells seem to fulfill that as well, as there the signals from the BSs within cooperation sets are usually strong. In the rural macro-cells, this is not the case anymore. Accordingly, we have seen that the block ZF approach does not lead to significant improvements in the rural scenario.

A further restriction on the block ZF approach as we applied it, is that closed BS clusters are required that have to be fixed during a transmission period. Moreover, clusters cannot overlap and each has to serve a disjoint set of MSs. Cooperation with dynamic clusters can therefore provide better performance, as e.g. proposed in [37]. For this, other transmission schemes than block ZF that are more flexible are potentially better suited. E.g with SLNR precoding, each MS could choose its own BS cluster it is served from and each BS can be part of multiple different BS groups. Such a dynamic clustering approach with individual and optimally chosen BS sets for each user is considered in Chapter 8.

Another source of suboptimality of the precoding design that we have applied arises from the approximation of the OCI for the optimization process. While this assumption allows to formulate convex optimization problems that can efficiently be solved, the true interference seen by the MSs leads to achievable rates that are worse than expected by the optimization. This leads to performance degradations especially on the cell edges where this OCI is strong. Nevertheless, modeling the OCI as spatially white transmit signals is reasonable. When many out of cluster BS antennas transmit signals whose precoding is calculated independently from the channels to the MSs in the cooperation area, these signals do not appear much different from spatially white signals. For the evaluations in the simulations, however, the fading on these channels has to be taken into account with pathloss and shadowing.

In order the get an estimate of the performance loss of the approximated OCI for the optimization under the block ZF conditions, we compare this approach to a direct optimization in which the ICI is not nulled and the true OCI can be taken into account. As the resulting optimization is not convex and more difficult to solve, we only apply it to a small simplified network.

3.6.2   Direct Optimization

An alternative to the block ZF approach is to maximize the minimal rate directly without the additional ZF conditions to be fulfilled and without approximating the OCI. Even though an optimization problem without the restriction of nulling interference is non-convex, we can apply a convex optimization algorithm with different initializations to see to which solutions it converges. To this end, we introduce such a direct optimization scheme that can lead to higher rates. Due to the higher computational complexity, it is however not well suited for application in realistic scenarios.

The transmit signal of BS $b \in \mathcal{M}_c$ is given by

$$\mathbf{x}_b = \sum_{j=1}^{K} \mathbf{Q}_{j,b} \cdot \mathbf{s}_j, \tag{3.27}$$

where the precoding matrices $\mathbf{Q}_{j,b}$ are here not restricted to be designed such that the ICI is forced to zero. Instead, we attempt to solve the following optimization problem directly:

$$\mathbf{Q}_{j,b}^\star = \arg \max_{\{\mathbf{Q}_{j,b}\}_{j,b}} \min \{R_1, \ldots, R_K\}$$

$$\text{s.t. } \mathrm{Tr}\left\{\sum_{j \in \mathcal{K}_c} \mathbf{Q}_{j,b} \mathbf{Q}_{j,b}^{\mathsf{H}}\right\} \leq P_{\mathrm{B}}, \quad \forall b \in \mathcal{M}_c. \tag{3.28}$$

To this end, we apply a gradient based optimization algorithm that converges to a local optimum. Note that here we do not have to make the additional assumption that the OCI is spatially white or fixed, but can include the true interference covariance matrix in the rate R_k.

Equivalent Optimization Problem

In order to get rid of the power constraint, the precoding matrices can be scaled according to the variable substitution

$$\mathbf{Q}_{k,b} = \frac{\sqrt{P_{\mathrm{B}}}}{\sqrt{\mathrm{Tr}\left\{\sum_{j \in \mathcal{K}_c} \tilde{\mathbf{Q}}_{j,b} \tilde{\mathbf{Q}}_{j,b}^{\mathsf{H}}\right\}}} \cdot \tilde{\mathbf{Q}}_{k,b}, \qquad \forall k \in \mathcal{K}_c, \forall b \in \mathcal{M}_c, \tag{3.29}$$

where $\tilde{\mathbf{Q}}_{j,b}$ are the initial choices of the precoding matrices that possibly violate the constraint. By this normalization, it is ensured that the power constraint is always fulfilled. With this, the optimization problem can be reformulated into the equivalent problem [147]

$$\max_{\{\tilde{\mathbf{Q}}_{j,b}\}_{j,b}} \tau$$
$$\text{s.t. } R_i - \tau \geq 0, \quad \forall i \in \mathcal{K}_c. \tag{3.30}$$

For an $\epsilon > 0$ we define the ϵ-active set as

$$\mathcal{A}_\epsilon = \{i \in \mathcal{K}_c : R_i - \tau < \epsilon\}. \tag{3.31}$$

With this definition, we can design a gradient based optimization algorithm that solves (3.30) and thus also (3.28). This algorithm updates the coefficients in the precoding matrices corresponding to the rates in the ϵ-active set $\mathcal{A}_\epsilon$ iteratively with a step into the direction of the gradient of the respective achievable rates. To this end, the gradient of the rates R_i is derived in the following.

Gradient

Due to the variable substitution (3.29), the gradient of the achievable rate of user k with respect to the initial precoding coefficients can be obtained by applying the chain rule of differentiation

$$\nabla_{\tilde{\mathbf{Q}}^*} R_k = \left((\nabla_{\mathbf{Q}^*} R_k)^{\mathsf{T}} \cdot \mathbf{J}_{\mathbf{Q}^* \tilde{\mathbf{Q}}^*} + (\nabla_{\mathbf{Q}^*} R_k)^{\mathsf{H}} \cdot \mathbf{J}_{\mathbf{Q}\tilde{\mathbf{Q}}^*} \right)^{\mathsf{T}}, \tag{3.32}$$

where $(\cdot)^*$ means complex conjugate. The outer gradient is given by the vector of all partial derivatives of R_k with respect to all precoding coefficients [16]

$$\nabla_{\mathbf{Q}^*} R_k = 2 \cdot \left[\frac{\partial R_k}{\partial \mathbf{Q}_{1,1}^*[1,1]}, \ldots, \frac{\partial R_k}{\partial \mathbf{Q}_{j,b}^*[p,q]}, \ldots, \frac{\partial R_k}{\partial \mathbf{Q}_{K,M}^*[N_\mathrm{B}, N_\mathrm{B}]} \right]^{\mathsf{T}}. \tag{3.33}$$

with $\mathbf{Q}_{j,b}^*[p,q]$ being the entry in the p-th row and q-th column of the matrix $\mathbf{Q}_{j,b}^*$. These partial derivatives can be calculated according to [105] by

$$
\frac{\partial R_k}{\partial Q^*_{j,b}[p,q]} = \frac{1}{\ln(2)} \mathrm{Tr}\left\{ \left(\mathbf{K}^{(s)}_k + \mathbf{K}^{(i)}_k + \mathbf{K}^{(n)}_k \right)^{-1} \cdot \left(\frac{\partial \mathbf{K}^{(s)}_k}{\partial Q^*_{j,b}[p,q]} + \frac{\partial \mathbf{K}^{(i)}_k}{\partial Q^*_{j,b}[p,q]} \right) \right\}
$$
$$
- \frac{1}{\ln(2)} \mathrm{Tr}\left\{ \left(\mathbf{K}^{(i)}_k + \mathbf{K}^{(n)}_k \right)^{-1} \cdot \left(\frac{\partial \mathbf{K}^{(i)}_k}{\partial Q^*_{j,b}[p,q]} \right) \right\}
$$

with the inner derivatives

$$
\frac{\partial \mathbf{K}^{(s)}_k}{\partial Q^*_{j,b}[p,q]} = \frac{\partial}{\partial Q^*_{j,b}[p,q]} \sum_{m \in \mathcal{M}_c} \sum_{n \in \mathcal{M}_c} \mathbf{H}_{k,m} \mathbf{Q}_{k,m} \mathbf{Q}^H_{k,n} \mathbf{H}^H_{k,n}
$$
$$
= \sum_{m \in \mathcal{M}_c} \sum_{n \in \mathcal{M}_c} \mathbf{H}_{k,m} \mathbf{Q}_{k,m} \frac{\partial \mathbf{Q}^H_{k,n}}{\partial Q^*_{j,b}[p,q]} \mathbf{H}^H_{k,n} \tag{3.34}
$$
$$
= \sum_{m \in \mathcal{M}_c} \mathbf{H}_{k,m} \mathbf{Q}_{k,m} \mathbf{E}^H_{p,q} \mathbf{H}^H_{k,m} \tag{3.35}
$$

and

$$
\frac{\partial \mathbf{K}^{(s)}_k}{\partial Q^*_{j,b}[p,q]} = \frac{\partial}{\partial Q^*_{j,b}[p,q]} \sum_{m \in \mathcal{M}_c} \sum_{n \in \mathcal{M}_c} \sum_{\substack{i \in \mathcal{K}_c \\ i \neq k}} \mathbf{H}_{k,m} \mathbf{Q}_{i,m} \mathbf{Q}^H_{i,n} \mathbf{Q}^H_{k,n}
$$
$$
= \sum_{m \in \mathcal{M}_c} \sum_{n \in \mathcal{M}_c} \sum_{\substack{i \in \mathcal{K}_c \\ i \neq k}} \mathbf{H}_{k,m} \mathbf{Q}_{i,m} \frac{\partial \mathbf{Q}^H_{i,n}}{\partial Q^*_{j,b}[p,q]} \mathbf{H}^H_{k,n}
$$
$$
= \sum_{m \in \mathcal{M}_c} \mathbf{H}_{k,m} \mathbf{Q}_{j,m} \mathbf{E}^H_{p,q} \mathbf{H}^H_{k,m}, \tag{3.36}
$$

since

$$
\frac{\partial \mathbf{Q}^H_{i,n}}{\partial Q^*_{j,b}[p,q]} = \begin{cases} \mathbf{E}^H_{p,q}, & \text{if } i = j \text{ and } n = b \\ \mathbf{O}, & \text{otherwise}, \end{cases} \tag{3.37}
$$

where $\mathbf{E}_{p,q}$ is the matrix with all entries 0 except the one in row p and column q which is 1. Note that $\partial \mathbf{K}^{(n)}_k / \partial Q^*_{j,b}[p,q] = \mathbf{O}$, as the effective noise is not a function of the precoding coefficients.

Next, we need to compute the Jacobian matrices $\mathbf{J}_{\mathbf{Q}\tilde{\mathbf{Q}}^*}$ and $\mathbf{J}_{\mathbf{Q}^*\tilde{\mathbf{Q}}^*}$. These contain the partial derivatives of

$$
\mathbf{Q}_{k,b} = \sqrt{\frac{P_\mathrm{B}}{\tilde{P}_b}} \cdot \tilde{\mathbf{Q}}_{k,b} \text{ and } \mathbf{Q}^*_{k,b} = \sqrt{\frac{P_\mathrm{B}}{\tilde{P}_b}} \cdot \tilde{\mathbf{Q}}^*_{k,b}, \quad \forall k, b, \tag{3.38}
$$

where $\tilde{P}_b$ is the power that would result at BS b if the unscaled initial precoding matrices

$\tilde{\mathbf{Q}}_{j,b}$ were applied

$$\tilde{P}_b = \mathrm{Tr}\left\{ \sum_{i \in \mathcal{K}_c} \tilde{\mathbf{Q}}_{i,b} \tilde{\mathbf{Q}}_{i,b}^{\mathsf{H}} \right\}. \tag{3.39}$$

The entries of the Jacobians are given by

$$\frac{\partial \mathbf{Q}_{j,b}^{(*)}[p,q]}{\partial \tilde{\mathbf{Q}}_{m,n}^{*}[r,c]} = \begin{cases} 0, & \text{if } n \neq b \\ \dfrac{\xi_{m,r,c}^{(j,p,q)} - \frac{1}{2}\sqrt{P_{\mathrm{B}}} \tilde{\mathbf{Q}}_{p,q}^{(*)}[j,b] \tilde{P}_b^{-\frac{1}{2}} \partial P}{\tilde{P}_b}, & \text{if } n = b, \end{cases} \tag{3.40}$$

and

$$\partial P = \frac{\partial}{\partial \tilde{\mathbf{Q}}_{m,n}^{*}[r,c]} \mathrm{Tr}\left\{ \sum_{i \in \mathcal{K}_c} \tilde{\mathbf{Q}}_{i,b} \tilde{\mathbf{Q}}_{i,b}^{\mathsf{H}} \right\} = \mathrm{Tr}\left\{ \tilde{\mathbf{Q}}_{m,n} \cdot \mathbf{E}_{r,c}^{\mathsf{H}} \right\}. \tag{3.41}$$

The notation $(\cdot)^{(*)}$ means here that the argument is or is not complex conjugate, depending on whether the term corresponds to $\mathbf{J}_{\mathbf{Q} \cdot \tilde{\mathbf{Q}}^{*}}$ or $\mathbf{J}_{\mathbf{Q}^{*} \cdot \tilde{\mathbf{Q}}^{*}}$.

Optimization

In each step of the optimization algorithm, the precoding coefficients of the corresponding rates R_i with $i \in \mathcal{A}_\epsilon$ are updated with a step in the direction of the gradient that corresponds to these rates, until all rates are in the ϵ-active set and a further increase in the rates is not possible anymore. In order to find an appropriate step size μ for each iteration step, a line search is used which starts with an initial step size $\mu_0 = 1$ and decreases this value until the evaluation of the achievable rate with the corresponding update leads to a minimal rate that is smaller than with the previous choice. With this line search it is ensured that the minimal rate is increased in each iteration step until it converges eventually to a local minimum. In order to increase the chance that the algorithm converges to the global optimum or at least to a good local one, the algorithm can be restarted with different initializations.

Comparison to Block ZF

As the achievable rates considered in the direct optimization also contain the true interference terms, the optimization problem is not convex. This makes the optimization rather complex and requires many iterations in general. Moreover, a convergence to the global optimum cannot be guaranteed; the gradient search converges only to a local optimum.

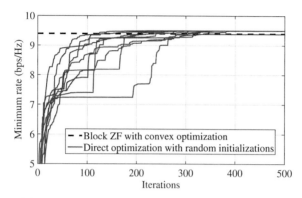

Figure 3.20.: Minimum achievable rates for a typical channel realization, optimized with the direct gradient based optimization compared with the convex optimization with the block ZF approach.

In Fig. 3.20, we compare the direct optimization with the block ZF approach where the power loading matrices $\mathbf{G}_{j,b}$ are optimized by the Yalmip optimization toolbox [88]. For the simulation, we consider a typical channel realization in a small network with only $M = K = 2$ BSs and MSs that lie in the same single cooperation area. The BSs are equipped with $N_B = 4$ and the MSs with $N_M = 2$ antennas and the per BS transmit power constraint and the noise variance at the MSs are set to $P_B = 1$ and $\sigma_w^2 = 0.1$, respectively. The elements of all channel matrices between BSs and MSs are i.i.d. $\mathcal{CN}(0,1)$, i.e. pathloss and other large scale fading effects are normalized to 1. We can observe that the direct gradient based optimization converges to at least two different local optima in this example, where the better one is higher than the convex optimization with the block ZF approach. The difference is however small. In higher SNR regimes or when more nodes or antennas are considered, the performance difference between the two schemes decreases even further. The block ZF approach thus achieves good results when the SNR is reasonably good and when the number of BS antennas is large enough. A similar observation is also reported in [82] or [39], where it is concluded that block ZF can achieve close to optimal rates when the number of BS antennas is large relative to the number of MS antennas and even asymptotically optimal in the high SNR regime.

3.6.3   Limitations of Simulations & Further Work

The applied block ZF approach leads to a performance that is close to optimal under certain conditions and the performance achieved in the simulation results is promising. For the simulation setups considered in this chapter, however, we made several assumptions that have to be taken into account by the interpretation of the results.

On one hand, all BSs (and thus the OCI) are assumed to be under full transmit power of $P_B = 80\,W$. This can lead to pessimistic interference scenarios, as real networks include power control which is not considered here. If BSs throttle down their transmit power to serve users that are e.g. in their close vicinity, the neighbor cells are less affected by interference. However, in the choice of the transmission schemes and optimization criteria, we were interested in achieving a high coverage and thus a good QoS that can be guaranteed with high probability in the entire network. In this sense, considering full interference corresponds to a worst case scenario. The observed 5% outage rates and outage probabilities are thus also guaranteed when a certain user is affected by the strong interference of neighboring BSs that transmit with full power. When power control is applied on top of the considered transmissions schemes, the performance for all users can potentially be improved significantly. So can e.g. the block ZF approach be adapted to transmit with a power that is sufficient to serve its users with a target rate. Alternatively, power control could be included in a way that multiple BSs transmit signals to a specific MS with a transmit power that is optimized such that this user gets the highest possible rate while other users are not affected by too strong interference. This can also be realized in a coordinated way between different BSs. Also for the BS selection, adapting transmit power can lead to benefits. In the macro diversity UB discussed here, each MS receives its signal only from a single BS that transmits with full power. This can be extended and improved in a way that multiple BSs transmit the same symbols (either precoded if CSIT is available or spatially white without CSIT) with a power allocation that is optimized for the current channel and interference condition. Such an optimization is introduced and discussed in Chapter 8.

A second assumption that differs from real networks and might lead to pessimistic interference scenarios in certain situations is that the networks considered here are built with regular cells that all have the same form and size. In reality, it is hardly possible to place all BSs with an equal distance to each other and to form perfectly hexagonal cells. If the distance between two adjacent BSs assumed in our simulation setups would be the minimum distance in a network where the BSs are located more randomly, situations can appear in which OCI generating BSs are further away than in the regular setup. This would then lead to situations in which this interference is lower than here. Regarding this interference, the setup with regularly placed interferers can again be considered as a worst-case assumption. On the other hand, if the cooperating BSs are located in a more random fashion, also the desired signal contributions might be weaker due to larger distances, which might reduce the performance again.

Additionally, we have only considered cooperation setups with 3 BSs as well as with 6 BSs. These configurations arise immediately from the geometry of the basic network. Other configurations in which more than 6 BSs cooperate with each other, e.g. 7, could lead to higher performance gains than the networks studied here. In order to extend the studies described here, one can consider also different, possibly more exotic, network configurations, which might lead to higher data rates or lead to additional insights. In Chapter 8, we also drop the assumption of regularly placed BSs and consider networks where all nodes are located in a random fashion. This provides further possibilities to optimize the MS-BS assignments as the different distances and corresponding signal strengths can be exploited.

Furthermore, also the max-min criterion can be reconsidered. In order to maximize coverage or 5% outage rates, other objective functions might lead to different results. An inherent drawback of the max-min optimization is that there are situations, in which one MS has so poor receive signals that optimizing its rate might lead to unacceptably low data rates at all other MSs as well. In order to avoid this, one could e.g. formulate target rates and if a user cannot achieve this despite the optimization, this user can be dropped completely such that the others achieve higher rates. One can even assign different target rates for different users that might wish to use different data services with differing performance requirements. Such considerations however lead to scheduling algorithms which are not in focus of this work. The potential of coordinated scheduling is however reflected in the macro diversity UB. In order to achieve a high performance in the entire network, joint scheduling should be combined with CoMP such that MSs are chosen that are suitable for the respective scheme.

Other limitations of the applied simulations lie in the signal and knowledge assumptions that we have made. For instance, we assumed perfect CSIT that can be shared with other BSs of a cluster in an unlimited way when the joint beamforming is applied. This is somewhat optimistic, as the channels are estimated with a finite precision and the channels have a certain coherence time after which the CSI might be outdated.

Also other imperfections such as synchronization errors or others are not taken into account here. Regarding imperfect CSIT, however, we will discuss its implications on the block ZF and the selection scheme in the following chapter.

For the time being, we stay with the block ZF approach and the assumptions made in this chapter and extend the framework of cooperative cells to the case where additional nodes such as relays, femto-cell BSs, or other low power nodes assist the communication between BSs and MSs. With a unified framework that allows to include different types of nodes, we study the influence of different nodes in such heterogeneous networks.

4

Small Cells and DF Relaying

In the last chapter, different concepts of cooperative multiuser communication in which multiple BSs are involved have been considered. These schemes try to enhance the coverage and/or achievable rates of cellular networks by serving MSs by multiple BSs instead of only one as in conventional non-cooperative networks. The general network topology, however, remains the same as in conventional architectures, only the sectorization and antenna orientations have been adapted to the cooperative transmission. Besides joint transmission, exploiting macro diversity with BS selection has also been identified as an efficient means to increase the performance. However, when locally restricted cooperation is applied, interference limited areas cannot be avoided in the entire network. Some areas with poor coverage and low data rates remain.

Besides CoMP, also the use of relays, femto-cell BSs (also known as home NodeBs – HNBs), RRHs, or other types of infrastructure nodes can help in improving the performance, especially in hot spots or locations where only poor rates can be provided. In order to fill these spots of poor coverage, the network can be extended by such nodes. As a consequence, the network is densified and diversified with different types of nodes that have to coexist. Thereby, their usage has to be designed such that they do not disturb the signals of other nodes or conventional BSs. For this, these additional nodes should transmit with low power. Moreover, as many small nodes are required to achieve a high degree of densification, they should of low cost. Their application can thus also range from relatively simple transmit node selection for which no explicit CSIT is required up to more sophisticated cooperative transmission between multiple nodes with joint beamforming.

The conceptual potential of heterogeneous networks has also been identified in the literature as promising solutions to enhance future networks. In the 4G standard [3], the use of relays is already foreseen. Thereby, the relays are considered as a tool to improve coverage of high data rates, temporary network deployment, cell-edge throughput, and/or to provide coverage into new areas. Thereby, these relays are considered as repeaters that are connected to the radio access network via a donor cell with a connection that can be in-band or out-band. From the point of view of the MSs, the relay cells can further appear as separate cells or they can be part of the same cell as the one of the donor BS. The opportunities and challenges that arise with such heterogeneous networks are discussed in [41,98]. Therein, the trend to densify the networks is motivated, as this provides a large potential to increase the capacity of the network and to serve more users concurrently. The additional interference that these nodes introduce has however to be controlled and limited such that no performance loss in other areas has to be accepted. A combination of using small cells and benefiting from cooperation can be seen in the concept of distributed antenna systems (DASs) [23]. In this approach, a cluster of cooperating BSs is assisted by additional antenna arrays distributed in space that contribute to the joint transmission. For this, however, in-band relays that are fed by the BSs on the same resources as the users are not suitable. A fast out-band connection is desirable when coherent signals with joint beamforming together with the BSs is applied. The assisting nodes can then be considered as RRHs or HNBs that are connected to the backhaul by a link of sufficient capacity.

These research efforts indicate that future networks will be heterogeneous and dense and that a multitude of nodes are coordinated with each other. As a large variety of possible transmission schemes exists, which all differ in their complexity, we limit ourselves to two signaling approaches that reflect the range of them: a simple selection scheme that does not require accurate CSIT but assigns each MS to a BS or a small cell depending on the current conditions and a CoMP scheme in which additional antennas support the joint transmission of BSs.

The performance of such schemes has been reported in the literature to be promising: In [146], a network that is assisted by DF relays is studied. Therein, significant performance gains are observed with a selection scheme that assigns the MSs to the nodes that offer the best instantaneous SINR. The additional interference that these relays impose is thereby not as severe as to impair the performance in other places of the network noticeably. A good tradeoff between node density and high performance could be achieved with four relays per cell in the setting the authors applied. For the evaluation, however, only a simplified network setup has been considered and all nodes were equipped with a single antenna.

A more general network setup with multiple cells and multiple antennas at the BSs and relays is studied in [101]. Therein, the concept of shared relays is introduced, in which relays are placed on corners of cells to support the MSs in the adjoining sectors, i.e. one MIMO relay on the corner of three sectors assists the communication to the MSs in these sectors. Different relaying strategies are compared with each other in this setup. A considerable benefit from these relays was however only reported when sophisticated signal processing is applied at the BSs as well as at the relays; otherwise, the potential performance gain due to the relays cannot be exploited due to the additional interference, which is especially high in the cell corners where the assistance of relays is needed most.

Higher gains can be achieved when joint transmission between a BS and multiple supporting antenna arrays is applied. Achievable rates in such a DAS are studied in [23]. Therein, a multi-antenna BS is assisted by multiple single antenna relays whose first hop is not considered (perfect BS-to-relay link with sufficient capacity) such that a single MS, also with a single antenna, is jointly served by all these nodes. In this case, a high performance can be achieved. However, only a single user is included in the evaluation and interference arising from signals to other users is not taken into account.

In order to see what benefits additional low power nodes can bring in our setup, we extend the framework from Chapter 3 to capture also heterogeneous networks with small cells. To this end, we introduce a simple decode-and-forward (DF) relaying scheme that can readily be applied for range extension and apply it then to dense networks where it can also achieve improvements in interference limited areas. As future networks are expected to be very heterogeneous, i.e. they can contain sophisticated BSs, HNBs, RRHs, or relays, a comprehensive discussion of future cellular networks has also to take different types of infrastructure nodes in the same network into account. To this end, we formulate a framework in which we can capture the different schemes in a unified way, ranging from a non-cooperative reference, relaying with transmit node selection, BS cooperation, up to a combination of relaying and cooperation. The different node types are thereby modeled in a way that they are distinguished only by parameters such as transmit power, backhaul connection, and antenna configuration. Combined with the channel and interference model from the last chapter, this allows to compare different communication strategies of high practical relevance and to study their performance and potential in practical scenarios. Thereby, we can see how the performance of a network depends on the network topology and cooperation complexity. By means of computer simulations, we evaluate various relevant scenarios with respect to performance, robustness, and complexity. Important aspects that we address include the influence of the node type and their transmit power. Also the impact of the number of cooperation nodes as well as the required amount and quality of CSI that limits the cooperation level from relaying to full CoMP is discussed. Furthermore, we also comment on the feedback and backhaul load required for the different schemes.

4.1 Network Model & Transmission Schemes

In order to capture networks with different types of nodes within a unified framework, we simplify the description of the nodes in a way that they are only distinguished by parameters such as transmit power, backhaul connectivity, and antenna configuration (including sectorization). This allows for considering and comparing different scenarios that include relaying and CoMP as well as combinations thereof. In our framework, we consider two types of infrastructure nodes, namely BSs and *supporting nodes* (SNs). While we consider the former as sophisticated high power nodes that are connected to the backhaul by high capacity links such as fiber optics, the SNs assist the communication between BSs and MSs with lower power and are connected to the backhaul via a donor BS, possibly with a connection of limited capacity. The SNs can represent relays, HNBs, or RRHs. Due to the abstraction of the infrastructure nodes, we can easily change SNs into BSs, or vice versa, by adjusting the corresponding parameters in our model. In the following, we focus on the DL and transmission on a single subcarrier as before.

Similar as in Chapter 3, the set of all BSs and SNs is divided into cooperation clusters that can comprise multiple BSs and/or SNs. The infrastructure nodes of one cooperation cluster can then jointly serve one or multiple MSs. The cooperation clusters are described by the index sets $\mathcal{M}_1, \ldots, \mathcal{M}_C$, where C is the number of clusters within a network and the elements of these sets correspond to the indices of BSs and SNs that are associated to these clusters. The cooperation clusters are assumed to be fixed during a transmission period and a specific resource (time/frequency) block. Note that different clusters can contain different numbers of infrastructure nodes; some can consist of only a single BS, while others can contain a plurality of BSs and SNs. Each element of a cooperation cluster $\mathcal{M}_c$, $c \in \{1, \ldots, C\}$, transmits signals to a set of MSs described by the index set $\mathcal{K}_c$. Also these sets can contain multiple or only one active node, depending on the specific scenario.

Each infrastructure node within a cooperation cluster, say node $b \in \mathcal{M}_c$, transmits a sum of linearly precoded signals

$$\mathbf{x}_b = \sum_{k \in \mathcal{K}_c} \mathbf{Q}_{k,b} \cdot \mathbf{s}_k, \tag{4.1}$$

where each summand corresponds to the signal intended for one of the associated MSs. The matrix $\mathbf{Q}_{k,b}$ denotes the precoding matrix of the signal from infrastructure node b to MS k, and $\mathbf{s}_k$ is the corresponding data symbol vector. Note that $\mathbf{Q}_{k,b} \in \mathbb{C}^{N_b^{(\mathrm{T})} \times N_k^{(\mathrm{s})}}$ and $\mathbf{s}_k \in \mathbb{C}^{N_k^{(\mathrm{s})}}$, with $N_b^{(\mathrm{T})}$ and $N_k^{(\mathrm{s})}$ the number of transmit antennas of node b (BS or SN) and data streams for MS k, respectively.

The receive signal $\mathbf{y}_k \in \mathbb{C}^{N_k^{(\mathrm{R})}}$, with $N_k^{(\mathrm{R})}$ the number of receive antennas of MS $k \in \mathcal{K}_c$ served by the cooperation cluster $\mathcal{M}_c$, can then be written as

$$\mathbf{y}_k = \sum_{b \in \mathcal{M}_c} \mathbf{H}_{k,b} \mathbf{Q}_{k,b} \mathbf{s}_k + \sum_{b \in \mathcal{M}_c} \sum_{j \neq k} \mathbf{H}_{k,b} \mathbf{Q}_{j,b} \mathbf{s}_j + \sum_{i \notin \mathcal{M}_c} \mathbf{H}_{k,i} \mathbf{x}_i + \mathbf{n}_k. \tag{4.2}$$

Therein, the first term captures all desired signals (transmitted by the nodes in $\mathcal{M}_c$). The second term contains the signals transmitted by nodes within $\mathcal{M}_c$ but intended for other MSs of $\mathcal{K}_c$ (the intra-cluster interference – ICI). The third and fourth term describe the interference caused by nodes outside cluster $\mathcal{M}_c$ (out-of-cluster interference – OCI) and noise. The matrix $\mathbf{H}_{k,b} \in \mathbb{C}^{N_k^{(\mathrm{R})} \times N_b^{(\mathrm{T})}}$ describes the channel from the b-th transmitting node to MS k, where we assume that the channel is frequency flat and constant for one transmission period, i.e., we look at a single sub-carrier of an OFDM based system with a block fading channel.

The precoding matrices $\mathbf{Q}_{k,b}$ can be chosen in many ways which differ in complexity, required knowledge (such as CSI), performance, and robustness. In order to reflect the different functionalities that different types of nodes can perform, we consider three different levels of transmission schemes with varying complexity:

- A simple non-cooperative reference scheme without any SNs where each BS transmits spatially white signals to a single MS. This reference corresponds to the conventional network already introduced in the last chapter.

- A transmit node selection scheme in which each MS can choose if it is served directly by a BS or by a SN. In this scheme, we distinguish two types of SNs, namely in-band DF relays and out-band femto-cell HNBs. The selection scheme resembles the macro diversity scheme from the previous chapter extended to include additional small cells but limited to a single cell/sector.

- A multiuser MIMO CoMP scheme where a set of multiple transmitting nodes (BSs and/or SNs) serve multiple MSs with block ZF with max-min precoding.

In the following, we describe these schemes in more detail and derive their achievable rates.

4.1.1 Non-Cooperative Reference

The non-cooperative reference scheme is of the lowest complexity. No SNs are present (or they are all shut off) and each BS independently serves a single user (per resource block) by a spatially white signal of transmit power P_B uniformly allocated across all antennas. The cooperation clusters thus only contain a single BS. Therefore, the precoding matrix of BS b is $\mathbf{Q}_k^{(b)} = \sqrt{P_B/N_b^{(T)}} \cdot \mathbf{I}_{N_b^{(T)}}$. Note that this transmission strategy is optimal in the absence of CSIT which is not required in this case. This scheme is of very low complexity as no CSI feedback and no complicated precoder calculation is required. The resulting achievable rate for MS k can be calculated as in (3.4) by

$$R_k = \log_2 \det \left(\mathbf{I} + \left(\mathbf{K}_k^{(i)} + \mathbf{K}_k^{(n)} \right)^{-1} \cdot \mathbf{K}_k^{(s)} \right), \tag{4.3}$$

where $\mathbf{K}_k^{(s)}$, $\mathbf{K}_k^{(i)}$, and $\mathbf{K}_k^{(n)}$ are the covariance matrices of the desired signal, interference, and noise.

4.1.2 Transmit Node Selection

The deployment of SNs is an approach to enhance the coverage of BSs in a simple way. When the network is densified by such additional nodes, the MSs can benefit from a macro diversity gain and the signals are stronger due to the lower pathloss the signals experience as the distance between a MS and its serving nodes becomes smaller. A simple way to benefit from such additional nodes is to apply a selection scheme. Each MS can choose its transmitting node from a set that contains BSs and SNs. For this selection scheme, we limit ourselves to two types of SNs: in-band DF relays and out-band femto-cell HNBs.

DF Relays

Relay stations (RSs) are nodes that do not transmit own information to/from the users but support the communication between BSs and MSs by forwarding signals that have been transmitted on the primary link. The use of relays promises better coverage and increased data rates than in conventional BS-MS communication due to the intermediate node that can boost the attenuated signal by additional power. In this chapter, we limit ourselves to DF relays that operate in half-duplex mode and use the same physical channel for both hops (in-band). If a single RS forwards its signal only to one MS and the direct channel between BS and MS is ignored or blocked, the data rate on the two-hop channel is limited by the worst link of the two hops. The second hop can thus not transmit more data than provided in the first hop. Accordingly, DF relays show a higher performance than with AF relaying because the relay noise is removed. AF relays on the other hand would lead to a worse performance in this setup as the relay noise is accumulated and the end-to-end SNR thus decreased [87]. When multiple relays serve the same MS, AF relays can however lead to a better performance. This case will be considered in the following chapters. In this chapter, we limit the discussion to DF relaying.

For the signaling protocol, we assume again that in each sector a single MS is active in one given resource block. For each transmission period, this MS can now choose over which link it receives the data it is interested in. This offers a diversity gain and, depending on where the relays are located, a range extension of the coverage of the corresponding BS. In order to describe the selection scheme applied here, we focus on a single BS that is assisted by N_R RSs, which together build a cooperation set $\mathcal{M}_c$, as depicted in Fig. 4.1.

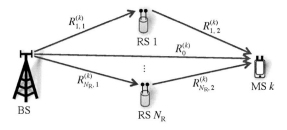

Figure 4.1.: DF relaying with transmit node selection.

When the BS transmits a spatially white signal with power P_B, i.e. the transmit signal of the BS intended for MS k is $\mathbf{x}_B = \sqrt{P_B/N_b^{(T)}} \cdot \mathbf{s}_k$, while the other nodes in

$\mathcal{M}_c$ remain silent, the achievable rate on the direct link $R_0^{(k)}$ can be calculated by

$$R_0^{(k)} = \log_2 \det \left(\mathbf{I}_{N_b^{(\mathrm{R})}} + \left(\sigma_n^2 \cdot \mathbf{I}_{N_b^{(\mathrm{R})}} + \mathbf{K}_k^{(i)} \right)^{-1} \cdot \frac{P_\mathrm{B}}{N_b^{(\mathrm{T})}} \mathbf{H}_k \mathbf{H}_k^\mathsf{H} \right), \qquad (4.4)$$

where $\mathbf{K}_k^{(i)}$ is the OCI generated by the nodes in the other cooperation clusters. Note that the relays within $\mathcal{M}_c$ do not cause interference if the direct link is chosen; the relays are turned off in this case.

In a similar way, the rate $R_{j,1}^{(k)}$ of the link between BS and RS j can be calculated. Thereby, it is assumed that the BS again transmits with full power P_B while the other RSs $i \neq j$ are silent. The transmission from the source to the relay takes place in an orthogonal resource (either different time slot or frequency band, depending on whether a TDD or FDD protocol is used). The achievable rate on the second hop is denoted by $R_{j,2}^{(k)}$ and we assume that RS k transmits with power P_R while the other nodes in $\mathcal{M}_c$ remain silent in this resource block. Assuming equal time slots or frequency bands on each hop, the resulting total achievable rate on the link from the BS to MS k via RS j is given by

$$R_j^{(k)} = \frac{1}{2} \min \left\{ R_{j,1}^{(k)}, R_{j,2}^{(k)} \right\}, \qquad (4.5)$$

where the prelog factor $\frac{1}{2}$ stems from the fact that two resource blocks are required for the transmission of a single symbol vector. However, equal time slot or frequency band allocation for both hops is not optimal. Better performance can be achieved if the resource allocation can be chosen in a more flexible way. To this end, we can optimize the time in which the two hops operate. In this way, less time can be assigned for the better and more time for the worse hop in order to balance the two rates. With t_1 and t_2 denoting the normalized durations of the first and second hop transmission, this can be formulated as the following optimization problem

$$R_j^{(k)} = \max_{t_1, t_2} \min \left\{ t_1 \cdot R_{j,1}^{(k)}, t_2 \cdot R_{j,2}^{(k)} \right\}, \quad \text{s.t. } t_1 + t_2 = 1. \qquad (4.6)$$

By solving the linear equation system

$$t_1 \cdot R_{j,1}^{(k)} = t_2 \cdot R_{j,2}^{(k)} \qquad (4.7)$$

$$t_1 + t_2 = 1, \qquad (4.8)$$

this leads to the solution

$$R_j^{(k)} = \frac{R_{j,1}^{(k)} \cdot R_{j,2}^{(k)}}{R_{j,1}^{(k)} + R_{j,2}^{(k)}}. \tag{4.9}$$

Equivalently, two fractions b_1 and b_2 of the bandwidth allocated to the transmission for both hops can be used by FDD relays instead of the two time slots.

For each transmission block, an active MS associated to cluster $\mathcal{K}_c$ now chooses the link to the best infrastructure node from its serving set $\mathcal{M}_c$. When the best link is chosen, the corresponding infrastructure node (BS or RS) transmits again spatially white (with transmit power P_{B} if the BS is chosen or with power P_{R} in the case a RS is chosen), while the other nodes within $\mathcal{M}_c$ are silent. The resulting achievable rate is then

$$R_k = \max\left\{ R_0^{(k)}, R_1^{(k)}, \ldots, R_{N_{\mathrm{R}}}^{(k)} \right\}. \tag{4.10}$$

The selection of the best link is based on a measurement of the link quality. Therefore, a rate feedback from the MS is required. However, as the precoding is spatially white, no CSIT is required also in this case and the increase of complexity as compared to the reference is only small. Additionally, relaying seems therefore to be well suited also in case of (fast) moving mobile users.

Femto-Cells

A similar communication scheme with transmit node selection can also be applied if other nodes than in-band relays are used. When these nodes are connected to the BS by a wired connection or a wireless out-band link that is of sufficient capacity, the first hop rate does not have to be considered for the calculation of the end-to-end rate. To this end, we use HNBs to reflect this case. The HNBs are small cell (femto-cell) BSs of reduced complexity and with lower power than the more sophisticated full BSs. Even though such a HNB might have a backhaul connection of limited capacity (e.g. via a Digital Subscriber Line – DSL), we assume that this link is of sufficient capacity and small delay such that only the links to the MSs have to be taken into account for the rate calculation. In this way, we can compare in-band relays, that have to share the same physical channel for the communication with the BS as well as the MS, with a selection scheme with perfect data delivery to the small cell.

In the case of femto-cells, the scheme described above can be adopted and the achievable rate for user k is given as in (4.10), with the exception that the involved rates $R_1^{(k)}, \ldots, R_{N_{\mathrm{R}}}^{(k)}$ are the achievable rates on the links between the HNBs and the MSs directly. In contrast to the relaying scheme, there is no time or frequency band allocation between BS-to-RS and RS-to-MS link required. The HNB receives its required information via the out-band link that is here assumed to be of a capacity that is sufficiently large to support the rate achievable to the MS. Therefore, the entire time slot or frequency band can be used for the transmission to the MS and the prelog factor can be omitted. Otherwise, the HNBs and RSs are very similar from an abstract point of view.

4.1.3 Distributed Antenna Systems with Block ZF

A more sophisticated transmission scheme can be realized if the SNs jointly transmit together with one or multiple BSs. To this end, we apply a multiuser CoMP scheme that is based on block ZF and optimized power allocation across all transmitted data streams as in the previous chapter. For this case, we assume that the SNs are connected to their associated BSs via a backhaul link of sufficient capacity. The SNs can correspond to HNBs or RRHs and the cooperation cluster with BSs and SNs becomes a DAS.

All mobiles within a cooperation area, that can comprise multiple sectors, are served jointly by the corresponding transmit nodes (BSs and SNs) where the ICI is nulled and the power is allocated to each stream such that the minimum rate is maximized. To this end, the precoding matrices are decomposed to $\mathbf{Q}_{k,b} = \mathbf{Z}_{k,b} \cdot \mathbf{G}_{k,b}$, where $\mathbf{Z}_{k,b}$ is the block ZF matrix and the power allocation for the different streams is handled in $\mathbf{G}_{k,b}$. The ZF matrices are obtained by components of the null space of all undesired links within the cooperation set, i.e. of null $\left\{ \left[\bar{\mathbf{H}}_{1,c}^T, \ldots, \bar{\mathbf{H}}_{k-1,c}^T, \bar{\mathbf{H}}_{k+1,c}^T, \ldots, \bar{\mathbf{H}}_{|\mathcal{K}_c|,c}^T \right]^T \right\}$, where $\bar{\mathbf{H}}_{i,c}$ is the collocated channel matrix from all transmitting nodes within a cooperation cluster $\mathcal{M}_c$ to MS $i \in \mathcal{K}_c$.

Once the ZF matrices are calculated, the power loading matrices $\mathbf{G}_{k,b}$ need to be found. As we assume that the transmitting nodes have only CSI from links within the cooperation area, the OCI is ignored for the calculation of $\mathbf{G}_{k,b}$. This allows to formulate the optimization problem

$$\max_{\left\{ \mathbf{G}_{k,j} \right\}_{\substack{j \in \mathcal{M}_c \\ k \in \mathcal{K}_c}}} \min \left\{ \tilde{R}_k \right\}_{k \in \mathcal{K}_c} \tag{4.11}$$

$$\text{s.t. } \mathrm{Tr} \left\{ \sum_{k \in \mathcal{K}_c} \mathbf{Q}_{k,b} \mathbf{Q}_{k,b}^H \right\} \leq P_x, \ \forall b \in \mathcal{M}_c, \tag{4.12}$$

where $\tilde{R}_k$ is the achievable rate that would result without OCI (the ICI is already nulled) and $P_x = P_B$ or P_S is the per node power constraint depending on whether the corresponding node is a BS or SN. Note that as there is no interference present, this optimization problem is convex and can thus efficiently be solved by standard optimization tools. However, in the evaluation of the rates, the OCI is taken into account. This form of cooperation requires accurate CSI for all links within the corresponding cooperation area. It is therefore of relatively high complexity.

4.1.4   Unified Framework

Depending on how the cooperation clusters are chosen and which infrastructure node is set to a BS or SN, different network configurations can be realized. A collection of example networks can be seen in Fig. 4.2. The network topologies that have already been discussed in Chapter 3 can be built as shown in Figs. 4.2a - 4.2c. The conventional network that acts as a reference is shown in Fig. 4.2a where the triangles with arrows correspond to BSs. There, each sector is served by a single BS[1] that operates independently of other BSs and with a transmit power of P_B. Consequently, the cooperation clusters contain only a single BS and no SNs (the nodes shown as gray circles are turned off). By turning off the triangles and considering the three circles in the corners of a cell as BSs that cooperate with each other, the 3-BS CoMP scenario with $30°$ orientation can be formed (Fig. 4.2b). Three sectorized BS arrays form a cooperation cluster that serves three adjacent sectors. Fig. 4.2c shows the somewhat more exotic 6 BS super-cell network configuration: six BSs placed on a ring around a center cell form a cooperation cluster. In this case, six BS antenna arrays can serve nine sectors, while the BS that would be located in the center cell of a conventional network is not in operation. Such a configuration can be used to save BSs and/or to compensate BS failures.

A network in which a BS is supported by relays or HNBs for range extension is depicted in Fig. 4.2d where the SNs are marked in light colors. These nodes can have omnidirectional or directed antennas. Fig. 4.2e shows a relaying scenario where each sector is assisted by additional relay nodes. In this case, a cooperation cluster consists of a BS (triangles in dark colors) and two sectorized relays (circles in light colors) located on the cell corners. The relays assist the communication within a sector with

---

[1]As in the previous chapter, we consider multiple BS arrays located on the same site as different BSs when they serve different independent sectors.

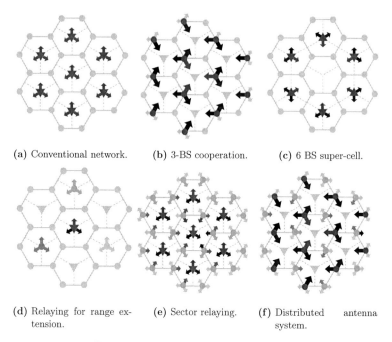

(a) Conventional network. (b) 3-BS cooperation. (c) 6 BS super-cell.

(d) Relaying for range ex- (e) Sector relaying. (f) Distributed antenna
tension.   system.

Figure 4.2.: Example network configurations.

a transmit power of P_S. Instead of relays, also HNBs can be used in this setup. A combination of BS cooperation and SNs can be seen in Fig. 4.2f, where each cell is served by three BSs and three additional low power nodes that can be HNBs or RRHs (in light colors) on the six cell corners. Such a DAS scenario resembles the six BS super-cell in a smaller format.

4.2 Simulation Results

In the following, the described transmission schemes are compared with each other in various network settings. We are particularly interested in the performance gain offered by the additional SNs, the impact the network topology has on the performance of the different schemes, but also in the robustness of the schemes with respect to CSI imperfections and the overhead they introduce in a practical system. To this end, we first consider a single remote BS whose coverage range is extended by an SN. In a later step, we then consider the different schemes in denser networks consisting of a multiplicity of nodes.

4.2.1 Relaying for Range Extension

In this section, we discuss the use of the selection scheme with SNs to enlarge the coverage of a single BS in a remote rural environment. In situations without (significant) interference, the transmit power is the main limiting factor of the communication between a BS and an MS. In such situations, simple relaying of the signals can be used to enhance the performance considerably. The environment is modeled by the WINNER II based channel model for the rural scenario D2 [76] as described in Section 2.4. The BS and the SN are equipped with $N_B = N_S = 4$ antennas, directed with the antenna pattern given in (3.25) and we assume a LOS component in the channels. As before, a single MS with $N_M = 2$ omnidirectional antennas is selected to be served by a BS sector in one resource block. The reference scenario without additional nodes is compared to networks where a single SN assists the communication of the BS. Different distances between BS and SN are chosen: 1, 2, 3, and 4 km. In the simulations, the BS transmits with full power of $P_B = 80$ W and the SNs can transmit with $P_S = 6$, 20, or 80 W. The noise induced in the MS and RS has a variance (power) of $\sigma_n^2 = 5 \cdot 10^{-12}$ W.

Fig. 4.3 shows the 5% outage rates and the outage probabilities for a target rate of 1 bps/Hz. The setting corresponds to Fig. 4.2d with FDMA frequency allocation, i.e. only one BS sector is considered. It can be seen that in the reference case without an SN, the outage probability drops below 5% at a distance of about 3 km from the BS and the 5% outage rate is below 2 bps/Hz after a distance of about 2 km from the BS. If an in-band relay is deployed that transmits with $P_S = 6$ W, the coverage is increased significantly. When a relay is placed at a distance of 2 or 3 km, the area with good performance (5% outage rate above 4 bps/Hz and outage probability smaller than 5%) can be doubled. If the relay is too close to the BS (1 km), the coverage increase is small and if the relay is too far away (4 km), the data rates start to drop between the BS and the relay. In this setting, placing a relay at 3 km from the BS seems therefore to be a good choice regarding coverage.

The performance of the selection scheme with in-band relays and femto-cells with out-band connection to the BS are compared in Fig. 4.4, where the 5% outage rates as well as the average user rates are plotted. The SN (relay or HNB) is located at a distance of 3 km from the BS. Regarding the coverage (area with high 5% outage

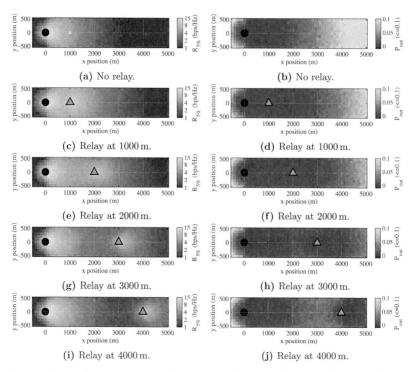

Figure 4.3.: A relay assists a single BS in a rural environment without interference. Different relay locations are compared. The BS transmits with $P_B = 80\,\mathrm{W}$, the relay with $P_S = 6\,\mathrm{W}$. 5% outage rates (on the left) and outage probabilities for a target rate of 1 bps/Hz (on the right) are shown.

rates), the two types of nodes do not show a considerable difference, even though the femto-cell is not affected by a prelog factor while the relay has to share resources for the two wireless links. The first hop is in our setup usually much stronger than the second one. Due to the 4×4 MIMO gain, the data rate can potentially be doubled as compared to the link to the MS with only two antennas. Therefore, a much smaller time or frequency slot can be allocated to this links than to the second one. The end-to-end performance is thus mostly limited by the RS-to-MS link, which is the same as with a femto-cell. A difference in the two schemes arises only close to the SN, where also a strong second hop link is present. There, the rate of the RS-to-MS link is in the same range as on the BS-to-RS link and the resources have to be shared when a relay

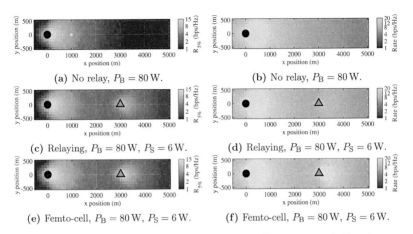

Figure 4.4.: Rural macro-cell with assisting small cell. 5% outage rates (left) and average user rates (right) for different types of supporting nodes.

is used. This is reflected in larger average user rates close the the HNB on the right hand side of the figure. Apart from this small area, however, the influence of the first hop is small and relays and femto-cells perform very similar.

So far, the BS transmit power was set to $P_B = 80\,\text{W}$, while the SN transmits with $P_S = 6\,\text{W}$. In Fig. 4.5, we look at the influence the transmit power has on the selection scheme. Thereby, we focus on the femto-cell case, as the results are very similar when in-band relays are applied. For the simulation, the femto-cell is treated as a full BS and both nodes transmit with the same power $P_B = P_S = 6, 20, 80\,\text{W}$. We show the 5% outage rates as a function of the distance between the MS and the BS. The values are averaged over the y coordinate and the plot thus resembles a cross section of the area plots shown before. It can be seen that the 5% outage rates gradually increase when more power is applied, as there is no interference in this network. If only the BS transmits with full power of 80 W and the femto-cell only with $P_S = 6\,\text{W}$, the performance is almost as good as when both nodes have 80 W. Also the use of an in-band relay with only 6 W leads to almost the same performance. Installing a node that transmits only with low power has already a large impact and coverage can be extended. Due to the transmit node selection, a MS located around 4 km from the BS can achieve almost the same 5% outage rate when the BS transmits with 80 W and the SN with 6 W as when both nodes transmit with $P_B = P_S = 20\,\text{W}$. Relaying is thus an efficient means for performance enhancement in remote areas with limited coverage

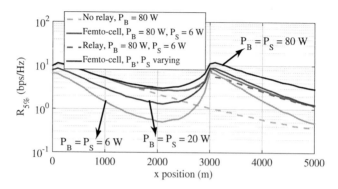

Figure 4.5.: DF relaying/femto-cells: average 5% outage rates for different transmit powers.

where installing new full BSs is difficult due to financial or other reasons. With relay and/or femto-cells, expensive BSs can be saved and the coverage of an existing BS can be enlarged considerably.

4.2.2 Heterogeneous Networks

In contrast to the scenario without interference discussed for the remote rural case, simple relaying in dense urban environments might not necessarily increase the performance since the relays cause additional interference to other users. Poor reception is thereby not caused by the weak receive power of the signals, but mostly due to strong interference. In this section, we study the potential of applying SNs in such interference limited situations. For the simulations, we again apply the WINNER II based channel model and consider the urban environment, i.e. the urban micro-cells as they are introduced in the chapter before. If not stated otherwise, the basic network model consists of 19 regularly arranged hexagonal cells, all divided into three sectors. The diameter of each cell is 700 m and three BSs (one for each sector) are placed in the center of the corresponding cell, each equipped with $N_B = 4$ antennas. Assuming a total bandwidth of 100 MHz, the total transmit power of each BS array is $P_B = 80$ W if not stated otherwise. The basic model is extended by additional SNs placed on each cell corner, i.e. two SNs for each sector, as shown in Fig. 4.2e. The transmit power of these nodes is denoted by P_S and is usually set to 6 W. The BSs and SNs are both equipped with $N_B = N_S = 4$ sectorized antennas directed to the corresponding sector. The antenna patterns are the same as introduced in (3.25). The MSs are equipped with $N_M = 2$

omnidirectional antennas. The noise power in the MSs and RSs is $\sigma_n^2 = 5 \cdot 10^{-12}$ W. We assume the relays to be dedicated infrastructure nodes which are intentionally placed at locations with good connection to their associated donor BS. As a consequence, we assume the channels between BSs and their associated relays to be with LOS with a pathloss given by (2.25). All other channels (particularly those to MSs and between nodes belonging to different cells) are NLOS with pathloss as in (2.24).

Modeling the interference scenario is more difficult with active SNs than in the scenarios considered in the last chapter. If in-band relays are deployed, which operate in half-duplex mode, they receive during a certain time interval and transmit in another, in which they also cause interference to other listening nodes. As the time slot assignment is optimized for each relay link, this results in differently long time intervals of transmission for different nodes. In order to avoid the optimization of the time slots for all links in the entire network and to predict the potential performance that can be guaranteed also for cell edge users with high probability, we again make a worst case assumption and assume that all nodes in cells other than the cell of interest transmit with full power of P_B and P_S, respectively. With this, also the interference from nodes which are closest to the test MS, whose data rates are evaluated, are considered. Again, no power control is applied and the performance is pessimistic as the cell of interest might be less affected by interference when other nodes throttle down their transmit power. In the cell of interest, only the nodes that are chosen by the selection scheme are active and transmitting while the others are idle for the current channel realization.

Relay Selection

First, we again focus on the selection scheme without beamforming. All nodes transmit spatially white and do not require any CSIT. In Fig. 4.6, we compare the selection scheme for in-band relays and out-band femto-cells to the conventional reference without additional SNs. The network corresponds to the setup depicted in Fig. 4.2e and the FDMA frequency allocation across the different cells is applied, i.e. each neighboring sector uses a different third of the available spectrum than the other two. Both the outage probabilities for a target rate of 1 bps/Hz as well as the 5% outage rates are plotted. We can see that the SNs clearly enhance the performance in the cell edges and the coverage of each sector is enlarged (the small squares on the cell corners where relays are located are not simulated). Close to the cell corners where an SN is located, the SINR and thus the rates are improved significantly. The data rates in the center of the cell close to the BSs, however, are not reduced noticeably as compared to the case without relays. The small transmit power of the SNs does thus not add much interference, even though the worst case scenario is considered in which all nodes (BSs and SNs) of neighboring sectors transmit with full power. The coverage, however, is not as homogeneous as with the BS cooperation discussed in the last chapter; areas

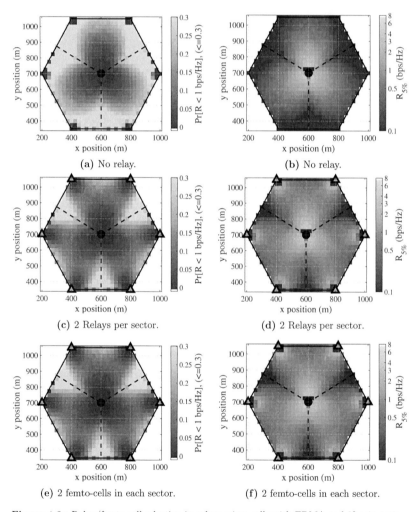

(a) No relay.

(b) No relay.

(c) 2 Relays per sector.

(d) 2 Relays per sector.

(e) 2 femto-cells in each sector.

(f) 2 femto-cells in each sector.

Figure 4.6.: Relay/femto-cell selection in urban micro-cells with FDMA and 0° orientation.

with low rates and high outage probability remain. The performance in the triangles on the cell edges between two SNs is not improved. In order to further enlarge the coverage also to these areas, additional SNs could be placed in these locations. As already observed in the rural case discussed before, in-band relaying and femto-cells with out-band HNBs show only a small difference. If the time slot assignment in the case of relaying is optimized, most of the transmission time is again allocated to the second hop, while the one for the first hop is small. The four spatial streams on the link between BS and SN and the higher signal powers due to the LOS pathloss on these links lead to a higher data rate on the first hop.

In the next figure (Fig. 4.7), we compare the FDMA frequency allocation (in the left column) with reuse 1 (right column) for the same setup. Here, the average user rates for each position are shown. This measure emphasizes the higher data rates that can be achieved close to the transmitting nodes in the reuse 1 case. Especially in the main beam of the BS, much higher data rates are achieved than with FDMA as no prelog factor of $1/3$ is applied. Also the data rates around the SNs are higher. The areas with poor performance are however larger than with FDMA, as it can be seen by the larger triangles on the cell edge between the SNs. In these places, the higher interference from the neighboring cells kicks in and decreases the SINR. Also here, the performance could be improved by installing additional SNs in these spots. In the reuse 1 case, in-band relays are not as good as the selection scheme with femto-cells. Close to the relay, higher second hop rates can be achieved, while the first hop rates are reduced due to the increased interference in this case. The rates on the two hops are thus a bit more balanced and a smaller fraction of the time can be allocated for the BS-to-RS link. Due to the time sharing, the femto-cells thus show some higher rates close to the SNs, but the difference is small.

In Fig. 4.8, the empirical CDFs of the SN selection scheme for both the FDMA and reuse 1 frequency allocation are shown. The curves confirm the better coverage of FDMA. Except in the high rate regime, the dark solid lines of the FDMA case show higher rates. On the 5% outage line, the data rates are almost doubled as compared to the reuse cases. Reuse 1 only leads to a bit higher peak rates. The difference between in-band relays and out-band femto-cells is almost not visible. When the SN power is increased to $P_S = 80\,\mathrm{W}$, the data rates are further improved, with a larger difference with FDMA than with reuse 1. In the latter case, the network is more in the interference limited regime and a higher transmit power has thus not much impact as also the interference is increased by the same amount.

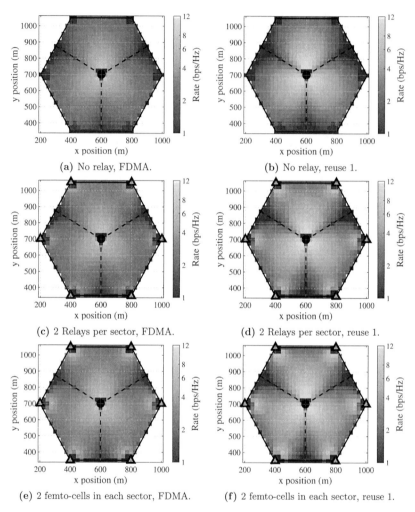

Figure 4.7.: Average user rates for relay/femto-cell selection in urban micro-cells. FDMA frequency allocation (left) is compared to reuse 1 (right).

Rural Macro-Cell

The SN selection scheme for the case of rural macro-cells is shown in Fig. 4.9. For these simulations, the same network architecture as before is applied, but the distance

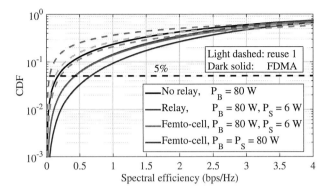

Figure 4.8.: Empirical CDFs of instantaneous user rates in an urban micro-cell.

between adjacent BSs is enlarged to 1500 m and the channels are drawn from the model in the rural scenario. Over all, the performance gain with SNs is larger than in the urban case. As the BSs are more distant from each other, the network is less affected by interference (especially with FDMA) and the additional power of the SNs helps in enlarging the coverage. For this reason, also the difference between in-band relays and out-band femto-cells is larger. Because the interference is smaller and the desired signal components stronger due to the somewhat smaller pathloss, the second hop rates are higher than in the urban case relative to the ones of the first hop. The optimized time sharing leads thus to a larger fraction for the first hop. With out-band SNs where the first hop does not have to be taken into account, the end-to-end rates are higher than with in-band relays. The difference increases further in the high rate regime (above 1 bps/Hz in the CDF) when the FDMA frequency allocation is applied. In the reuse 1 case, this effect is less pronounced, as more interference is present and the network behaves similar to the FDMA case in urban networks. Nevertheless, installing additional SNs can also help here in order to enlarge coverage and provide better service to cell edge users.

Joint Beamforming with Supporting Nodes

If SNs are more sophisticated nodes or RRHs, they can also contribute in joint beam-forming with the BSs. If all involved nodes are connected together with a backhaul link of sufficient capacity, this scenario corresponds to a DAS. In this section, we evaluate

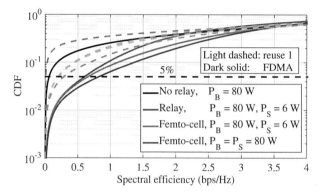

Figure 4.9.: Empirical CDFs of instantaneous user rates in a rural macro-cell.

the performance of DASs with block ZF and max-min optimization. To this end, we apply again the network architecture depicted in Fig. 4.2e, with three sectors where the BSs are located in the middle of the cell. The three BS sectors cooperate with each other and the transmission is assisted by six additional SNs in the cell corners, which are also sectorized and the three antenna arrays of each of them points towards the center of their respective cell. The transmit power of the BSs is fixed to $P_B = 80$ W, while the transmit power of the SNs is varying between $P_S = 1$ dBm and 49 dBm (1 mW to 80 W). With this, we can study the influence of the SNs and their transmit power on the CoMP scheme. If the SNs are turned off ($P_S = 0$ W), the scenario corresponds to sector cooperation as introduced in the last chapter, if the power is set to the maximal value of $P_S = 80$ W, the three MSs in the cell are served by one virtual antenna array that consists of 9 full-fledged individual antenna arrays that form one cooperation area. In all cases, we make again the worst case assumption for the interference and assume that all nodes in the neighboring cells transmit with the same power of P_B and P_S, respectively.

The empirical CDFs of this setup are shown in Fig. 4.10, for the reuse 1 frequency allocation. The CoMP case is also compared to the non-cooperative reference (the three BS arrays operate independently of each other and the SNs are turned off) and to the selection combining scheme with femto-cells (relays would lead to a similar performance), both with FDMA. It can be seen that the coherent cooperation with block ZF and optimized power allocation can profit more from increasing P_S than selection combining. Even with a very small transmit power of $P_S = 1$ dBm = 1 mW,

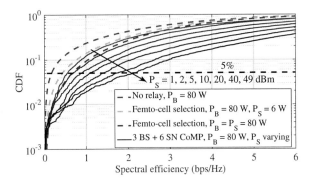

Figure 4.10.: CDFs of instantaneous user rate for 3 BS cooperation with 6 assisting SNs with varying transmit powers. The BSs transmit with $P_B = 80\,\text{W}$.

the CoMP scheme achieves the same performance as the selection scheme when the SNs transmit with $P_S = 80\,\text{W}$. When the SN power is increased, the performance steadily improves until the 5% outage rate is approximately tripled as compared to the selection scheme. With increasing SN power, however, also the interference in the network increases.

The selection combining scheme, on the other hand, already achieves notable improvements with much lower complexity, while the full CoMP scheme transmits with 9 antenna arrays to 3 MSs and a correspondingly large precoding matrix has to be calculated. The high performance with the DAS requires thus high computational complexity and backhaul access of high capacity to all involved antenna arrays in the cooperation set. Moreover, the selection scheme has further advantages regarding robustness, as we will see in Section 4.2.4.

While the CoMP scheme with SNs as studied before comprises nine antenna arrays for the transmission to only three MSs, a high performance can be achieved. Many excess antennas are however used for this setup. In the following, we look at the same cooperation strategy with block ZF and max-min optimization but with an adapted network geometry. We apply the CoMP scenario with three cooperating BSs with the 30° orientation and place only three additional SNs into the corners which have no BS. This CoMP setup with three BSs and three SNs corresponds to Fig. 4.2f and resembles the six BS super-cell in a smaller dimension. With this setup, we hope to maintain much of the performance gain of the DAS but with smaller complexity as less antenna arrays are involved. By placing one antenna array in each of the corners of the cooperation area, we expect a similar behavior as the six BS super-cells with homogeneously distributed data rates in the entire area.

The area plot of the 5% outage rates for this setup, when the SNs transmit with $P_S = 6\,\text{W}$ and the BSs with $P_B = 80\,\text{W}$, is shown in Fig. 4.11d. For this simulation, we applied the FDMA frequency allocation, i.e. the three cooperating sectors use a different frequency than the direct neighbors. As with the other max-min schemes discussed in the previous chapter, a very homogeneous rate distribution with high 5% outage rates can be observed. Even more homogeneous and better than with three BS cooperation or the six BS super-cells. With the FDMA frequency allocation, the interference is small and the additional three SNs can also provide a significant contribution to the performance. By transmitting jointly from all sectors, with four times the number of antennas than the MSs, data rates of more than 2 bps/Hz can be

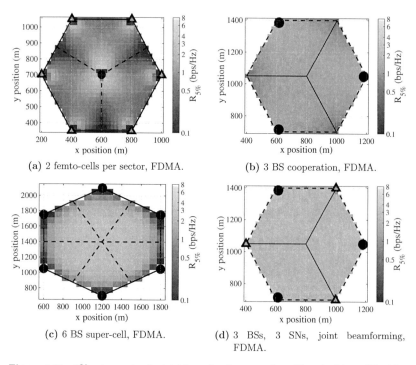

(a) 2 femto-cells per sector, FDMA.

(b) 3 BS cooperation, FDMA.

(c) 6 BS super-cell, FDMA.

(d) 3 BSs, 3 SNs, joint beamforming, FDMA.

Figure 4.11.: 5% outage rates for joint beamforming scenarios with and without SNs. The femto-cell selection scheme is also shown.

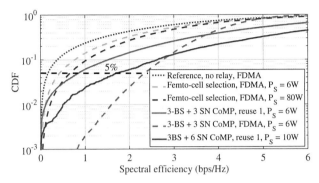

Figure 4.12.: CDFs of instantaneous user rate for BS and SN cooperation with varying transmit powers. The BS arrays transmit with $P_B = 80\,\text{W}$.

achieved with high probability almost everywhere in the network. For comparison, the femto-cell selection scheme with two SNs per sector as well as the cooperation scheme with three BSs and six BS super-cells are also shown in Fig. 4.11. In comparison to the three BS case, the average 5% outage rates can significantly be boosted by the SNs. Also compared to the six BS super-cells, the coverage is better, especially at the cell edges. This is due to the smaller cell size which leads to more balanced signal strengths in the cell and the higher number of excess antennas which leave more room for optimization. The data rates of the SN assisted cooperation scheme are however smaller than the peak rates in the selection scheme. This is due to the max-min optimization.

The empirical CDFs of the three BS plus three SN setup is shown in Fig. 4.12, where we compare this setup to the CoMP setup with three BSs and six SNs, as well as with the reference and the SN selection scheme. The cooperation with three BSs and three SNs is shown for both the FDMA as well as the reuse 1 case. In the regime above the 5% outage line, the smaller CoMP setup with only three SNs clearly outperforms the SN selection and the reference, for both frequency allocations. In the reuse 1 case, it shows a similar behavior as the CoMP scheme with six SNs but shifted by about 1 bps/Hz to the left. With less antennas, the cooperation scheme is thus less effective, but the SNs still improve the coverage as well as the individual user rates as compared to CoMP without the additional SNs. In the FDMA case, the cooperation scheme shows a significant improvement in the low rate regime. In the high rate regime, however, the rates are smaller than with the selection scheme due to the max-min optimization as already observed in previous simulations; the data rates are much more homogeneously distributed. Hence, the peak rates are smaller in this case than with the same scheme in the reuse 1 frequency allocation. With the latter, no FDMA prelog loss of 1/3 affects the rates and close to the BSs or SNs higher rates can be achieved.

4.2.3 Comparison of Different Schemes

In this section, we compare the most promising SN assisted communication scenarios with BS cooperation discussed in the previous chapter. Thereby, we focus on the dense urban scenario with micro-cells. Fig. 4.13 compares the various schemes in their best cell layout and sector orientation. The non-cooperative reference forms the baseline of a conventional network in the configuration of Fig. 4.2a with FDMA frequency allocation. This reference is clearly outperformed by all other schemes. The selection scheme with femto-cells in the network layout of Fig. 4.2e (and similarly with in-band relays) leads to improvements in the entire network (in all rate regimes shown in the CDF) with low complexity as no CSIT is required in this case and each MS selects the best transmitting node (either the BS or one of the two SNs in the corresponding sector) for which only rate feedback is required. This selection scheme can be improved if the MS of interest has a wider choice of transmitting nodes, as shown by the macro diversity UB (here with reuse 1 and the network layout of Fig. 4.2a), which leads to significantly higher rates on the 5% outage line. This UB could be further increased, when additional SNs were placed in the network, preferably towards the cell edges where no BS is located.

When three BSs cooperate to perform joint beamforming with max-min optimization, the performance above the 5% outage line can be further improved (green solid line); the three BS CoMP scheme with FDMA and in the 30° orientation (Fig. 4.2b) shows a more homogeneous rate distribution. When this joint beamforming scheme is extended to also include additional SNs, the high number of antennas in the virtual antenna arrays leads to a higher performance. In the reuse 1 frequency allocation, the data rates with six SNs (the setup of Fig. 4.2e) are the highest above 10% with the best performance among all considered scenarios. In the low rate regime, the performance is a bit lower than three BS CoMP with FDMA due to the higher interference the additional SNs cause. When FDMA is applied, the SN assisted CoMP scheme leads to a more homogeneous rate distribution as it can be seen by the steeper slope of the CDF, which has a similar behavior as the 3 BS CoMP. The additional antennas of

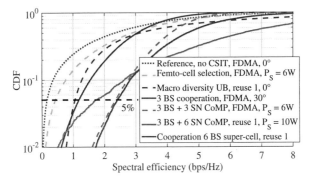

Figure 4.13.: CDFs of instantaneous user rate for different schemes with and without SNs.

the three SNs however, shift the curve to higher rates; the 5% outage rate is further improved, even with only three SNs instead of six. The cooperation area in this setup leads to smaller interference between different cooperation sets due to FDMA, while the six SNs increase the interference in the network which affects especially the cell edge users. The six BS super-cells achieve a very similar behavior as the CoMP with three BSs and three SNs, even though the super-cells comprises a larger area in one cell. Due to the larger distances between the BSs, a similar interference scenario can be observed as with FDMA in the smaller cells. The evenly distributed antenna arrays in all corners of the cooperation area lead to balanced signal strengths in most of the area and high data rates can be achieved almost everywhere.

The main results of BS cooperation, with and without the assistance of SNs, are summarized in Table 4.1, where the key performance indicators defined in Chapter 2 are given. In can be seen that also here the selection scheme with in-band relays as well as out-band femto-cells achieve a very similar coverage and average 5% outage rates: 27% and 0.89 bps/Hz for femto-cells and 25% and 0.87 bps/Hz for in-band relays. The performance loss due to sharing the same physical channel for both hops is small when the time sharing is optimized. When the relay/femto-cell power is increased from 6 W to 80 W, the performance is with a coverage of 50% and an average outage rate of 1.08 bps/Hz already comparable to the macro diversity UB which is a bit better. The SN selection scheme is however of lower complexity as less nodes are involved and the scheme is limited to a single sector. The joint beamforming with three BSs achieves a coverage of 86% when no SNs are used. When three additional SNs are deployed that can also contribute in the joint transmission, this can be extended to

Scheme	Configuration	Coverage	Av. $R_{5\%}$ (bps/Hz)
Reference	FDMA, 0°	15 %	0.53
Beamforming	FDMA, 0°	24 %	0.77
Macro diversity UB	reuse, 0°	61 %	1.19
3 BS cooperation	FDMA, 30°	86 %	1.26
6 BS super-cell	reuse, super-cell	77 %	1.88
Relay selection	FDMA, 0°	25 %	0.87
Femto-cell selection	FDMA, 0°	27 %	0.89
Femto-cell 80 W	FDMA, 0°	50 %	1.08
3 BS + 3 SN cooperation	FDMA, 30°	99 %	2.38

Table 4.1.: Key performance indicators for the different schemes in dense urban micro-cells.

almost full coverage (99%) and the average 5% outage rate can almost be doubled to 2.38 bps/Hz. This is considerably higher than with the six BS super-cells. In the SN case, however, the double amount of antennas is applied relative to the number of MSs. The main observations from the previous chapter thus carry over also to the case when additional SNs are deployed: The selection scheme achieves a good performance with low complexity when a large number of nodes is present from which the MSs can choose from. The block ZF with max-min optimization leads to very homogeneously distributed date rates with a high coverage and the performance improves with an increasing number of transmit antennas. In order to get low outage probabilities, however, the interference between the locally restricted cooperation areas has to be kept low. To this end, the sector orientations have to be adapted to the geometry of the virtual cells and neighboring cells should be separated, e.g. with sectorization and different frequency bands. The best coverage can be achieved with FDMA when the transmitting antenna arrays are located around the cooperation area and point towards its center. The 30° orientation, preferably with antenna arrays in all six corners, as well as the six BS super-cells are thereby particularly beneficial.

4.2.4 Robustness to Imperfect CSI

So far, perfect CSIT was assumed for all transmission schemes. For the CoMP with or without additional SNs, full channel knowledge of all links within a cooperation area is required to perform block ZF and max-min optimization. In the case of the selection scheme, perfect rate feedback from the MSs is assumed. As the latter is simpler to acquire than full CSIT and because the data rates required for the selection are mostly dominated by the pathloss, we expect that the selection scheme is more robust with respect to wrong or outdated channel knowledge. In the following, we briefly evaluate and discuss the performance loss of the different schemes when the CSI is affected by imperfections. To this end, we model imperfect CSI as follows. Instead of the true channel matrix $\mathbf{H}_{k,b}$, the transmitting nodes use an estimate

$$\hat{\mathbf{H}}_{k,b} = \sqrt{\frac{1}{L_{k,b}}} \left(\sqrt{1 - \vartheta^2} \mathbf{H}_{k,b} + \vartheta \mathbf{W}_{k,b} \right), \tag{4.13}$$

where $\mathbf{W}_{k,b}$ is the estimation error with elements i.i.d. $\mathcal{CN}(0,1)$ and $\vartheta^2 \in [0,1]$ the CSI noise scaling factor. This model captures effects as outdated CSI (if MSs are moving) as well as estimation errors. Note that the pathloss $L_{k,b}$ is not affected by CSI noise, as we assume that its estimation is much easier in practice than that of the actual fading coefficients and that it can be averaged over multiple transmission blocks.

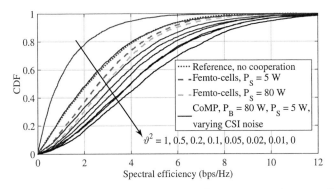

Figure 4.14.: CDFs of instantaneous user rate for BS cooperation with varying CSI noise scaling. The BS arrays transmit with $P_{\mathrm{B}} = 80\,\mathrm{W}$, the SNs with $P_{\mathrm{S}} = 5\,\mathrm{W}$.

Fig. 4.14 shows the behavior of the same setup as in Fig. 4.10 with three BSs and six SNs in the cell corners, but with fixed transmit power $P_{\mathrm{S}} = 5\,\mathrm{W}$ and varying CSI noise scaling factors ϑ^2. The curves show that for reasonable choices of ϑ^2, the CoMP scheme is quite robust and outperforms selection combining; only if $\vartheta^2 \geq 0.5$ ($\vartheta^2 = 1$ corresponds to no CSI at all), the performance is worse. In this case, block ZF fails completely and the max-min optimization allocates power to the wrong streams. Better performance can in this case be achieved with robust beamforming schemes such as described e.g. in [12]. A CSI estimation SNR, defined as $\mathrm{SNR}_{\mathbf{H}} = \frac{1-\vartheta^2}{\vartheta^2}$, of about 10 dB is however sufficient for a good performance. For higher inaccuracies, the performance of CoMP decreases but relaying does (almost) not vary with ϑ^2 (cf. also Fig. 4.15). The reason for this is that the link selection requires only knowledge of the link quality which is essentially given by the second order statistics of the channel, which corresponds to $L_{k,b}$. This estimation, in turn, can be assumed to be very robust due to the diversity offered by the MIMO channels.

Even though CoMP achieves higher gains than selection combining, at least for accurate CSI, relaying and/or femto-cells have still some advantages. Fig. 4.15 shows the 5% outage rates that are achieved for the different schemes. It can be seen that the selection scheme proves to be a better choice than CoMP when the CSI imperfections are too large. However, even with low transmit power and high CSI noise, the cooperation scheme exceeds the target rates for cell edge users in the LTE-Advanced. These targets are specified to be 0.07 bps/Hz for 2×2 MIMO transmission, 0.09 bps/Hz for 4×2, and 0.12 bps/Hz for 4×4 MIMO transmission [1].

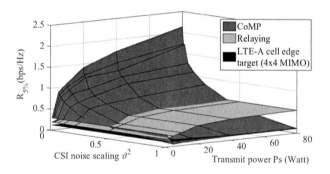

Figure 4.15.: CoMP with SNs vs. relay selection with varying SN transmit power and varying CSI noise scaling.

The spatial distribution of the 5%-outage rates of the network configurations from Fig. 4.2b with FDMA frequency allocation are shown in Fig. 4.16 and compared to the selection scheme with femto-cells (Fig. 4.2e). Plots for 3 BS cooperation with imperfect CSI are also shown. Here, we do not apply the CSI error model from (4.13), but assume perfectly estimated CSI that is quantized with 3 and 4 bits per real and imaginary dimension of each channel coefficient. To this end, a scalar linear and uniform quantizer is applied to the small scale fading, while the second order knowledge

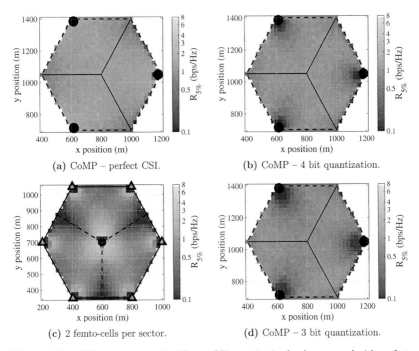

Figure 4.16.: 3 BS cooperation with different CSI quantization levels compared with perfect CSI and femto-cell selection. Urban area with FDMA frequency allocation.

(pathloss) remains unaffected.[2] It can be seen that the performance is only slightly decreased when the CSI is quantized with fewer bits, at least in large parts of the network. Close to the BSs, however, the performance drop is more severe. In these locations, the residual interference cannot be suppressed sufficiently due to the quantization noise that destroys the block ZF. In this case, the interference for MSs close to the BSs is strong, which limits the data rates. In these areas, however, the direct transmission from a single BS without cooperation can be applied, which would lead to even higher rates than with perfect cooperation. In the rest of the network area,

[2]The results of the cooperation with quantized CSI have been published by G. Psaltopoulos, R. Rolny, M. Kuhn, M. Kuhn, and A. Wittneben in "Future Cooperative Wireless Networks, Deliverable 4," ETHZ, Tech. Rep., Oct. 2011, available on request.

cooperation leads to a high performance increase with cooperation, also with coarse quantization. Relaying, on the other hand, offers an interesting alternative when no (sufficiently accurate) CSI is available.

4.3   Critical Discussion

In this Chapter, we have evaluated and compared the performance of SN assisted cells in various network settings. Based on the insights from the previous chapter, we extended the schemes introduced before to the case where additional nodes are deployed. These can comprise DF relays, femto-cells HNBs, or RRHs to form DASs. The selection scheme and the joint beamforming with block ZF and max-min optimization reflect therefore a simple and a sophisticated transmission strategy. While the transmit node selection benefits from a diversity gain within a single cell where the MSs can choose the best transmit node within a limited set, the joint beamforming scheme allows the MSs to be served by a plurality of nodes distributed to different locations. The former already leads to considerable performance gains with low complexity as no CSIT is required. With the latter scheme, higher data rates and better coverage can be achieved, but with the price of much higher complexity. As with BS cooperation discussed in the previous chapter, the transmitting nodes require accurate CSIT to calculate and optimize the precoding matrices. Moreover, the knowledge has to be delivered and shared with all contributing antenna arrays within the cooperation area. This can either be done in a distributed way such that all involved nodes exchange their CSI and data symbols and each node calculates the beamforming matrices individually, or in a distributed way. Thereby, the central unit (e.g. a BS) collects all the required CSI and calculates all precoding matrices and disseminates the transmit signals then to the other antenna arrays of the cooperation set. In this way, the SNs can be seen as RRHs that form a DAS together with the central BS. In the distributed approach where each node calculates and builds its transmit signal itself, the SNs are more attributed to HNBs or possibly relays. In either case, all nodes need to be connected with each other by a link of sufficient capacity and significant computational capabilities are required to perform the necessary calculations.

For the evaluation of the different schemes, different network setups are applied that arise as extensions from the basic setup of conventional cellular networks. In most setups, the SNs resemble the shared relay approach from [101]. The SNs are placed on the corner between three different cells and support the transmission in each of them simultaneously. While [101] reports good performance only when the relays apply sophisticated signal processing to cancel or mitigate interference that these relays cause to the different transmissions, we have achieved this with the static sectorization of the antenna arrays. Thereby, the interference caused by the SNs can be kept low and relays and femto-cells achieve significant improvements even without CSIT. Due to the sectorization of the SNs, the interference level in the network is not increased significantly. This observation however holds only for the considered network setups. Different performance and behavior could be observed when the SNs were placed at different locations or when different cell or sector configurations were applied. The chosen setups however seem to be a good choice, as the SNs are placed in locations that have poor performance when only BSs serve the users. Nevertheless, the performance can potentially be further improved by installing more relays or femto-cells. As reported in e.g. [146], a better tradeoff between number of nodes and performance can be achieved with four relays per MS additional to the direct link from the BS. With this, however, also the costs of the network increases, especially when the more sophisticated joint transmission is applied. Our studies are therefore limited to the cases discussed here. Networks where more relays are deployed, but made simpler so as to implement them with lower complexity and thus with lower costs, are discussed later in this work.

4.3.1 Limitations

Apart from the limited selection of network geometries considered in this chapter, also other limitations have to be considered in the appreciation of the results. The cooperation areas are still static. Therefore no flexible choice of transmitting nodes across different cells is possible, which might lead to additional diversity gains as indicated by the macro diversity UB. Also all network geometries discussed in this chapter consist of regular hexagonal cells, which cannot be realized in practice. A more random distribution of the different nodes might change the interference situation and thus the achievable rates. BS cooperation and the use of relays in irregular networks are considered in Chapter 8. There, however, with AF relays instead of DF relays or femto-cells, as they can potentially be implemented with lower complexity and offer benefits when users are served by multiple relays, as we will see.

For the calculation of the precoding matrices, the OCI is ignored in the cooperation process when the joint beamforming approach is applied. This simplification leads to a convex optimization problem that can be solved efficiently, but it is suboptimal with respect to the achievable rates. For the joint beamforming, only the max-min approach was applied here. With other objective functions such as sum rate maximization, individual user rates can be increased while users in locations of worse coverage would be further decreased. Further improvements can be achieved when the cooperation scheme also attempts to mitigate the interference the cooperation area causes to other cells. Especially when SNs are installed, a high number of excess antennas are available. While a larger number of transmit antennas improves the achievable rates within the cooperation area, excess antennas could also be used to cancel or at least reduce the OCI that is generated by them. This would lead to a tradeoff, as less OCI would improve the achievable rates, while canceling this OCI leads to less room for optimization of the performance within the cooperation area. In Chapter 8, we will apply a different precoding strategy, the SLNR approach, which attempts to solve this tradeoff. With this scheme, the ratio of the desired signal strength and the interference to other, also out-of-cell, users is maximized. With this, also OCI can be managed and no conditions of the number of antennas have to be fulfilled as it is the case with block ZF.

In the selection scheme, on the other hand, only a single node is selected for the transmission to each user. With other transmission schemes and relaying architectures, the direct link of the BS and the links via SNs can be combined. DF relaying has however a disadvantage when signals from multiple relays are combined for a single user, as the data rate of the relays is limited to the one with the weakest link. Only then, all relays can decode successfully and contribute to joint signaling. To this end, we will also consider AF relays which can also be used in multiple numbers but without applying a joint transmission scheme as the block ZF used here.

As in the chapter before, no power control is applied to the transmission schemes. The selected nodes in the cooperation area of focus transmit with full power, as well as all nodes of cooperation areas adjacent to it. With this worst case assumption, the potential of the considered schemes can also be utilized for cell edge users which might be affected by such high interference from neighboring cells. In practice, however, the interference situation would generally look less severe when not all nodes transmit with full power. In Chapter 7, we also look at networks with relays in which transmit power control at the different nodes is applied. With this, further performance gains can be achieved.

An additional limitation that arises from the considered scenarios is that the transmission schemes are designed for perfect CSIT at all involved nodes. Even though the block ZF with max-min optimization is quite robust with respect to CSI imperfections, more robust schemes can be applied. When CSI inaccuracies are already taken into account by the derivation of the precoding matrices as e.g. in [12], better results are possible.

4.4 Conclusions

In Chapter 3, we have seen that block ZF with optimized power loading on the different streams achieves data rates that are homogeneously distributed in the entire area of service. If complexity and the costs of installing more BSs is not an issue, the CoMP areas should be realized with as many BSs as possible. With a sufficient BS density, high data rates can be provided. Otherwise, the cooperation has to be restricted to clusters of limited size such that the joint transmission is locally focused.

In such locally restricted cooperation areas, we have seen in this chapter that the deployment of SNs can increase the data rates. Thereby, as many RRHs as possible should be installed into the cooperation area. With additional SNs, the data rates are still homogeneous with the max-min optimization, but increased significantly due to the additional antennas and the higher density of transmitting nodes. Surprisingly, CoMP is rather robust with CSIT imperfections and quantization. The simulations show that only a small rate loss has to be accepted if the CSI noise scaling factor ϑ^2 is not higher than 0.02 or 0.01, which leads to a similar performance as when the errorless CSI is quantized with a simple scalar uniform quantizer with 3 and 4 bits per dimension, respectively. For a frequency selective 2×4 MIMO channel of 100 MHz bandwidth with 20 relevant channel taps between infrastructure nodes and MSs, and assuming that the CSI is updated by the MS every 10 ms, the resulting feedback rate required for the CSI dissemination does not exceed 288 kbit/s for 3 bit quantization or 384 kbit/s if 4 bit quantization is applied. In either case, the LTE-Advanced uplink rates (or future versions of cellular network standards) will certainly be able to support these rates. Compared to the dissemination of user data, also the backhaul rates are not affected too much when CSI is exchanged between different BSs. The considered schemes with the shown gains thus seem possible for implementation in cellular networks of the upcoming generation.

CoMP combined with RRHs shows the best results. Such DASs with joint beamforming can thus solve the problem of interference limitedness to a large extent and quite high data rates can be achieved in the entire network. When SNs are used in the joint beamforming scheme, however, the computational complexity to acquire the CSI and to perform the precoding calculations might lead to implementations that are too complicated and expensive. A massive deployment of SNs might therefore be infeasible in certain scenarios due to the costs and the high computational complexity. Relay selection offers in this case a simpler alternative. Areas with poor reception can be improved by installing relays or HNBs. These additional nodes can provide coverage in areas where the BSs can offer only poor service due to weak signals or strong interference. The additional nodes however also increase the interference in other locations. Interference management is thus necessary in this case.

In practical networks, CoMP can also be combined with the selection scheme. So can CoMP be applied in nodes which have sufficient capabilities to perform joint precoding and when the channels to the corresponding users are only slowly fading or static. In areas where the sophisticated precoding is not feasible or when users move with higher velocity, transmit node selection can be applied. With a TDMA or FDMA scheme, the transmitting nodes can also operate simultaneously with the different schemes and serve MSs with both strategies.

5

Distributed Cooperation with AF Relays

In the previous chapters, we have studied the potential of cellular networks with simple cooperation and relaying schemes that we applied in a straight forward way. Thereby, the DF relays offer a conceptually simple way to enhance the performance in locations with poor reception. Due to the additional signal power the relays offer, higher data rates can be achieved when the direct signal from the BS is weak at the corresponding MS due to high pathloss or when the MS is affected by strong interference.

DF relays are however of relatively high complexity, as they need to decode the entire signals before they can re-encode (possibly with a new codebook) and retransmit them. AF relays that only amplify the received signal (or form linear combinations of their receive signals if they are equipped with multiple antennas) might therefore be an interesting alternative. They do not only promise gains in terms of coverage, but when combined with more sophisticated signal processing, might also turn out to be valuable means for enhancing the efficiency of MIMO communication. For example, a mobile device in communication range of RSs can exploit them in order to enforce well conditioned MIMO channel matrices even though the environment provides only little scattering. This potential of relay nodes to act as "active scatterers" has been identified in [111]. Another advantage of AF relays as compared to the DF counterpart is that multiple RSs can act together and cooperate with each other. If DF relaying would be applied, they all have to be able to decode the signals, which limits the data rate to the RS with the weakest signal. This is not the case with AF relays, as they all can contribute to an effective two-hop channel as they do not distinguish between desired signal, interference, or noise. This is particularly interesting in networks of continuously growing node densities, as there the performance is severely affected by interference. Sophisticated transmission schemes that can exploit interference to mitigate these impairments and allow for an efficient use of the available resources are thus required. Furthermore, AF relays are easy to implement, especially if they are realized as FDD devices, they avoid coding delays, and are fully transparent to the modulation alphabet used by the terminal nodes (BSs and MSs). This is particularly interesting in heterogeneous networks with several nodes of different complexity.

In this chapter, we look at more sophisticated relaying schemes that can exploit the available resources in a more efficient way and can contribute to an effective interference management. To this end, we apply MIMO AF relays that do not decode but forward optimized linear combinations of their receive signals. Wile promising gains can be expected by the use of coherent AF relays (see e.g. [10]), it is known to achieve the capacity of two-hop networks only asymptotically in the number of nodes [14]. In networks with a finite number of relays, however, AF relaying is suboptimal. An inherent drawback of non-regenerative relaying systems (as AF) is the accumulation of relay and destination noise. In AF-based systems, this accumulation arises through amplification and forwarding of the noise components of the receive signals of the RSs. With an appropriate choice of relay transformation matrices, however, the performance can be optimized. In [97], it is shown how the relay gain matrix can be calculated such that the end-to-end rate of a two-hop link is maximized. Also multiple users can be served in parallel with relays that shape the effective channel between the terminals in a way that the relays can cancel interference or that other criteria as e.g. the sum rate or the MMSE of the receive signals are optimized [10]. AF relays are thus an efficient means to bring the effective channel between sources and destinations into a beneficial form. With this, the full spatial degrees of freedom of a relay network can be exploited, even without cooperation or interference management at the terminals [64]. To this end, it is important to efficiently exploit the CSI to the extent it is available at the relay nodes in order to benefit from them.

In the following, we consider two different AF relaying approaches. First, we develop a sum rate maximization scheme that allows each node to calculate its transformation matrix with local CSI. This distributed gain allocation scheme is applied to a quite general multihop network setup that can comprise multiple relay stages between the terminals and also includes precoding at the source nodes that can be identified with BSs. In a second approach, we attempt to make the optimization more efficient and consider a block ZF approach in which the relays completely cancel the interference between the different terminals. We apply this block ZF approach to relay networks that are generalized to two-way relaying and are able to form linear combinations over multiple subcarriers of a wideband system. By this scheme, we can show that such a subcarrier cooperation can offer further gains than when each subcarrier is treated independently. Moreover, we outline how such a scheme can be applied to cellular networks.

Both schemes show a high potential for large performance gains, as we will see. The optimization of the relay gain matrices is however rather complex. We will therefore limit ourselves to maximize the sum rate in simplified network setups where network geometries or distance dependent channel models are not included. Nevertheless, the insights provided by the studies allow to draw conclusions on how AF relays can beneficially be applied in practical mobile communication networks. In the following chapters, we will thus develop simpler schemes that can achieve similarly good performance with lower complexity and apply them in large numbers to more practical networks.

5.1 Multihop Networks

Different research results suggest that the use of relays such as in multihop communication via multiple relay stages can be a cost effective solution to provide ubiquitous access to high data rates. In such multihop networks, source nodes (which can be identified with BSs if the DL is considered) transmit signals which are received and forwarded by multiple relays. Each relay stage thereby forwards its signals to the next stage until they reach the destinations, as depicted in Fig. 5.1. With this, long distances between the terminals can be overcome and the signal strengths that would be very weak at the destinations due to the high pathloss can be boosted significantly by the relays. In [78], the potential gains and the challenges that have to be solved for practical applications such as in cellular networks are summarized. Among others, finding an optimal path through the network for each signal from the source to the intended destination, possibly via multiple hops, is a difficult problem. Thereby, the delays that each intermediate stage (relay) introduces has to be traded with the power gain that this node can offer due to smaller distances of the wireless links and the interference that the different signals cause to each other. Such a routing protocol that attempts to optimize the resources is provided e.g. in [20]. Particularly in environments that are affected by strong shadowing, the use of multiple relays can enhance the performance considerably. Therein, however, DF relays were applied to the multihop network. With

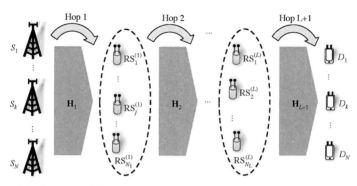

Figure 5.1.: Structure of the considered multihop network. Destination node D_k is interested in the message sent by source node S_k, while other sources are interfering. The communication is assisted by intermediate stages of relays.

AF relays, the signals can implicitly be routed to the destinations, when the relay gains are chosen appropriately. As all relays forward all signals they receive in this case, no explicit signal path has to be found. The interference terms can thereby still be reduced or completely canceled by the transformations in the relays, while the desired signal components can be amplified. A way to optimize the signals over multiple AF relays is shown in [119]. Therein, the AF relays diagonalize the effective multihop channel and maximize the sum rate between a single source-destination pair. When multiple user pairs are considered, the interference between them has also be taken into account. In [147], a scheme is introduced that forces all interference terms to zero. With this, the full spatial degrees of freedom of the network can be achieved when enough relays are present. Regarding sum rate or individual data rates, the zero-forcing approach is however not optimal and finding gain coefficients that maximize the performance is generally difficult, as polynomial equation systems have to be solved or non-convex optimization problems arise when more than two hops are considered.

To maximize the sum rate of a multihop network, an iterative optimization scheme is thus required. This, however, requires global CSI in general, which introduces large overheads, since all nodes have to disseminate their local CSI to all other nodes or to a central processing unit. In order to reduce this overhead, a distributed gain alloca-tion scheme that allows each node to compute its own transformation matrix based on locally available CSI is therefore desirable. A gain allocation scheme that allows for optimizing the gain coefficients based on local CSI and very limited feedback from the destination nodes is inspired by [31], which studies gain allocation in two-hop networks with single antenna nodes. In this chapter, we consider a much more general framework of multihop networks, where an arbitrary number of source-destination pairs communicate with each other via an arbitrary number of relay stages (hops) and where all nodes are equipped with multiple antennas. The optimization of such general networks is therefore fundamentally different from [31]. The generalization to multiuser multihop MIMO networks has already been considered in our prior work [117], where a distributed gradient based optimization algorithm attempts to find optimal relay gain coefficients in such a multihop network. The optimization is however restricted to the relays, the source nodes transmit spatially white. Here, we extend these results to the case where also source precoding (beamforming at the BSs) is included and put the results into the cellular context and discuss how the algorithm can be used for channel tracking. For the gradient calculation, the extension to source precoding can readily be included into the framework. The additional source precoding has however implications on the distributed version of the algorithm. The transmit signals of the relays are functions of the source precoding coefficients. In order to optimize the precoding, the impact of these coefficients on the transmit signals of subsequent stages has thus also to be considered in the optimization of them. Therefore, additional terms have to be acquired and some terms need to be approximated for the scheme to work in a distributed way without introducing overhead that scales with the number of nodes.

After introducing the general setup of multiuser multihop networks, we adopt the distributed sum rate maximization algorithm from [117] that is based on a gradient search. This gain allocation scheme requires only local CSI and very limited feedback, also for the case when the relay gains are jointly optimized together with the source precoding. The resulting scheme is thus particularly interesting for channel tracking in slow fading situations. We show that the gain allocation does (almost) not scale with the number of involved nodes and that it is quite robust with respect to channel changes.

5.2 System Model

The multihop networks under consideration consist of N_0 source-destination pairs that wish to communicate concurrently over the same physical channel. The sources can be identified as BSs and the destinations accordingly as MSs if the DL is considered or vice versa for the UL. The source nodes as well as the destination nodes are all equipped with an arbitrary number of antennas and we identify the number of antennas in source node S_k, $k = 1, \ldots, N_0$, with $n_k^{(0)}$ and those of destination node D_k with $n_k^{(L+1)}$. The communication between sources and destinations is assisted by L intermediate relay stages, where stage l contains N_l RSs. The relays are also equipped with multiple antennas and we denote the number of antennas at relay $\mathrm{RS}_j^{(l)}$ by $n_j^{(l)}$.

The communication is divided into $L + 1$ time or frequency slots and is initiated by the source nodes which transmit their signals simultaneously in the first time slot[1]. Each source node S_k transmits a symbol vector $\mathbf{s}_k \in \mathbb{C}^{n_k^{(0)}}$ which is linearly precoded by the beamforming matrix $\mathbf{Q}_k \in \mathbb{C}^{n_k^{(0)} \times n_k^{(0)}}$, i.e. the transmit signal of S_k is

$$\mathbf{x}_k^{(0)} = \mathbf{Q}_k \cdot \mathbf{s}_k. \tag{5.1}$$

Note that each source transmits only the data to its intended destination, i.e. the source nodes are not able to exchange information with each other.

In time slot l, the relays in stage l receive signals from stage $l - 1$ (which corresponds to the source stage if $l = 1$). Each relay $\mathrm{RS}_j^{(l)}$ performs then a linear transformation of its receive signals with a complex gain matrix $\mathbf{G}_j^{(l)} \in \mathbb{C}^{n_j^{(l)} \times n_j^{(l)}}$ before retransmission in time slot $l + 1$. The transmit signal of $\mathrm{RS}_j^{(l)}$ can thus be written as

$$\mathbf{x}_j^{(l)} = \mathbf{G}_j^{(l)} \cdot \left(\mathbf{y}_j^{(l)} + \mathbf{n}_k^{(l)} \right), \tag{5.2}$$

with $\mathbf{y}_j^{(l)}$ being the receive signal and $\mathbf{n}_j^{(l)}$ the noise induced in $\mathrm{RS}_j^{(l)}$. The relay noise is thus also amplified and accumulated over the different stages. In time slot $L + 1$, the destination stage receives the transmission of the relays in stage L, affected by the local destination noise.

In the following, we assume slow and frequency flat fading channels with coefficients that are i.i.d. $\mathcal{CN}(0, 1)$, if nodes are located in adjacent stages, and zero otherwise. The channel matrix of the l-th hop is denoted by $\mathbf{H}_l$, for $l = 1, \ldots, L + 1$. The overall effective channel from source stage to the destinations follows as the concatenated product

$$\mathbf{H}_{\mathrm{eff}} = \mathbf{H}_{L+1} \cdot \mathbf{G}_L \cdot \mathbf{H}_L \cdots \mathbf{H}_2 \cdot \mathbf{G}_1 \cdot \mathbf{H}_1, \tag{5.3}$$

where $\mathbf{G}_l$ is the block diagonal matrix consisting of all transformation matrices of the l-th stage as its diagonal blocks. Setting the channels between non-adjacent stages to

---

[1] In the following, we assume that the relays operate in a TDD mode for the sake of conceptual simplicity. The same schemes can also be applied to FDD relays, which would lead to a simpler relay architecture.

zero follows from the underlying assumption that the sources wait for $L+1$ time slots until they inject new messages into the network and each receiving stage l ignores all signals except the one in the l-th slot or that all channels over more than a single hop are blocked by strong shadowing. The more general case in which also a direct link is present between source and destination stage, but only for a two-hop network, is discussed in [118].

Node D_k of the destination stage is only interested in the signal $\mathbf{s}_k$ transmitted by S_k, while the signals from all other nodes are interference and treated as noise. The receive signal of D_k can thus be written as

$$\mathbf{d}_k = \mathbf{H}_{\text{eff},k,k} \cdot \mathbf{Q}_k \cdot \mathbf{s}_k + \mathbf{H}_{\text{eff},k,-k} \cdot \mathbf{Q}_{-k} \cdot \mathbf{s}_{-k} + \sum_{l=1}^{L} \mathbf{H}_{\text{rd},k}^{(l)} \cdot \mathbf{G}_l \cdot \mathbf{n}_l + \mathbf{w}_k, \qquad (5.4)$$

with

- $\mathbf{H}_{\text{eff},k,k}$ the components of the effective channel $\mathbf{H}_{\text{eff}}$ that correspond to the k-th source-destination pair,
- $\mathbf{H}_{\text{eff},k,-k}$ the components of $\mathbf{H}_{\text{eff}}$ to D_k originating from all sources $j \neq k$,
- $\mathbf{s}_k$ the data symbol vector from source k intended to D_k,
- $\mathbf{s}_{-k} = \left[\mathbf{s}_1^\mathsf{T}, \ldots, \mathbf{s}_{k-1}^\mathsf{T}, \mathbf{s}_{k+1}^\mathsf{T}, \ldots, \mathbf{s}_{N_0}^\mathsf{T}\right]^\mathsf{T}$ the data symbols of all other sources $j \neq k$,
- $\mathbf{Q}_{-k}$ the block diagonal beamforming matrix from all sources $j \neq k$,
- $\mathbf{H}_{\text{rd},k}^{(l)}$ the components of the concatenated channel $\mathbf{H}_{L+1}\mathbf{G}_L\mathbf{H}_L \cdots \mathbf{G}_{l+1}\mathbf{H}_{l+1}$ to the k-th destination node, i.e. the effective channel from stage l to D_k,
- $\mathbf{n}_l$ and $\mathbf{w}_k$ the noise induced in stage l and D_k with elements i.i.d $\mathcal{CN}(0, \sigma_n^2)$ and $\mathcal{CN}(0, \sigma_w^2)$, respectively.

On each source node as well as each relay, we impose an instantaneous transmit power constraint

$$P_j^{(l)} \leq P, \quad \forall j \in \{1, \ldots, N_l\}, \quad \forall l \in \{0, 1, \ldots, L\}, \qquad (5.5)$$

where $P_j^{(l)}$ is the instantaneous transmit power of $\text{RS}_j^{(l)}$ or source S_j if $l = 0$. Note that we impose the same transmit power constraint to all relays as well as sources. A generalization to different power constraints would be straight forward. An achievable sum rate of the considered multihop network can then be given in terms of the covariance matrices of the desired signal, interference, and noise denoted by $\mathbf{K}_k^{(s)}$, $\mathbf{K}_k^{(i)}$, and

$\mathbf{K}_k^{(n)}$, as

$$
R_\Sigma = \sum_{k=1}^{N_0} \log_2 \det \left(\mathbf{I} + \left(\mathbf{K}_k^{(i)} + \mathbf{K}_k^{(n)} \right)^{-1} \cdot \mathbf{K}_k^{(s)} \right)
$$

$$
= \sum_{k=1}^{N_0} \left(\log_2 \det \left(\mathbf{K}_k^{(s)} + \mathbf{K}_k^{(i)} + \mathbf{K}_k^{(n)} \right) - \log_2 \det \left(\mathbf{K}_k^{(i)} + \mathbf{K}_k^{(n)} \right) \right). \tag{5.6}
$$

The covariance matrices for user k are given by

$$
\mathbf{K}_k^{(s)} = \mathbf{H}_{\text{eff},k,k} \mathbf{Q}_k \cdot \mathbf{Q}_k^{\mathsf{H}} \mathbf{H}_{\text{eff},k,k}^{\mathsf{H}} \tag{5.7}
$$

$$
\mathbf{K}_k^{(i)} = \mathbf{H}_{\text{eff},k,-k} \mathbf{Q}_{-k} \cdot \mathbf{Q}_{-k}^{\mathsf{H}} \mathbf{H}_{\text{eff},k,-k}^{\mathsf{H}} \tag{5.8}
$$

$$
\mathbf{K}_k^{(s)} = \sigma_n^2 \sum_{l=1}^{L} \mathbf{H}_{\text{rd},k}^{(l)} \mathbf{G}_l \cdot \mathbf{G}_l^{\mathsf{H}} \mathbf{H}_{\text{rd},k}^{(l)\mathsf{H}} + \sigma_w^2 \cdot \mathbf{I}, \tag{5.9}
$$

since $\mathsf{E}\left[\mathbf{s}_k \mathbf{s}_k^{\mathsf{H}} \right] = \mathbf{I}$. Note that we dropped the prelog factor in (5.6). Due to the use of multiple time slots for the transmission of one symbol vector, the achievable rate needs to be multiplied by a prelog factor. Since its value, however, is not immediately clear without additional assumptions, we drop this factor here. Depending on the pathloss and shadowing effects, new signals can be injected into the network more often than every $L+1$ time slots without causing (significant) interference to previously transmitted signals.

5.3 Distributed Optimization

In the following, we aim to find transformation matrices for the relays and source nodes that maximize the achievable sum rate (5.6) of the network. However, finding optimal precoding and relay gain coefficients requires global CSI in general, which is not readily available in practice. Therefore, we wish to distribute the gain allocation scheme such that each node is able to optimize its gain coefficients based on locally available CSI and limited feedback from the destinations only. In order to include the source precoding into the framework from the previous work in [117], we can identify the beamforming matrices of the source stage as an additional first relay stage, i.e. we extend the effective channel (5.3) to

$$
\mathbf{H}_{\text{eff}} = \mathbf{H}_{L+1} \cdot \mathbf{G}_L \cdot \mathbf{H}_L \cdots \mathbf{H}_2 \cdot \mathbf{G}_1 \cdot \mathbf{H}_1 \cdot \mathbf{G}_0 \cdot \mathbf{I}, \tag{5.10}
$$

where $\mathbf{G}_0$ is a block diagonal matrix that contains the source precoding matrices $\mathbf{Q}_k$ on its diagonal. With this, the network with source precoding is equivalent to a network with $L + 1$ relay stages where the first hop channel is described by the unity matrix.

For the distributed optimization inspired by [31], the extension from the two-hop case without source precoding fails, because several terms that are required to compute the gradient are not locally available. One of the issues arises due to the power constraint (5.5). The transmit power of $\text{RS}_j^{(l)}$ depends on the power of the receive signals, which is a function of the source precoding and the gain coefficients of the preceding relay stages. This implies that the transmit powers of all subsequent stages $i > l$ are dependent on the gain coefficients in stage i. For the gradient calculation, $\text{RS}_j^{(l)}$ would thus have to know also the local CSI of these subsequent stages. Therefore, the different stages have to be decoupled. To this end, we relax the optimization problem by imposing a norm constraint on each node instead of the power constraint:

$$\left\| \mathbf{G}_j^{(l)} \right\|_{\text{F}}^2 = \text{Tr}\left\{ \mathbf{G}_j^{(l)} \cdot \mathbf{G}_j^{(l)\text{H}} \right\} \le \rho, \qquad \forall j, l. \tag{5.11}$$

Note that (5.11) is equivalent to the true power constraint (5.5) for the source stage as

$$\text{Tr}\left\{ \mathbf{Q}_k \cdot \text{E}\left[\mathbf{s}_k \cdot \mathbf{s}_k^{\text{H}} \right] \cdot \mathbf{Q}_k^{\text{H}} \right\} = \text{Tr}\left\{ \mathbf{Q}_k \mathbf{Q}_k^{\text{H}} \right\}, \tag{5.12}$$

when we identify $\mathbf{G}_k^{(0)}$ with $\mathbf{Q}_k$. The optimization problem that we wish to solve can now be described by

$$\max_{\left\{ \mathbf{G}_j^{(l)} \right\}_{j,l}} R_\Sigma \quad \text{s.t.} \quad \left\| \mathbf{G}_j^{(l)} \right\|_{\text{F}}^2 \le \rho, \quad \forall j, l. \tag{5.13}$$

In first instance, we turn the constrained optimization problem (5.13) into an unconstrained one by fulfilling the norm constraint with equality through the variable substitution

$$\mathbf{G}_j^{(l)} = \sqrt{\rho} \cdot \tilde{\mathbf{G}}_j^{(l)} / \sqrt{\text{Tr}\left\{ \tilde{\mathbf{G}}_j^{(l)} \cdot \tilde{\mathbf{G}}_j^{(l)\text{H}} \right\}}. \tag{5.14}$$

Note that fulfilling the constraint in each node with equality is not optimal in general. The algorithm described later in this chapter, however, can be adapted for optimization without this additional restriction. In the following, the gradient with respect to the gain coefficients in the transformation matrices is computed.

5.3.1 Distributed Gradient Computation

The complex gradient of the sum rate (5.6) is calculated similar to the one in Chapter 3 through the chain rule of differentiation

$$\nabla_{\tilde{G}^*} R_\Sigma = \left((\nabla_{\mathbf{G}^*} R_\Sigma)^{\mathsf{T}} \cdot \hat{\mathbf{J}} + (\nabla_{\mathbf{G}^*} R_\Sigma)^{\mathsf{H}} \cdot \mathbf{J} \right)^{\mathsf{T}}, \tag{5.15}$$

with $\mathbf{J}$ and $\hat{\mathbf{J}}$ the Jacobians of the variable substitution (5.14) and its complex conjugate.

The "outer" gradient is defined as [16]

$$\nabla_{\tilde{\mathbf{G}}^*} R_\Sigma = 2 \cdot \left[\frac{\partial R_\Sigma}{\partial \tilde{g}_{1,1}^{(1,0)*}}, \frac{\partial R_\Sigma}{\partial \tilde{g}_{1,2}^{(1,0)*}}, \cdots, \frac{\partial R_\Sigma}{\partial \tilde{g}_{1,n_1^{(0)}}^{(1,0)*}}, \cdots, \frac{\partial R_\Sigma}{\partial \tilde{g}_{p,q}^{(j,l)*}}, \cdots \right]^{\mathsf{T}}, \tag{5.16}$$

with $\tilde{g}_{p,q}^{(j,l)*}$ being the complex conjugate of the coefficient in the p-th row and q-th column of $\tilde{\mathbf{G}}_j^{(l)}$. Its elements are given by [105]

$$\frac{\partial R_\Sigma}{\partial \tilde{g}_{p,q}^{(j,l)*}} = \frac{1}{\ln(2)} \sum_{k=1}^{N_0} \left(\mathrm{Tr} \left\{ \left(\mathbf{K}_k^{(s)} + \mathbf{K}_k^{(i)} + \mathbf{K}_k^{(n)} \right)^{-1} \cdot \left(\frac{\partial \mathbf{K}_k^{(s)}}{\partial \tilde{g}_{p,q}^{(j,l)*}} + \frac{\partial \mathbf{K}_k^{(i)}}{\partial \tilde{g}_{p,q}^{(j,l)*}} + \frac{\partial \mathbf{K}_k^{(n)}}{\partial \tilde{g}_{p,q}^{(j,l)*}} \right) \right\} \right.$$
$$\left. - \mathrm{Tr} \left\{ \left(\mathbf{K}_k^{(i)} + \mathbf{K}_k^{(n)} \right)^{-1} \cdot \left(\frac{\partial \mathbf{K}_k^{(i)}}{\partial \tilde{g}_{p,q}^{(j,l)*}} + \frac{\partial \mathbf{K}_k^{(n)}}{\partial \tilde{g}_{p,q}^{(j,l)*}} \right) \right\} \right). \tag{5.17}$$

In order to compute the inner derivatives in (5.17), we factorize the effective channel (5.3) as

$$\mathbf{H}_{\text{eff}}^{(l)} = \mathbf{H}_{\text{eff}} \cdot \mathbf{I} = \underbrace{\mathbf{H}_{L+1}\mathbf{G}_L \cdots \mathbf{H}_{l+1}}_{\triangleq \mathbf{H}_{\text{rd}}^{(l)}} \cdot \mathbf{G}_l \cdot \underbrace{\mathbf{H}_l \cdots \mathbf{H}_1 \mathbf{G}_0 \mathbf{I}}_{\triangleq \mathbf{H}_{\text{sr}}^{(l)}}$$
$$= \mathbf{H}_{\text{rd}}^{(l)} \cdot \mathbf{G}_l \cdot \mathbf{H}_{\text{sr}}^{(l)}, \qquad l = 0, 1, \ldots, L, \tag{5.18}$$

and obtain $L+1$ different notations for $\mathbf{H}_{\text{eff}}$, each with respect to the gain matrix of a particular stage. We further define $\mathbf{H}_{\text{rd},k}^{(l)}$ as the components of $\mathbf{H}_{\text{rd}}^{(l)}$ that correspond to D_k, $\mathbf{H}_{\text{sr},k}^{(l)}$ as the components of $\mathbf{H}_{\text{sr}}^{(l)}$ that originate from source S_k, and $\mathbf{H}_{\text{sr},-k}^{(l)}$ as the components of $\mathbf{H}_{\text{sr}}^{(l)}$ that correspond to all other source nodes $j \neq k$. Then, we can rewrite the derivatives of the covariance matrices in (5.17) as

$$\frac{\partial \mathbf{K}_k^{(\mathrm{s})}}{\partial \tilde{g}_{p,q}^{(j,l)*}} = \mathbf{H}_{\mathrm{eff},k,k} \cdot \mathbf{H}_{\mathrm{sr},k}^{(l)\mathsf{H}} \cdot \mathbf{E}_{p,q}^{(j,l)\mathsf{H}} \cdot \mathbf{H}_{\mathrm{rd},k}^{(l)\mathsf{H}} \tag{5.19}$$

$$\frac{\partial \mathbf{K}_k^{(\mathrm{i})}}{\partial \tilde{g}_{p,q}^{(j,l)*}} = \mathbf{H}_{\mathrm{eff},k,-k} \cdot \mathbf{H}_{\mathrm{sr},-k}^{(l)\mathsf{H}} \cdot \mathbf{E}_{p,q}^{(j,l)\mathsf{H}} \cdot \mathbf{H}_{\mathrm{rd},k}^{(l)\mathsf{H}} \tag{5.20}$$

$$\frac{\partial \mathbf{K}_k^{(\mathrm{n})}}{\partial \tilde{g}_{p,q}^{(j,l)*}} = \sigma_n^2 \cdot \left(\mathbf{H}_{\mathrm{rd},k}^{(l)} \mathbf{G}_l \cdot \mathbf{E}_{p,q}^{(j,l)\mathsf{H}} \mathbf{H}_{\mathrm{rd},k}^{(l)\mathsf{H}} + \sum_{i=1}^{l-1} \mathbf{H}_{\mathrm{rd},k}^{(i)} \mathbf{G}_i \cdot \mathbf{G}_i^{\mathsf{H}} \cdot \frac{\partial \mathbf{H}_{\mathrm{rd},k}^{(i)\mathsf{H}}}{\partial \tilde{g}_{p,q}^{(j,l)*}} \right), \tag{5.21}$$

where $\mathbf{E}_{p,q}^{(j,l)}$ is the single-entry matrix of the same size as $\mathbf{G}_j^{(l)}$ with all entries zero except the entry in the p-th row and q-th column, which is one. The last term in (5.21) is given by

$$\frac{\partial \mathbf{H}_{\mathrm{rd},k}^{(i)\mathsf{H}}}{\partial \tilde{g}_{p,q}^{(j,l)*}} = \begin{cases} \mathbf{H}_{i+1}^{\mathsf{H}} \mathbf{G}_{i+1}^{\mathsf{H}} \cdots \mathbf{H}_l^{\mathsf{H}} \cdot \mathbf{E}_{p,q}^{(j,l)\mathsf{H}} \cdot \mathbf{H}_{l+1}^{\mathsf{H}} \cdots \mathbf{G}_L^{\mathsf{H}} \mathbf{H}_{L+1}^{\mathsf{H}}, & \text{if } j < l \\ \mathbf{O}, & \text{otherwise.} \end{cases} \tag{5.22}$$

The Jacobians $\mathbf{J}$ and $\hat{\mathbf{J}}$ contain the partial derivatives of (5.14) and of its complex conjugates, both with respect to the entries of $\mathbf{G}_j^{(l)*}$. Entries of $\mathbf{J}$ are given by

$$\frac{\partial g_{p,q}^{(j,l)}}{\partial \tilde{g}_{p',q'}^{(j',l')*}} = \frac{-\frac{1}{2}\sqrt{\rho}\tilde{g}_{p,q}^{(j,l)} \left\| \tilde{\mathbf{G}}_j^{(l)} \right\|_{\mathrm{F}}^{-\frac{1}{2}} \mathrm{Tr}\left\{ \tilde{\mathbf{G}}_j^{(l)} \mathbf{E}_{p',q'}^{(j',l')\mathsf{H}} \right\}}{\left\| \tilde{\mathbf{G}}_j^{(l)} \right\|_{\mathrm{F}}^2} \tag{5.23}$$

and those of $\hat{\mathbf{J}}$ by

$$\frac{\partial g_{p,q}^{(j,l)*}}{\partial \tilde{g}_{p',q'}^{(j',l')*}} = \frac{\xi_{j',l',p',q'}^{(p,q,j,l)} - \frac{1}{2}\sqrt{\rho}\tilde{g}_{p,q}^{(j,l)*} \left\| \tilde{\mathbf{G}}_j^{(l)} \right\|_{\mathrm{F}}^{-\frac{1}{2}} \mathrm{Tr}\left\{ \tilde{\mathbf{G}}_j^{(l)} \mathbf{E}_{p',q'}^{(j',l')\mathsf{H}} \right\}}{\left\| \tilde{\mathbf{G}}_j^{(l)} \right\|_{\mathrm{F}}^2}, \tag{5.24}$$

with

$$\xi_{j',l',p',q'}^{(p,q,j,l)} = \begin{cases} \sqrt{\rho} \cdot \left\| \tilde{\mathbf{G}}_j^{(l)} \right\|_{\mathrm{F}}, & \text{if } j = j', l = l', p = p', q = q' \\ 0, & \text{otherwise.} \end{cases} \tag{5.25}$$

Note that the Jacobians are block diagonal matrices, since the variable substitution is applied to each node separately. Due to the applied norm constraint instead of the power constraint, the scaling of the gain coefficients in one relay does not depend on the values of other relays. The overall gradient can thus be decomposed into components that correspond to one relay or source node. These components are given by

$$(\nabla R_\Sigma)_j^{(l)} = \left((\nabla_{\mathbf{G}^\star} R_\Sigma)_j^{(l)\mathsf{T}} \cdot \hat{\mathbf{J}}_j^{(l)} + (\nabla_{\mathbf{G}^\star} R_\Sigma)^{\mathsf{H}} \cdot \mathbf{J}_j^{(l)} \right)^\mathsf{T}, \tag{5.26}$$

where $(\nabla_{\mathbf{G}^\star} R_\Sigma)_j^{(l)}$, $\mathbf{J}_j^{(l)}$ and $\hat{\mathbf{J}}_j^{(l)}$ are the components of the "outer" gradient (5.17) and the blocks of the Jacobians that correspond to $\mathrm{RS}_j^{(l)}$ or, if $l = 0$, to S_j.

Each of these components shall now be computed in each node separately with local CSI and limited feedback from the destinations. In the following we refer to *local* CSI at $\mathrm{RS}_j^{(l)}$ as the channel coefficients that can be estimated locally at $\mathrm{RS}_j^{(l)}$, i.e. the channel coefficients from the sources to $\mathrm{RS}_j^{(l)}$ and the channel coefficients from $\mathrm{RS}_j^{(l)}$ to the destinations. With this, the overhead of the feedback to acquire the necessary information does not scale with the number of relays in the different stages. This is shown in the following.

While it is immediately clear that each node can compute its corresponding blocks of the Jacobians (they depend only on the own transformation matrix), the computation of the terms in $(\nabla_{\mathbf{G}^\star} R_\Sigma)_j^{(l)}$ requires some feedback from the destination. In order to compute (5.17), the effective channel $\mathbf{H}_{\mathrm{eff}}$ as well as the noise covariance matrices $\mathbf{K}_k^{(n)}$ need to be known. These can be estimated at the destinations and fed back to the other nodes. In order to compute (5.19) and (5.20), local CSI and the knowledge of $\mathbf{H}_{\mathrm{eff}}$ is sufficient, since

$$\mathbf{H}_{\mathrm{sr},k}^{(l)\mathsf{H}} \cdot \mathbf{E}_{p,q}^{(j,l)\mathsf{H}} \cdot \mathbf{H}_{\mathrm{rd},k}^{(l)\mathsf{H}} = \left(\mathbf{H}_{\mathrm{rd},k}^{(l)}[:,p] \cdot \mathbf{H}_{\mathrm{sr},k}^{(l)}[q,:] \right)^\mathsf{H} \tag{5.27}$$

$$\mathbf{H}_{\mathrm{sr},-k}^{(l)\mathsf{H}} \cdot \mathbf{E}_{p,q}^{(j,l)\mathsf{H}} \cdot \mathbf{H}_{\mathrm{rd},k}^{(l)\mathsf{H}} = \left(\mathbf{H}_{\mathrm{rd},k}^{(l)}[:,p] \cdot \mathbf{H}_{\mathrm{sr},-k}^{(l)}[q,:] \right)^\mathsf{H}, \tag{5.28}$$

where $\mathbf{A}[:,p]$ and $\mathbf{A}[q,:]$ denotes the p-th column and q-th row of the matrix $\mathbf{A}$, respectively. The matrices on the right hand side of (5.27) and (5.28) can be estimated locally at the respective nodes. For the computation of (5.21), however, not all terms are locally accessible as the relay noise of $\mathrm{RS}_j^{(l)}$ also depends on the gain coefficients of the preceding stages $i < l$. We therefore approximate the derivative of the noise covariance matrix as

$$\frac{\partial \mathbf{K}_k^{(n)}}{\partial g_{p,q}^{(j^l)*}} \approx \sigma_n^2 \cdot \mathbf{H}_{\mathrm{rd},k}^{(l)} \mathbf{G}_l \cdot \mathbf{E}_{p,q}^{(j,l)\mathsf{H}} \mathbf{H}_{\mathrm{rd},k}^{(l)\mathsf{H}}, \tag{5.29}$$

for which local CSI is also sufficient since $\mathbf{G}_l$ is block-diagonal:

$$\mathbf{H}_{\mathrm{rd},k}^{(l)} \mathbf{G}_l \mathbf{E}_{p,q}^{(j,l)\mathsf{H}} \mathbf{H}_{\mathrm{rd},k}^{(l)\mathsf{H}} = \left(\mathbf{H}_{\mathrm{rd},k}^{(l)} \mathbf{E}_{p,q}^{(jl)} \mathbf{G}_l^{\mathsf{H}} \mathbf{H}_{\mathrm{rd},k}^{(l)\mathsf{H}} \right)^\mathsf{H}$$

$$
\begin{aligned}
&= \left(\mathbf{H}_{\mathrm{rd},k}^{(l)} \mathbf{E}_{p,q}^{(jl)} \underbrace{\begin{bmatrix} \mathbf{G}_1^{(l)\mathsf{H}} \left(\mathbf{H}_{\mathrm{rd},k}[:,\mathcal{K}_1^{(l)}] \right)^{\mathsf{H}} \\ \vdots \\ \mathbf{G}_j^{(l)\mathsf{H}} \left(\mathbf{H}_{\mathrm{rd},k}[:,\mathcal{K}_j^{(l)}] \right)^{\mathsf{H}} \\ \vdots \end{bmatrix}}_{\triangleq \mathbf{X}} \right)^{\mathsf{H}} \\
&= \left(\mathbf{H}_{\mathrm{rd},k}^{(l)}[:,p] \cdot \mathbf{X}[q,:] \right)^{\mathsf{H}},
\end{aligned}
\tag{5.30}
$$

with $\mathcal{K}_j^{(l)}$ the index set corresponding to the antennas of $R_j^{(l)}$. The single entry matrix $\mathbf{E}_{p,q}^{(j,l)}$ selects a single row of $\mathbf{X}$ that contains only channel coefficients that are locally available at $R_j^{(l)}$. With the approximation (5.29), each relay considers only its "own" noise (and the destination noise) for the computation of the derivative (5.21), while the noise terms induced in the other stages are ignored. With only two hops, the derivative of (5.29) is exact, as there is no noise forwarded by the source nodes. Even though a two-hop network with source precoding resembles a three-hop network with the extension in (5.10), the exact gradient can thus be calculated locally.

5.3.2 Sum Rate Maximization

In order to compute the gain coefficients in each node, we apply a gradient based optimization algorithm that updates the transformation matrices according to

$$
\tilde{\mathbf{G}}_j^{(l)}[m+1] = \mathbf{G}_j^{(l)}[m] + \mu[m] \cdot \Delta_j^{(l)}[m], \quad \forall j,l,
\tag{5.31}
$$

where $\mu[m]$ is the step-size and $\Delta_j^{(l)}$ is the search direction of the j-th node in the l-th stage in iteration m. If the norm constraint is enforced with equality, the search direction can be chosen as

$$
\Delta_j^{(l)}[m] = \left(\nabla_{\tilde{\mathbf{G}}^*} R_\Sigma \left(\tilde{\mathbf{G}}_j^{(l)}[m] \right) \right)_j^{(l)}.
\tag{5.32}
$$

But as already mentioned, enforcing the constraint with equality is not optimal in general. A relay node that receives a weak signal due to small channel coefficients from the previous stage retransmits mainly noise and little desired signal. Forcing this relay to transmit with high power can, in this case, possibly reduce the achievable rate. In order to drop this additional restriction, the variable substitution (5.14) can be modified to

$$
\mathbf{G}_j^{(l)} = \begin{cases} \sqrt{\rho} \cdot \tilde{\mathbf{G}}_j^{(l)} / \left\| \tilde{\mathbf{G}}_j^{(l)} \right\|_{\mathrm{F}}, & \text{if } \left\| \tilde{\mathbf{G}}_j^{(l)} \right\|_{\mathrm{F}} > \sqrt{\rho} \\ \tilde{\mathbf{G}}_j^{(l)}, & \text{if } \left\| \tilde{\mathbf{G}}_j^{(l)} \right\|_{\mathrm{F}} \le \sqrt{\rho}, \end{cases} \tag{5.33}
$$

where the transformation matrix is only normalized, if the constraint is violated. If a specific $\tilde{\mathbf{G}}_j^{(l)}$ does not violate the constraint, $\mathbf{G}_j^{(l)} = \tilde{\mathbf{G}}_j^{(l)}$ and the search direction can be chosen as

$$
\Delta_j^{(l)} = \left(\nabla_{\mathbf{G}^\star} R_\Sigma \left(\mathbf{G}_j^{(l)}[m] \right) \right)_j^{(l)}, \tag{5.34}
$$

i.e., only the "outer" gradient is considered and the Jacobians are dropped. In the other case, i.e. if $\mathbf{G}_j^{(l)}$ is a normalized version of $\tilde{\mathbf{G}}_j^{(l)}$, two different search directions have to be considered:

- If an update with $\Delta_j^{(l)}[m] = \left(\nabla_{\mathbf{G}^\star} R_\Sigma \left(\mathbf{G}_j^{(l)}[m] \right) \right)_j^{(l)}$ with a small step-size $\mu_{\min}$ leads to a transformation matrix that does *not* violate the constraint, this search direction is used,

- if this update leads to a transformation matrix that *does* violate the constraint, the norm constraint is enforced with equality and the total gradient including the Jacobians has to be considered, i.e., the search direction is in this case given by $\Delta_j^{(l)} = \left(\nabla_{\tilde{\mathbf{G}}^\star} R_\Sigma \left(\tilde{\mathbf{G}}_j^{(l)}[m] \right) \right)_j^{(l)}$.

After the search direction for iteration step m is obtained, we optimize the step-size $\mu[m]$ by a line search method that is designed to achieve low complexity. To this end, we start with some small initial step-size $\mu_0 = \mu_{\min}$ and increase the step-size according to $\mu_{t+1} = \alpha \cdot \mu_t$, for some $\alpha > 1$, as long as the sum rate with instantaneous choice of μ_{t+1} is larger as the sum-rate with the previous choice. As soon as μ_{t+1} leads to a smaller sum rate than with the previous step-size, the line search terminates and $\mu[m] = \mu_t$ is chosen for the step-size of iteration step m.

In order to achieve a faster convergence behavior than with the simple gradient search method outlined above, we can refine the algorithm by utilizing conjugate gradients (CG) [130]. To this end, the search direction is modified according to

$$
\Delta_j^{(l)}[m] = \left(\nabla_{\tilde{\mathbf{G}}^\star} R_\Sigma \left(\tilde{\mathbf{G}}_j^{(l)}[m] \right) \right)_j^{(l)} + \beta_{\mathrm{cg}}[m] \cdot \Delta_j^{(l)}[m-1]. \tag{5.35}
$$

The parameter β_{cg} is a design parameter. We choose

$$\beta_{\text{cg}}[m] = \frac{\left(\nabla_{\tilde{\mathbf{G}}^*} R_\Sigma\left(\tilde{\mathbf{G}}_j^{(l)}[m]\right)\right)^{\mathsf{H}} \cdot \left(\nabla_{\tilde{\mathbf{G}}^*} R_\Sigma\left(\tilde{\mathbf{G}}_j^{(l)}[m]\right) - \nabla_{\tilde{\mathbf{G}}^*} R_\Sigma\left(\tilde{\mathbf{G}}_j^{(l)}[m-1]\right)\right)}{\left(\nabla_{\tilde{\mathbf{G}}^*} R_\Sigma\left(\tilde{\mathbf{G}}_j^{(l)}[m-1]\right)\right)^{\mathsf{H}} \cdot \left(\nabla_{\tilde{\mathbf{G}}^*} R_\Sigma\left(\tilde{\mathbf{G}}_j^{(l)}[m-1]\right)\right)} \tag{5.36}$$

according to Polak-Ribière [130]. In this case, however, inner products of the current and previous search direction need to be computed. These can be obtained by *over-the-air additions* (cf. [31]). Thereby, each $\text{RS}_j^{(l)}$ computes a value

$$\gamma_j^{(l)}[m] = \frac{\left(\nabla_{\tilde{\mathbf{G}}^*} R_\Sigma\left(\tilde{\mathbf{G}}_j^{(l)}[m]\right)\right)_j^{(l)\mathsf{H}} \cdot \left(\left(\nabla_{\tilde{\mathbf{G}}^*} R_\Sigma\left(\tilde{\mathbf{G}}_j^{(l)}[m]\right)\right)_j^{(l)} - \left(\nabla_{\tilde{\mathbf{G}}^*} R_\Sigma\left(\tilde{\mathbf{G}}_j^{(l)}[m-1]\right)\right)_j^{(l)}\right)}{\mathbf{H}_{\text{rd},k}^{(l)}[p_j, q]} \tag{5.37}$$

with p_j denoting an arbitrary antenna of $\text{RS}_j^{(l)}$ to an arbitrary destination antenna q. When all relays simultaneously transmit these analog values, the receive signal at destination antenna q becomes

$$d_q^{(l)}[m] = \sum_{j=1}^{(N_l)} \mathbf{H}_{\text{rd},k}^{(l)}[p_j, q] \cdot \gamma_j^{(l)}[m] \tag{5.38}$$

$$= \left(\nabla_{\tilde{\mathbf{G}}^*} R_\Sigma\left(\tilde{\mathbf{G}}_j^{(l)}[m]\right)\right)^{(l)\mathsf{H}} \cdot \left(\left(\nabla_{\tilde{\mathbf{G}}^*} R_\Sigma\left(\tilde{\mathbf{G}}_j^{(l)}[m]\right)\right)^{(l)} - \left(\nabla_{\tilde{\mathbf{G}}^*} R_\Sigma\left(\tilde{\mathbf{G}}_j^{(l)}[m-1]\right)\right)^{(l)}\right),$$

which is (by neglecting the noise) equivalent to the numerator of $\beta_{\text{cg}}[m]$. In order to obtain the denominator, each $\text{RS}_j^{(l)}$ transmits another value

$$\delta_k^{(l)}[m] = \frac{\left(\nabla_{\tilde{\mathbf{G}}^*} R_\Sigma\left(\tilde{\mathbf{G}}_j^{(l)}[m-1]\right)\right)_j^{(l)\mathsf{H}} \cdot \left(\nabla_{\tilde{\mathbf{G}}^*} R_\Sigma\left(\tilde{\mathbf{G}}_j^{(l)}[m-1]\right)\right)_j^{(l)}}{\mathbf{H}_{\text{rd},k}^{(l)}[p_i, q]}. \tag{5.39}$$

With this, the denominator can be calculated in the same way. The destination then calculates the fraction and broadcasts the resulting $\beta_{\text{cg}}[m]$ back to the relays. Note that in each of these *measurement cycles*, all relays of the same stage transmit simultaneously. Thus, the overhead does not grow with the number of nodes in each stage.

5.3.3 Overhead

The knowledge of $\mathbf{H}_{\text{eff}}$ as well as $\mathbf{K}_n^{(k)}$ is required by all source and relay nodes for the computation of the search directions. Estimates of these matrices can be obtained at the destination stage by the use of orthogonal pilot sequences transmitted simultaneously by the sources. The matrices can then be fed back through the network to the relay and source nodes. However, the dimensions of these matrices are not dependent on the number of relays or antennas therein and remain fixed, even if N_l grows large. The pilots transmitted by the sources and additional sequences transmitted simultaneously by the destinations in the feedback cycle can be used for estimates of the local CSI in the relays. The overhead due to such an estimation and feedback cycle does thus not depend on the relays per stage. More relay stages, however, imply a larger overhead, since all signals have to traverse the entire network.

An additional overhead is introduced by the line search as well as for the computation of the inner products for the CG algorithm, if this refinement is used. But also this overhead does not scale with the number of relays per stage. Moreover, numerical results show that also the number of iterations required to achieve close to optimal sum-rates does not scale at all, or if so, very slowly with the number of relays per stage. Accordingly, our gain allocation scheme proves to be particularly useful for networks with a large number of relays that are grouped into few stages.

5.4 Simulation Results

For the computer simulations, we choose a basic network with $L = N_0 = N_1 = N_2 = 2$ and two antennas at each node, if not stated otherwise. The noise variance is set to $\sigma_n^2 = \sigma_w^2 = 1$ and the transmit power P is variable. The channels between two consecutive stages have elements i.i.d. $\mathcal{CN}(0, 1)$ and stay constant for the whole transmission period. The channels over more stages are zero.

First, we compare the results of the proposed distributed scheme with the approximated derivative of the noise covariance matrix (5.29) with the achievable rates obtained by an optimization algorithm that has access to global CSI and applies the exact gradient. The performance for a typical channel realization is shown in Fig. 5.2. In this simulation, the transmit power at the sources and the relay norm constraints are fixed to $P = \rho = 100$, i.e. the average transmit SNR is 20 dB. We can observe that the scheme based on limited CSI achieves a sum rate that is very close to the one obtained

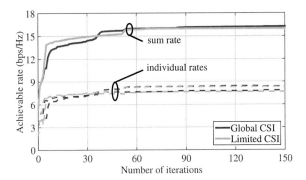

Figure 5.2.: Optimization with limited CSI (approximated derivative of noise covariance matrix) achieves rates close to the ones achievable with global CSI.

by the search direction with the correct derivative. For this choice of SNR and number of stages and nodes, the noise approximation does thus not have a significant impact. In the low SNR regime and when the network is long, i.e. for large L, the performance difference between the local algorithm and the exact one with global CSI is expected to increase as the noise has more impact.

The other relaxation that allows the gain allocation scheme to be distributed is the norm constraint that is applied instead of a per node transmit power constraint. The influence of this simplification is studied next. For a fair comparison, the transformation matrices after optimization with the norm constraint (5.11) have to be scaled such that the power constraint (5.5) is fulfilled. To this end, two different scaling methods are used. In the first method, each node scales its gain coefficients individually if the power constraint is violated. In the second method, the scaling is performed stage-wise, where all transformation matrices of the same stage are normalized by the same factor such that the node with the highest transmit power fulfills the power constraint exactly. Note that this scaling method requires an additional over-the-air addition cycle in order to distribute the scaling factor to all relays in a stage.

Since the optimization at hand is highly non-convex, the optimization algorithm can converge to different local optima, depending on the initialization. While some of the solutions achieve full multiplexing gain, other initializations can converge to local optima that do not suppress the interference of all users and hence achieve only a fraction of the achievable multiplexing gain. Starting the gain allocation scheme with a "good" initial guess is therefore crucial for achieving good performance. In Fig.

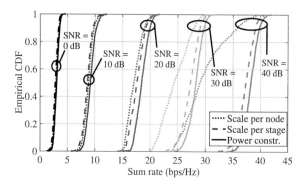

Figure 5.3.: CDF of achievable sum rate after optimization for different SNR values. For each channel realization, only the best solution is considered.

5.3, the empirical CDF of the sum-rates are plotted for different values of SNR $= P$, where the distributed algorithm is initialized with the best solution of the optimizer with global CSI. The optimum that can be achieved with per node transmit power constraints is compared to the distributed case in which the enforcement of the power constraint is achieved by the two scaling approaches discussed above. We observe that the distributed scheme achieves close to optimal solutions for all SNR values if the power scaling is performed stagewise. The performance loss due to the per node scaling is large in the high SNR regime. With this scaling approach, the coherent additions of the transmit signals and the interference reduction is destroyed as all relays scale their signals with different factors. For lower SNR's however, the per node scaling achieves better results relative to the per stage scaling and the exact results with the true per node power constraint. In this regime, the noise is dominant, and transmitting with full power is more important than adjusting the gain coefficients in order to suppress the interference as much as possible. Nevertheless, the approximated gradient scheme with norm constraint and scaling each stage with the same factor is justified in the high SNR regime, since its calculation requires only limited CSI and the algorithm can be distributed. In this case, close to optimal results are achievable and the AF relays can be used to maximize the data rates in a network. The proposed algorithm is thereby particularly useful for channel tracking in slow fading environments, where a global algorithm initiates the communication between sources and destinations and the distributed scheme tracks then the changes of the channel.

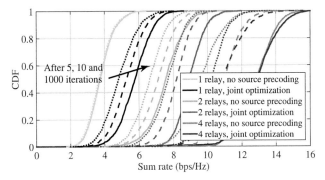

Figure 5.4.: Empirical CDF of the achievable sum rate after 5, 10, and 1000 iterations with $L = 1$ and $N_1 = 1, 2, 4$ relays.

In the following, we evaluate the impact of source precoding additional to the relay gain optimization. In Fig. 5.4, we consider a network with two source-destination pairs and a single relay stage, i.e. $L = 1$, with $N_1 = 1, 2, 4$ relays. All nodes are again equipped with 2 antennas. The noise variance and the transmit power of the nodes are set to $\sigma_n^2 = \sigma_w^2 = 1$ and $P = 10$. The achievable sum rate of the joint optimization of relay and source gain coefficients (dark colors) is compared with the case when only the relay gains are optimized and the sources transmit spatially white with $\mathbf{G}_k^{(0)} = \sqrt{P/n_k^{(0)}} \cdot \mathbf{I}$ (light colors). The achievable sum rates are shown after 5, 10, and 1000 iterations of the optimization algorithm (dotted, dashed, and solid lines). It can clearly be seen that the performance is considerably improved when the source precoding is also optimized and the number of relays is small. The performance gain becomes however smaller when 4 or more relays are used. In this case, the relay gain matrices provide already a large room for improvement of the rates and the additional source precoding cannot increase the rates much. Nevertheless, the rates increase significantly with more relays. But more relays also require more iteration steps to achieve close to optimal results. A similar behavior can also be expected when more relay stages are considered. When the number of relays per stage increases, room for optimization is enlarged and the source precoding has less impact. When the number of relays is small or when the functionalities of the relays are limited, such that they cannot fully optimize their signals, the source precoding is an important tool to improve the network wide performance. As the proposed algorithm is of rather high complexity as many iterations are required, we thus investigate different gain allocation schemes that are potentially of lower complexity and will then develop suboptimal but much simpler AF relaying schemes. These can however still achieve good performance when they are combined with an effective source precoding, as we will see in the next chapter.

Nevertheless, the gradient based algorithm described here can still be applied efficiently when it is used for channel tracking in slowly fading environments. Thereby, the proposed algorithm can be initialized with the solution of a global algorithm and the distributed scheme can then be updated to follow the changes of the channel. In Fig. 5.5, we show the tracking performance of the distributed algorithm for SNR = 20 dB with a time varying channel $\mathbf{H}_l[m+1] = \vartheta\mathbf{H}_l[m] + \sqrt{1-\vartheta^2}\mathbf{W}_l$, $\vartheta \in [0,1]$, that is fixed until the optimization has terminated and changes by an additive i.i.d. component $\mathbf{W}_l$ afterwards. The parameter ϑ is a measure for the correlation of the channel compared to the previous realization. In each optimization (for each channel realization), the distributed algorithm is initialized with the preceding solution (or with the global optimum at the beginning). If the channel increment is small ($\vartheta = 0.99$), the distributed scheme shows similarly good performance results as before. When the correlation of the actual channel realization with the initial one decreases (due to a smaller ϑ or increasing m), the performance decreases to a constant value. However, we can conclude that the proposed algorithm is an efficient scheme that achieves close to optimal solutions, if the channel changes sufficiently slowly, while the introduced overhead is reduced considerably with respect to a global optimization scheme.

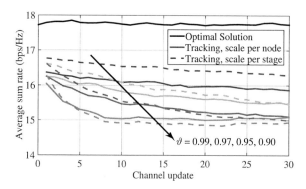

Figure 5.5.: Average sum rate after channel tracking in a slow fading environment.

5.5 Block Zero-Forcing for Two-Way Relaying

With appropriate relay gain coefficients, AF relays are an effective means to shape the effective channel between sources and destinations in a beneficial way. In the previous sections we have seen that optimized gains can maximize the achievable rate. Thereby, the interference is reduced to a extent that all spatial streams are "open" and the spatial degrees of freedom can be exploited when enough relays are deployed. The proposed gain allocation algorithm is however of high complexity as many iterations are required to achieve close to optimal results, especially with many relays. Even though the algorithm can be distributed, more iterations are required when the number of relays increases. Finding optimal relay gain matrices is thus a very difficult task, as the corresponding optimization problem is non-convex. Therefore, we apply a block ZF approach in the following that allows us to orthogonalize different user pairs in closed-form. With this approach, the sum rate maximization under the ZF conditions is simplified as the search space is drastically reduced. While the direct optimization with the gradient search optimizes each individual relay gain coefficient, the problem is reduced to null space selection where only few variables need to be optimized.

The generalization to multihop networks as considered in the last section is however difficult. With two-hop networks, the ZF conditions can be turned into a linear equation system and the ZF solution can be found in closed form as we will see in the next section. If more than one relay stage is present, this is no longer the case and fulfilling the ZF conditions requires the solution of a set of polynomial equations, which is much more involving, even in the case of single antenna nodes [148]. To this end, we limit ourselves to networks with only a single stage of relays but extend them to two-way relaying. Since the communication in cellular networks is usually bidirectional, i.e. they consist of a DL as well as an UL, two-way relaying has been identified as an efficient concept to combine both directions without a loss in spectral efficiency, even with half-duplex relays [112]. Also in this case, relays can perform distributed optimization of the network performance or can help to achieve the degrees of freedom [36].

Most research in this field, however, focuses on narrow-band systems in which the channels are frequency flat and can be described by single channel matrices. Exceptions are e.g. filter-and-forward relays [7, 21]. Since current (and probably also future) systems are OFDM based, the assumption of frequency flat narrow-band channels is certainly valid, as such systems can be described by parallel channels that operate on orthogonal subcarriers. In the literature, most models therefore treat only the single-carrier case, in which the system is optimized for a single subcarrier. An extension to the wide-band case can then be done by treating each subcarrier independently and combining them with a power allocation across the subcarriers. While this treatment is comfortable, recent research has shown that this is suboptimal in certain cases. If a capacity achieving transmission scheme is applied (e.g. DPC for the BC), treating the orthogonal subcarriers separately still achieves capacity when the transmission scheme is combined with power loading across the different subcarriers. Reference [54] proves however for the BC that separate treatment can be strictly suboptimal for certain channel realizations, when *linear* precoding is used. In [53], the same authors show that in a BC with linear precoding and cooperation among different subcarriers, i.e. when arbitrary linear combinations of all symbols across different channels are allowed, a performance gain to the carrier-non-cooperative case can indeed be achieved. Because a two-way relay network inherently contains a MAC as well as a BC phase and since AF relays are restricted to perform linear operations on their signals, we can expect a similar behavior of subcarrier cooperation as in the original BC. To this end, we consider a two-way AF relay network with multiple orthogonal subcarriers and allow the multi-antenna relays to forward arbitrary linear combinations of their receive signals in the spatial as well as frequency domain.

The ZF approach is similar to [84]. There, however, only a single relay and single antenna terminals are considered. The ZF conditions are therein applied for the BC and the MAC phase individually and the study is based on a single subcarrier. In the following, we apply two-way relaying to a more general network with multi antenna nodes, multiple relays, and multiple orthogonal subcarriers. Thereby, the relays can forward arbitrary linear combinations of their receive signals in the spatial as well as frequency domain. On the example of this network, we discuss how narrow band transmission schemes can be extended to the wideband case and study the gains that can be achieved by subcarrier cooperation. The discussion is however limited to two-way relaying for which the achievable performance gains can be expected to be particularly good. In one directional communication setups, these gains are less pronounced.

5.6   Subcarrier Cooperative Two-Way Relay Network

The network under consideration consists of K transceiver pairs that wish to communicate in a bidirectional way. The pair of terminals T_k and T_j that can be attributed to BSs or MSs which exchange information with each other is represented by (k, j) such

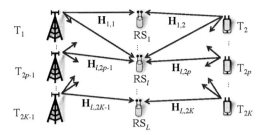

Figure 5.6.: Two-way relay network with multiple AF relays.

that the pth pair is denoted by $(2p-1, 2p)$. The communication is assisted by L relays while it is assumed that there is no direct link between the terminals. The terminal and relay nodes are equipped with an arbitrary number of antennas. The number of antennas at T_k is denoted by M_k and N_l is the number of antennas at relay station RS_l, $l \in \{1, \ldots, L\}$. A sketch of the system model can be seen in Fig. 5.6. Note that this network model is quite general in the sense that it also captures the special cases where all relay antennas cooperate with each other or where the network corresponds to a cellular network with relays.

In this work, we consider AF relays that operate in a half-duplex mode and allow for two-way relaying. Furthermore, the communication can use orthogonal subcarriers $c \in \{1, \ldots, C\}$ that can e.g. correspond to OFDM subcarriers. On each subcarrier c, the channel is assumed to be frequency flat. We describe the channel between T_k and RS_l on subcarrier c by $\mathbf{H}_{l,k}^{(c)}$ and assume reciprocity, i.e. the channel from RS_l to T_k is $\overleftarrow{\mathbf{H}}_{k,l}^{(c)} = \mathbf{H}_{l,k}^{(c)\mathsf{T}}$. For the two-way relaying protocol, we apply two time slots for each transmission cycle. In the first time slot, all terminals transmit simultaneously to the relays. The receive signal of RS_l on subcarrier c can thus be written as

$$\mathbf{r}_l^{(c)} = \sum_{k=1}^{2K} \mathbf{H}_{l,k}^{(c)} \cdot \mathbf{x}_k^{(c)} + \mathbf{n}_l^{(c)}, \quad c = 1, \ldots, C, \tag{5.40}$$

with $\mathbf{x}_k^{(c)} \in \mathbb{C}^{M_k}$ the transmit signal of T_k and $\mathbf{n}_l^{(c)} \in \mathbb{C}^{N_l}$ the noise induced in RS_l.

In order to get a more compact notation over *all* subcarriers, we rewrite the channels into the equivalent block-diagonal form

$$\mathbf{H}_{l,k} = \mathrm{blkdiag}\left(\mathbf{H}_{l,k}^{(1)}, \ldots, \mathbf{H}_{l,k}^{(C)}\right). \tag{5.41}$$

151

The corresponding form of (5.40) over all subcarriers is thus

$$\mathbf{r}_l = \sum_{k=1}^{2K} \mathbf{H}_{l,k} \cdot \mathbf{x}_k + \mathbf{n}_l, \tag{5.42}$$

where $\mathbf{x}_k \in \mathbb{C}^{CM_k}$ and $\mathbf{n}_l \in \mathbb{C}^{CN_l}$ are the transmit and noise signals of *all* subcarriers.

The AF relays multiply their received signal with a gain matrix $\mathbf{G}_l$ and broadcast the resulting signal $\mathbf{G}_l \cdot \mathbf{r}_l$ back to the terminals. In conventional approaches, the relays process their input separately for all subcarriers, i.e. a separate gain matrix $\mathbf{G}_l^{(c)} \in \mathbb{C}^{N_l \times N_l}$ is applied to each subcarrier c. The resulting gain matrix over all subcarriers is thus a block-diagonal matrix $\mathbf{G}_l = \text{blkdiag}\left(\mathbf{G}_l^{(1)}, \ldots, \mathbf{G}_l^{(C)}\right)$. This block-diagonal structure ensures that the transmission on the different subcarriers remain orthogonal. We will refer to this case as to the conventional, subcarrier *non-cooperative* case.

Subcarrier *cooperation*, on the other hand, allows the relays to form arbitrary linear combinations across all subcarriers, i.e., the gain matrices are no longer restricted to be block-diagonal but can have an arbitrary form $\mathbf{G}_l \in \mathbb{C}^{CN_l \times CN_l}$.

The receive signal over all subcarriers at terminal T_k is in either case (subcarrier cooperative or non-cooperative)

$$\mathbf{y}_k = \sum_{l=1}^{L} \mathbf{H}_{l,k}^{\mathsf{T}} \cdot \mathbf{G}_l \left(\sum_{n=1}^{2K} \mathbf{H}_{l,n} \cdot \mathbf{x}_n + \mathbf{n}_l \right) + \mathbf{w}_k, \tag{5.43}$$

where $\mathbf{w}_k \in \mathbb{C}^{CM_k}$ is the noise induced in terminal T_k.

One component thereof is self-interference, i.e. the signal of the terminal itself that has been retransmitted back to it. This self-interference can be cancelled at the terminal when the effective channel via the relays is known [112]. Hence, the receive signal (5.43) becomes

$$\tilde{\mathbf{y}}_k = \sum_{l=1}^{L} \mathbf{H}_{l,k}^{\mathsf{T}} \cdot \mathbf{G}_l \cdot \left(\mathbf{H}_{l,j} \cdot \mathbf{x}_j + \sum_{\substack{n=1 \\ n \notin \{k,j\}}}^{2K} \mathbf{H}_{l,n} \cdot \mathbf{x}_n \right) + \sum_{l=1}^{L} \mathbf{H}_{l,k}^{\mathsf{T}} \cdot \mathbf{G}_l \cdot \mathbf{n}_l + \mathbf{w}_k. \tag{5.44}$$

In the following, we focus on the design of the relay gain matrices $\mathbf{G}_l$ and assume that the terminals do not apply any beamforming or precoding. To this end, the transmit signals $\mathbf{x}_k$ of the terminals are assumed to be i.i.d $\mathcal{CN}(\mathbf{0}, \frac{P_s}{M_k} \cdot \mathbf{I})$, where P_s is the total allowed transmit power of T_k. For the relays we impose the transmit power constraint

$$\tilde{P}_l = \mathrm{Tr}\left\{ \mathbf{G}_l \cdot \left(\sum_{k=1}^{2K} \frac{P_{\mathrm{s}}}{M_k} \mathbf{H}_{l,k}\mathbf{H}_{l,k}^{\mathsf{H}} + \sigma_n^2\mathbf{I} \right) \cdot \mathbf{G}_l^{\mathsf{H}} \right\} \leq P_{\mathrm{r}}, \quad \forall l. \tag{5.45}$$

In this case, the achievable rate of user k can be stated as

$$R_k = \frac{1}{2}\log_2\det\left\{ \mathbf{I} + \left(\mathbf{K}_k^{(\mathrm{i})} + \mathbf{K}_k^{(\mathrm{n})} \right)^{-1} \cdot \mathbf{K}_k^{(\mathrm{s})} \right\}, \tag{5.46}$$

with the covariance matrices of the desired signal, interference, and noise given by

$$\mathbf{K}_k^{(\mathrm{s})} = \frac{P_{\mathrm{s}}}{M_k} \sum_{l=1}^{L}\sum_{i=1}^{L} \mathbf{H}_{l,k}^{\mathsf{T}}\mathbf{G}_l\mathbf{H}_{l,j}\mathbf{H}_{i,j}^{\mathsf{H}}\mathbf{G}_i^{\mathsf{H}}\mathbf{H}_{i,k}^{*} \tag{5.47}$$

$$\mathbf{K}_k^{(\mathrm{i})} = \sum_{\substack{n=1 \\ n \notin \{k,j\}}}^{2K} \sum_{l=1}^{L}\sum_{i=1}^{L} \frac{P_{\mathrm{s}}}{M_n} \mathbf{H}_{l,k}^{\mathsf{T}}\mathbf{G}_l\mathbf{H}_{l,n}\mathbf{H}_{i,n}^{\mathsf{H}}\mathbf{G}_i^{\mathsf{H}}\mathbf{H}_{i,k}^{*} \tag{5.48}$$

$$\mathbf{K}_k^{(\mathrm{n})} = \sigma_n^2 \sum_{l=1}^{L} \mathbf{H}_{l,k}^{\mathsf{T}}\mathbf{G}_l\mathbf{G}_l^{\mathsf{H}}\mathbf{H}_{l,k}^{*} + \sigma_w^2\mathbf{I}. \tag{5.49}$$

Therein, σ_n^2 and σ_w^2 are the variances of the noise induced in the relays and terminals, respectively.

5.6.1 Design of Relay Gain Matrix

Relay gain matrices that optimize the performance of the network, e.g. sum rate, are generally hard to find. Possible approaches to optimize the objective function by iterative algorithms are e.g. the gradient based algorithm presented in Section 5.3. This is now generalized to relay gain matrices that can form arbitrary linear combinations of the relay input signals in the spatial as well as frequency domain. This additionally increases the complexity of the optimization problem, especially when many subcarriers are involved. To this end, we apply a block ZF approach in which the relay gains cancel the interference between the terminal pairs completely. Note that ZF is not optimal with respect to achievable rate. However, it allows us to reduce the optimization problem to only finding combination weights of null space vectors.

The ZF conditions can be stated as follows

$$\sum_{l=1}^{L} \mathbf{H}_{l,k}^{\mathsf{T}} \mathbf{G}_l \mathbf{H}_{l,n} = \mathbf{O}, \quad \forall k, n, \quad n \notin \{k, j\} \tag{5.50}$$

$$\mathrm{rank}\left(\sum_{l=1}^{L} \mathbf{H}_{l,k}^{\mathsf{T}} \mathbf{G}_l \mathbf{H}_{l,j}\right) = d, \tag{5.51}$$

where $d \leq \min\{M_k, M_j\}$ is the number of desired spatial streams user pair (k, j) wishes to exchange. Note that self-interference can be cancelled at the terminals and is therefore not part of these conditions [112].

In order to solve the ZF conditions in closed form, we rewrite the left hand side of (5.50) into the product $\mathbf{H}_k^{\mathsf{T}} \mathbf{G} \mathbf{H}_n$, in which the matrices are with respect to all relays, i.e. $\mathbf{H}_k^{\mathsf{T}} = [\mathbf{H}_{1,k}^{\mathsf{T}}, \ldots, \mathbf{H}_{L,k}^{\mathsf{T}}]$ and $\mathbf{G} = \mathrm{blkdiag}(\mathbf{G}_1, \ldots, \mathbf{G}_L)$. The equation system $\mathbf{H}_k^{\mathsf{T}} \mathbf{G} \mathbf{H}_n = \mathbf{O}$ is linear in the entries of $\mathbf{G}$. It can be solved by rewriting it into the equivalent form

$$\mathbf{A}_{k,n} \cdot \mathbf{g} = \mathbf{O}, \quad \forall k, n, \quad n \notin \{k, j\}, \tag{5.52}$$

where $\mathbf{g} = \mathrm{vec}(\mathbf{G})$ is the vector that contains all relay gain coefficients. The matrix $\mathbf{A}_{k,n}$ is constructed by copying and permuting combinations of entries of $\mathbf{H}_k$ and $\mathbf{H}_n$ as follows

$$\mathbf{A}_{k,n} = \left[(\mathbf{1}_{M_n \times 1} \otimes \mathbf{I}_{M_k}) \cdot \mathbf{H}_k^{\mathsf{T}} \cdot \mathbf{E}\right] \odot \left[(\mathbf{H}_n^{\mathsf{T}} \cdot \mathbf{F}) \otimes \mathbf{1}_{M_k \times 1}\right], \tag{5.53}$$

where $\odot$, $\otimes$, and $\mathbf{1}_{n \times m}$ denote the element wise product, Kronecker product, and the all ones matrix of size $n \times m$ and the matrices $\mathbf{E}$ and $\mathbf{F}$ are given by

$$\mathbf{E} = \mathrm{blkdiag}(\mathbf{1}_{1 \times N_1} \otimes \mathbf{I}_{N_1}, \ldots, \mathbf{1}_{1 \times N_L} \otimes \mathbf{I}_{N_L}) \tag{5.54}$$

$$\mathbf{F} = \mathrm{blkdiag}(\underbrace{\mathbf{1}_{1 \times N_1}, \ldots, \mathbf{1}_{1 \times N_1}}_{N_1 \text{ times}}, \ldots, \underbrace{\mathbf{1}_{1 \times N_L}, \ldots, \mathbf{1}_{1 \times N_L}}_{N_L \text{ times}}). \tag{5.55}$$

Once the matrices $\mathbf{A}_{k,n}$ are constructed, they can be stacked on top of each other into the matrix $\mathbf{A} = [\mathbf{A}_{1,3}^{\mathsf{T}}, \ldots, \mathbf{A}_{1,2K}^{\mathsf{T}}, \ldots, \mathbf{A}_{2K,1}^{\mathsf{T}}, \ldots, \mathbf{A}_{2K,2K-2}^{\mathsf{T}}]^{\mathsf{T}}$ and the relay gains can be chosen to lie in the null space of $\mathbf{A}$. To this end, the SVD $\mathbf{A} = \mathbf{U} \cdot \mathbf{\Sigma} \cdot [\mathbf{V}_1 \mathbf{V}_0]^{\mathsf{H}}$ can be applied, where the columns of $\mathbf{V}_0$ form a basis of $\mathrm{null}\{\mathbf{A}\}$, as in Chapter 3. The relay gain matrices that fulfill the ZF conditions can then be stated as a linear combination

$$\mathbf{G}_l = \alpha_1 \cdot \mathbf{V}_1^{(l)} + \ldots + \alpha_D \cdot \mathbf{V}_D^{(l)}, \tag{5.56}$$

where α_i and $\mathbf{V}_i^{(l)}$ are the combination weights and the null space vectors turned back into matrices according to the dimensions of $\mathbf{G}_l$. The combination weights need thereby to be scaled such that the power constraint (5.45) is fulfilled in all relays.

The dimension D of the null space depends on the network topology and the number of relay antennas. In a network with a single relay, the number of relay antennas required to fulfill the ZF conditions (5.50) and (5.51) is

$$N = 2\sum_{p=1}^{K} \min\left\{M_{2p-1}, M_{2p}\right\} - \min\left\{M_1, \ldots, M_{2K}\right\}, \qquad (5.57)$$

irrespective of the number of subcarriers. If the network contains multiple relays, the total number of relay gain coefficients needs to be at least the same as in the single relay case.

5.6.2 Sum Rate Maximization

Solving the block ZF conditions, the relay gain matrices are found to lie in the null space spanned by the basis vectors from (5.56). If the null space consists of $D \geq 2$ dimensions, relay gain matrices can be formed as arbitrary linear combinations of all null space vectors. Multiple null space vectors are able to improve the diversity gain [84]. However, the linear combination of the null space vectors needs to be appropriately designed to benefit from this gain. A trivial choice of combination weights might destroy the diversity. It is therefore essential to choose the different combination weights α_i in a beneficial way.

A plot of the sum rate for different combination weights of a sample network can be seen in Fig. 5.7. Therein, a typical channel realization with $C = 2$ independent subcarriers for a network with $K = 2$ user pairs, $L = 2$ relays, and $M_k = 2$ and $N_l = 4$ antennas at the different nodes is considered. The noise variances are $\sigma_n^2 = \sigma_w^2 = 0.001$, the transmit power at the terminals as well as the relay transmit power constraints are set to $P_s = P_r = 1$. The sample network has $D = 8$ null space vectors, but only the real parts of the first two are varied; the others are set to zero. The achievable rate is set to zero if a specific weight combination violates the power constraint. It can be seen that there are multiple local optima. As a resulting optimization problem is not convex, it is difficult to find "good" linear combinations. Nevertheless, the complexity is drastically reduced as compared to a direct optimization of the relay gains without ZF,

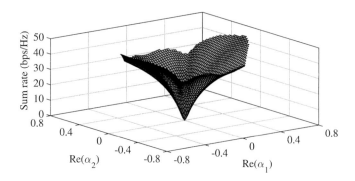

Figure 5.7.: Achievable sum rate for different combination weights α_1 and α_2.

as only combination weights have to be found. In order to find combination weights that achieve the expected gains of subcarrier cooperation in two-way relaying, we apply two methods that aim to maximize the sum rate of the network.

Null Space Selection

The first approach is to select the best null space vector and to scale it such that the power constraint at each relay is met. To this end, the null space basis is chosen according to

$$\arg \max_{i=1,\ldots,D} \sum_{k=1}^{2K} R_k(i),$$
(5.58)

where $R_k(i)$ is the achievable rate of user k when the relays choose the ith null space vector, i.e. $\mathbf{G}_l = \beta \cdot \mathbf{V}_i^{(l)}$. The scaling factor β is then chosen such that all relays fulfill the per-node sum power constraint $\tilde{P}_l \leq P_\mathrm{r}$, $\forall l$. Note that all relays have to apply the same scaling factor β as otherwise the ZF conditions would be violated. This is achieved by setting $\beta = P_\mathrm{r}/\max_l \tilde{P}_l$. Hence, in general, not all relays will transmit with full transmit power.

Numerical Optimization

The second, more complex, approach is to numerically find combination weights α_i that attempt to maximize the sum rate under the per-relay power constraint (5.45).

The resulting optimization problem can be formulated as

$$\max_{\alpha_1,\dots,\alpha_D} \sum_{k=1}^{2K} R_k \quad \text{s.t.} \quad \tilde{P}_l \le P_r, \quad \forall l. \tag{5.59}$$

As this optimization problem is not convex, it is generally difficult to find the global optimum. There are, however, standard optimization tools that converge to local optima. To this end, we can apply such tools as e.g. gradient based optimization algorithms [15] that are designed for convex problems, but also converge to local optima in non-convex problems. We therefore apply an interior-point method with random initialization in Matlab in order to numerically find a local optimum of (5.59).

5.7 Performance Evaluation

The performance of the weight allocation schemes and the improvements due to sub-carrier cooperation are assessed by means of computer simulations. A comparison of the weight allocation schemes is shown in Fig. 5.8. There, the numerically optimized combination and the null space vector selection are compared with choosing a random null space vector and using all vectors equally weighted. For the simulation, a network with $K = L = 2$ terminal pairs and relays has been used. All terminals are equipped with $M_k = 2$ antennas, the relays with $N_l = 4$ antennas, and $C = 4$ subcarriers are considered. The elements of the channel matrices on all subcarriers are drawn i.i.d. according to $\mathcal{CN}(0,1)$. The transmit powers are set to $P_s = P_r = C$ for all terminals and relays and the noise variances are chosen as $\sigma_n^2 = \sigma_w^2 = 0.01$. The numerically optimized weight allocation clearly outperforms all other schemes. Selecting the best null space vector leads to lower sum rates, but achieves the same diversity gain as indicated by the slope of the empirical CDF for small values. Using all null space vectors in an equally weighted way leads to a further decrease in performance. Also the slope of the CDF is somewhat less steep, which indicates that not all diversity can be exploited. Using a random null space vector performs poorly and achieves a much smaller diversity gain. In this case, it can also be observed that the CDF follows a stair like behavior which is due to some null space vectors that are much worse than others and that different null space vectors have different importance. Even though the block ZF approach cancels all the interference in the network, it is crucial to chose the null

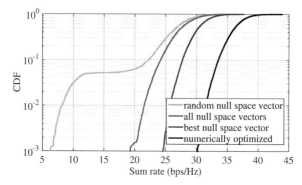

Figure 5.8.: Empirical CDF of the achievable sum rates for different weight allocation schemes. The rates are normalized by C.

space vectors appropriately. The numerically more complex optimization is justified by the higher performance. The simpler null space selection scheme achieves however a good tradeoff between complexity and performance.

The impact of subcarrier cooperation and the influence of the number of subcarriers is studied in the following. The sum rate of the two-way relaying schemes for different numbers of subcarriers $C = 1, 2, 4, 6$ is shown in Fig. 5.9. The sum rates are normalized by C. For the simulation, the same network as before is considered. The average SNR, however, is chosen to be SNR $= 30\,\mathrm{dB}$, irrespective of the number C of subcarriers, i.e. $P_\mathrm{s} = P_\mathrm{r} = 1$ and $\sigma_n^2 = \sigma_w^2 = 0.001$ for any choice of C. The CDFs of the sum rates for the numerically optimized and null space selection method are plotted. It can be seen that the performance as well as the diversity increases with the number of subcarriers for both the optimized scheme and the null space selection. Again, the numerically optimized scheme leads to considerably higher sum rates. Also the performance gain with increasing subcarriers is somewhat larger than with the null space selection scheme. With $C > 6$, however, the performance increase starts to saturate, at least in the shown regime of outage probabilities larger than 10^{-3}. The crossings of the null space selection CDFs in the high rate regime show a disadvantage of the simpler scheme. When a null space vector is selected, it is scaled such that the power constraint is met at all relays. As the scaling has to be the same in all relays, only one relay exploits the power constraint with equality, the others transmit with lower power. When more subcarriers are used, the probability increases that some channels are strong which results in a lower power at the weak relays. This is reflected in the

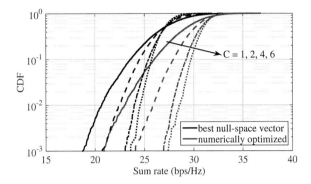

Figure 5.9.: Empirical CDFs of sum rates for different numbers of subcarriers. The rates are normalized by C.

peak rates of the selection scheme that slightly decrease when more subcarriers are applied. Nevertheless, the simpler null space selection scheme can exploit the diversity offered by the channel with multiple subcarriers.

The performance gain due to subcarrier cooperation over subcarrier non-cooperative schemes is studied in Fig. 5.10. Thereby, $C = 4$ subcarriers are considered and the proposed subcarrier-cooperative scheme is compared to two conventional subcarrier non-cooperative schemes. In both conventional schemes, the relays perform block ZF separately on each of the independent subcarriers. In the first one, the relay power $P_r = C$ is divided equally to all subcarriers, i.e. the transmit power on each subcarrier is $P_r^{(c)} = 1$. In the second subcarrier non-cooperative scheme, the power allocation on the different subcarriers is numerically optimized according to

$$\max_{P_r^{(1)}, \ldots, P_r^{(C)}} \sum_{c=1}^{C} \sum_{k=1}^{2K} R_k^{(c)} \left(P_r^{(c)} \right), \quad \text{s.t.} \quad \sum_{c=1}^{C} P_r^{(c)} = P_r, \tag{5.60}$$

where $R_k^{(c)} \left(P_r^{(c)} \right)$ is the achievable rate of terminal T_k in subcarrier c when a power $P_r^{(c)}$ is allocated on this subcarrier. The CDFs of the sum rates normalized to C show a significant gain in the subcarrier-cooperative case. It can also be observed that the optimized power allocation does not lead to much improvement compared to the equal power allocation. These results show that a notable portion of the achievable rate is lost when the relays operate independently on the different subcarriers and that subcarrier cooperation can indeed lead to a significant gain.

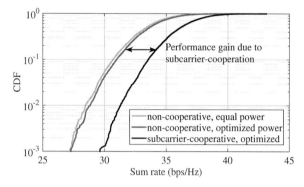

Figure 5.10.: Comparison between subcarrier cooperation and non-cooperation for $C = 4$ subcarriers. The rates are normalized by C.

Even though it seems counterintuitive at first glance to combine signals on different subcarriers that are orthogonal, we have shown that two-way AF relaying can be improved by subcarrier cooperation. Thereby, a part of the performance gain can be attributed to channel pairing [48], i.e. an advantageous mapping of the different subcarriers for the two hops that is inherently included in the subcarrier cooperation. In this regard, subcarrier-cooperative relaying differs to the setup in [53], where only a conventional BC is considered. The performance in two-way relaying with extension to multiple subcarriers is therefore particularly good. If only one directional communication is considered, the improvements are smaller, as indicated in [53] where the performance gain due to subcarrier cooperation is visible but small. The results however indicate that the performance gap between linear schemes and DPC is smaller than expected when subcarrier cooperation can be applied. For practical networks, on the other hand, the significant increase of complexity due to precoding across all subcarriers is hardly justified, especially when many subcarriers are considered as in practical OFDM based systems. Nevertheless, exemplarily for the case of two-way relaying we see how we can deal with wideband systems that contain multiple subcarriers. The results with the significant performance gains can however not be generalized to other setups. But we can expect that also schemes other than the block ZF approach might benefit from subcarrier cooperation.

5.8   Application to Cellular Networks

The relaying schemes discussed in this chapter are only studied in simplified network setups where no network geometry or channel models with distant dependent pathloss and other real-world considerations were included. Applying the multihop networks and the relaying scheme with block ZF into the framework provided in the last chapters is difficult as the complexity of the relay gain allocation schemes is rather high, especially when large networks with many nodes are considered. Simulations in which the gain coefficients of many relays are optimized are therefore hardly feasible. In order to get an insight into how such AF relaying schemes can be applied to cellular networks anyway and what potential gains we can expect, we consider the two-way AF relaying approach with block ZF in a small cellular setup. Thereby, we deploy relays within a cooperation area between neighboring cells. Relays that are spread in the cells can thereby cancel interference between different users. In this way, the relaying scheme can act as an alternative or complement to CoMP as discussed in the previous chapters. However, a potential drawback of the proposed scheme is the large number of relay antennas required for nulling the interference, especially when two-way relaying is applied.

This number, however, can be reduced when the two-way relaying scheme is combined with a simple form of BS cooperation. If neighboring BSs are connected with each other via the backhaul, they can share their user data. This allows the BSs not only to cancel self-interference, but also interference caused by adjacent BSs. If the BS signals contain training or pilot sequences, these can be used to estimate the effective channels via the relays to allow the cancellation of the BS interference without disseminating CSI among the BSs. In this way, some ZF conditions can be dropped, namely the ones of the signals received by the BSs that contain interference from other BSs, and the relaying scheme works with fewer antennas. This number can be further reduced if the interference the different MSs cause to each other is also ignored by the ZF. This can be justified by the assumption that MS signals are weak compared to BS signals. Ignoring them for cancellation would therefore only increase the noise floor by a small amount. The required number of relay antennas for the block ZF between three BSs and MSs, each equipped with two antennas (thus two data streams for each user in each direction), is shown in the following table:

Number of relays L	1	3	6	9
N_l full antennas	10	6	4	3
N_l reduced antennas	7	4	3	2

For the evaluation, we consider a network with three BSs that serve three sectors in the 30° orientation as depicted in Fig. 3.3d. The MSs as well as the BSs are equipped with two antennas each, i.e. no excess antennas and no beamforming is applied at the BSs. The same urban environment with the channel model with NLOS propagation condition as described in the previous chapters is applied for the simulations. The BSs, however transmit with a reduced power of 40 W. As we apply two-way relaying, the MSs also transmit signals, also without beamforming, with a power of 200 mW. The noise variance at the MSs as well as the BSs is set to $\sigma_w^2 = 5 \cdot 10^{-12}$ W as before. In order to keep the simulation complexity manageable, we consider only the three sectors as described, OCI injected by other nodes is not taken into account here. The communication between the BSs and MSs is assisted by three relays with $N_l = 3$ or 2 antennas in each sector. The relays are located on 2/3 of the distance between the BSs and the sector borders as depicted in Fig. 5.11.

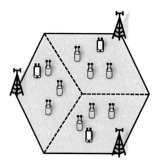

Figure 5.11.: The communication in three sectors is assisted by additional two-way relays.

The average user rates (UL plus DL) of this network are shown in Fig. 5.12. Therein, two-way relaying, once with the full number of antennas (three per relay) allowing to cancel all interference and once with a reduced number of antennas (two per relay) as explained above, are compared to a conventional network without relays where the three users are served on three orthogonal frequency bands (FDMA). It can be seen that a significant gain can be achieved by the relaying approach, especially when subcarrier cooperation is applied. With the reduced antenna setup where the MS interference is dropped in the ZF conditions and subcarrier cooperation is applied (yellow solid line), the performance is almost identical to the case with the full number of antennas but without subcarrier cooperation. The performance that comes without canceling the MS

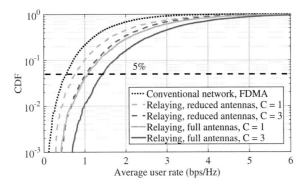

Figure 5.12.: Application of two-way relaying to a cellular network with 3 BS-MS pairs. Relaying with $L = 9$ and $N_l = 3$ and 2, respectively, is compared to a conventional network with FDMA. The rates are normalized by C.

interference can thus be compensated with subcarrier cooperation when a conventional system is considered. With subcarrier cooperation, the 5% outage rate can be shifted above 1 bps/Hz. The data rates shown here are however a bit optimistic as OCI is not taken into account. Nevertheless, the performance gain is comparable to the one that can be achieved with cooperation between three BSs and even better than with DF relays in the selection scheme described in the previous chapter.

The application of AF relaying in cellular networks thus shows to be an interesting alternative to the DF relays discussed in Chapter 4. Such an approach is especially interesting when e.g. the backhaul of the BSs is not sufficient to allow for sophisticated BS cooperation. For the reduced antenna setup at the relays, however, the BSs still need to exchange their user data in order to cancel the BS interference that is transmitted back from the relays. In the following, we investigate the potential of including relays into cellular networks. To this end, however, we apply simpler relaying schemes that do not require the solution of computationally complex optimization problems. If many relays are deployed, the network can still benefit from a higher coverage due to the relays and also from distributed cooperation as explained in the next chapters.

6

Ubiquitous Relaying

In the preceding chapters, the potential of PHY layer cooperation between multiple BSs and the use of SNs has been studied and evaluated. When several BSs share their information and precode their signals in an appropriate manner, interference can be mitigated and turned into useful signal contributions. Also the deployment of relays or other types of SNs offers interesting benefits. They can increase the coverage range of cells and can provide additional diversity gains. AF relays on the other hand can cancel interference and shape the effective channel between BSs and MSs for better receive signal qualities and higher data rates. AF relaying is thereby especially beneficial when multiple relays can jointly contribute to optimize the overall performance. Even though the considered techniques offer considerable performance gains, the boundaries of cellular systems have to be stretched further such that networks of future generations can provide data rates that are higher by orders of magnitude than today's systems.

From a fundamental perspective, there are different ways to achieve this: expanding the resources such as bandwidth or power, make use of a large number of antennas to increase the spatial degrees of freedom, and to apply cooperation to overcome the impairments of interference. While the former approach is limited by the scarcity or costs of the resources, spectral efficiency can be increased by an extension of the networks in the spatial domain. In order to enable the required gains, the BSs can be equipped with (many) more antennas, eventually leading to massive MIMO [94]. Such (very) large antenna arrays allow to serve many users at the same time, for instance using multiuser-MIMO beamforming methods, and to mitigate interference in adjacent cells.

An alternative is to increase the BS density and to reduce the cell sizes such that the network consists of pico- or femto-cells [98]. Such small cells can also coexist with micro- or macro-cells in heterogeneous networks [41]. The fundamental advantage is that the adaptation to the user position can be achieved by handovers between cells or sectors as indicated by the macro diversity scheme discussed earlier. Such selection schemes are easy to implement and require little overhead. In practice, however, this approach is, among others, limited by the difficulty to identify new BS sites, e.g. due to social acceptance, availability of backbone access etc., and by the cost of deployment. Besides this, it is also foreseen that future networks are supported by RRHs [23] or by wireless relays [146]. These, however, are so far primarily intended for range extension, or, as discussed in Chapters 4 and 5, require high complexity to cooperate with other nodes.

Either of these approaches can increase the total throughput of the entire network, while individual user rates remain limited when the MSs do not have more antennas. As much of the available resources (e.g. bandwidth) as possible should therefore be allocated to each user, up to a reuse factor of one. Due to the interference-limited nature of cellular networks, this can only be achieved by efficient interference management. To this end, BS cooperation (Chapter 3, [39,68]) can be applied to mitigate or cancel out-of-cell interference. With cooperation of limited complexity, however, the fundamental problems remain, as the cooperation areas (virtual cells) cause interference to adjacent areas, as in the case of conventional networks, just that the virtual cells contain multiple BSs instead of only one.

Moreover, if cooperation is applied on a larger scale, it suffers from severe challenges and difficulties. BSs that perform joint beamforming require very high backhaul rates, not only to support the data rates of their users, but also to exchange user data and CSI with their cooperation partners. Especially if BSs with large arrays are considered, the number of channel coefficients that need to be estimated grows rapidly with the number of involved antennas. This leads to an increasing overhead, as more pilots have to be included in the signals. Achievable performance gains might therefore stagnate or even decrease [109]. Moreover, even when this overhead can be overcome, the performance of CoMP remains limited by residual interference [90].

First attempts to combine the aforementioned approaches are presented in [13,56], where a layer of small cells operates in parallel to a macro-cell tier with a large array BS. Therein, a TDD based network architecture is proposed which allows the BS to benefit from channel reciprocity to simplify channel estimation. While the small cells transmit, the BS can estimate these signals and use them for interference mitigating beamforming on the downlink. When the small cell BSs are at fixed positions, the channels between them and the large macro BS can be assumed to be quasi static and the estimation of this channel is not susceptible to instantaneous channel variations and stable over longer time scales. The BS can then design its precoding to transmit independent data streams to its users while being orthogonal to the subspace spanned by the strongest interference directions. The sum interference imposed on the small cells can thus be minimized. [56]. The small cells, however, need fully equipped small BSs connected to the wired backbone; their massive deployment might therefore be difficult and expensive.

In the following, we develop an alternative concept that attempts to profit from the advantages of the previously mentioned approaches while avoiding their disadvantages. Based on the insights from the previous chapters, we combine BS beamforming with large antenna arrays with many relays without a fixed connection to the backhaul. If the relays are of low cost and low power, they can be installed in massive numbers across the entire area of the network. When users communicate via the relay links, the BSs see the static relays as the nodes they communicate with, while the MSs receive their signals from the small relay nodes. Thereby, the BSs can profit from transmitting to static receivers, which allows to apply sophisticated beamforming, and the mobile users in turn can profit from the advantages of dense networks with small cells.

To this end, we compare different relaying architectures and their feasibility for this concept of ubiquitous relaying. As the operation of the relays should be of low complexity such that they can be installed in large numbers, we refrain from sophisticated iterative optimization procedures, but limit ourselves to transmission schemes that can be applied in each node with local CSI only and can be calculated in closed form. The carefully selected transmission schemes should then be able to contribute together to achieve a high network performance. With this, we can study the behavior in large networks in which many nodes are included. The different approaches that we compare include AF as well as DF relays in one-way and two-way protocols. We compare these approaches with respect to achievable rates and complexity and propose methods to cope with the interference in such networks, e.g. based on beamforming at the BSs, relay filtering, and a specific form of BS cooperation, that are of comparably low complexity. Due to the interference mitigation, high performance gains can be achieved. Particularly two-way relaying proves thereby to be very beneficial in contrast to rather pessimistic results of prior work (cf. e.g. [104]).

Furthermore, we investigate the influence of imperfect CSI on these approaches and show that especially simple relays of low complexity are very robust and thus well suited for massive deployment. In Chapter 7, we will then go one step further and propose a concept in which a massive number of relays with a particularly simple implementation is deployed throughout the entire network and the entire communication in the whole area is performed in a two-hop fashion. We argue then that this "relay carpet" can enable massive MIMO and can lead to the desired performance of future networks.

6.1   Network Model

The basic organization of the network is similar to a conventional one with micro- or macro-cells. The area is divided into geographically separated cells, each with one BS that is equipped with a large antenna array and multiple MSs that are served simultaneously. In this chapter, we assume that the BSs do not apply cooperative joint transmission or joint decoding across multiple cells. Each MSs is served only by the BS associated to the cell in which the MS is located in. The BSs serve however jointly all MSs within their cell, i.e. no sectorization is applied but all BS antennas are used jointly for the transmission and reception in a cell. Furthermore, we assume that all transmissions are synchronized in time and frequency and that the propagation delays in the different links are compensated.

The communication between BSs and MSs (DL) and vice versa (UL) is assisted by a large amount of relays. Different relays can thereby transmit in different frequency bands such that adjacent relay cells form a reuse pattern. Accordingly, the MSs can be served in different resource blocks and the BSs communicate with MSs by assigning an appropriately chosen set of relays. We consider the relays as dedicated infrastructure nodes. As such, they are intentionally mounted at fixed positions, e.g. on lamp posts, at bus stops, or on the wall of a building, and might therefore have a good connection to the BS. Additionally, these links have a long coherence time and fast fading is eliminated. The MSs, on the other hand, are served by small relay cells. If sufficiently many relays are deployed, shadowing effects can be avoided to a large extent.

The relays not only improve the connectivity for the MSs, but can also apply different signal processing tasks. These depend on the architecture of the relays and can range from simple active scattering [151] up to sophisticated filtering, interference cancellation [9,31], or decoding and encoding [72]. Different implementations can thereby affect the signal processing and the complexity at the other nodes. In the following, we consider AF and DF relays that each can operate in TDD or FDD mode, i.e. they are half-duplex relays that can transmit and receive in different time slots or in different frequency bands. The complexity of the relays does not only depend on the relaying strategy but also on further implementation aspects, for instance receive and transmit filters. To this end, we consider two different types of relay implementations: In their simplest form, the relays do not use any special receive or transmit filter; the signal is only scaled with a gain matrix given by a scaled identity matrix. We refer to these relays as *type A* relays. The more complex *type B* relays use spatial receive and transmit filters.

In order to reflect the bidirectional communication in cellular networks, we apply each of these relay types in a one-way and a two-way relaying protocol that does not make use of the direct link between BSs and MSs. In the former case, the UL and DL are separated and the relays either forward the BS signals to the MSs or vice versa. In two-way relaying, both directions of communication are combined such that the relays receive the superposition of all BS and MS signals and broadcast a processed version of these signals back to all terminals. This can double the spectral efficiency as compared to one-way relaying, but the signals are affected by additional interference the terminals have previously transmitted and is backscattered by the relays [158].

For the studies in this chapter, we thus compare a preselection of relaying approaches with each other that arise from the different combinations of these aspects:

1. duplex mode (TDD/FDD),
2. relaying strategy (AF/DF),
3. implementation (type A/B), and
4. protocol (one-way/ two-way relaying).

An especially simple class of relays is given by type A AF relays in an FDD system; such relays can be implemented by a frequency conversion of the received signal. They are not only of very low complexity, but also introduce no (or very small) delays, as the signals are immediately retransmitted. This is not the case for DF relays. Due to the decoding and encoding, the retransmission is delayed by at least a block length, even in FDD mode. Additionally, the DF relays also require the most complex implementation, not only because of the decoding and encoding functionality but also due to the required CSIR that has to be obtained.

6.1.1 Input-Output Relation & Achievable Rates

In order to describe this network mathematically, we consider C cells, each with one BS and multiple MSs. For notational simplicity, we assume that all cells have the same number M of active MSs and that all nodes of the same kind have the same number of antennas, although an extension to a more general case is straightforward. The number of antennas at the BSs is denoted by N_B, the one of the MSs by N_M. The

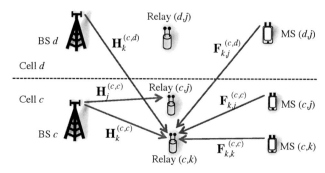

Figure 6.1.: Network model. The communication between BSs and MSs is assisted by relays.

considered communication is bidirectional, i.e. BS c, with $c \in \{1, \ldots, C\}$, wants to transmit $d_s \leq N_M$ data streams to MS (c, k) (the kth MS in cell c) in the DL and, in turn, each MS wishes to send d_s data streams to its BS in the UL.

As each BS simultaneously serves multiple MSs located in its corresponding cell, we assume $N_B \geq M \cdot N_M$ and write the DL signal of BS c as

$$\mathbf{x}_c^{(B)} = \sum_{k=1}^{M} \mathbf{Q}_{c,k}^{(B)} \cdot \mathbf{s}_{c,k}^{(B)}, \qquad (6.1)$$

where $\mathbf{s}_{c,k}^{(B)} \in \mathbb{C}^{d_s}$ is the transmit symbol vector from BS c intended for MS (c, k) and $\mathbf{Q}_{c,k}^{(B)} \in \mathbb{C}^{N_B \times d_s}$ the precoding matrix. In the UL, the MSs transmit

$$\mathbf{x}_{c,k}^{(M)} = \mathbf{Q}_{c,k}^{(M)} \cdot \mathbf{s}_{c,k}^{(M)}, \qquad (6.2)$$

with $\mathbf{s}_{c,k}^{(M)} \in \mathbb{C}^{d_s}$ and $\mathbf{Q}_{c,k}^{(M)} \in \mathbb{C}^{N_M \times d_s}$ being the transmit symbol vector and the precoding matrix of the signal from MS (c, k) intended for BS c.

The bidirectional communication between BSs and MSs is assisted by $K \geq M$ relays. We focus on a single resource block, i.e. all relays transmit in the same frequency band. Furthermore, each active MS is served by at least one relay and a relay cannot serve more than one MS.[1] The relays are equipped with N_R antennas, where $N_B \geq N_R \geq N_M$. A sketch of the network can be seen in Fig. 6.1. The narrow-band channel from BS d to relay (c, k) is denoted by $\mathbf{H}_k^{(c,d)} \in \mathbb{C}^{N_R \times N_B}$ and the reverse channel from relay (c, k)

---

[1] More MSs can be served in different frequency bands or by sharing the resources with a TDMA or FDMA scheme.

to BS d by $\overline{\mathbf{H}}_k^{(d,c)} \in \mathbb{C}^{N_\mathrm{B} \times N_\mathrm{R}}$. The channels from MS (d, j) to relay (c, k) and vice versa are denoted by $\mathbf{F}_{k,j}^{(c,d)} \in \mathbb{C}^{N_\mathrm{R} \times N_\mathrm{M}}$ and $\overline{\mathbf{F}}_{j,k}^{(d,c)} \in \mathbb{C}^{N_\mathrm{M} \times N_\mathrm{R}}$, respectively. When a TDD protocol is applied, the channels are assumed to be reciprocal, i.e. $\overline{\mathbf{H}}_k^{(d,c)} = \mathbf{H}_k^{(c,d)\mathsf{T}}$ and $\overline{\mathbf{F}}_{j,k}^{(d,c)} = \mathbf{F}_{k,j}^{(c,d)\mathsf{T}}$. If the system is operated in the FDD mode, the channels on the different directions are assumed to be independent. Direct channels between BS and MSs are not considered, the users are exclusively served via relays. In the following, we describe the end-to-end relations of the system for the different relaying strategies and derive their achievable rates.

AF One-Way Relaying

In one-way relaying, the UL and DL are separated, either by different time slots (TDD) or orthogonal frequency bands (FDD). Considering the DL, the BSs simultaneously transmit their signal (6.1) and relay (c, k) receives (in the forward direction)

$$\overrightarrow{\mathbf{r}}_{c,k} = \sum_{d=1}^{C} \mathbf{H}_k^{(c,d)} \cdot \sum_{j=1}^{M} \mathbf{Q}_{d,j}^{(\mathrm{B})} \cdot \mathbf{s}_{d,j}^{(\mathrm{B})} + \overrightarrow{\mathbf{n}}_{c,k}, \tag{6.3}$$

where $\overrightarrow{\mathbf{n}}_{c,k}$ is the noise induced in the relay. Assuming AF relaying, the relays multiply their receive signals (6.3) with a gain matrix $\mathbf{G}_{c,k} \in \mathbb{C}^{N_\mathrm{R} \times N_\mathrm{R}}$ and, after a possible frequency conversion in FDD, retransmit $\overrightarrow{\mathbf{t}}_{c,k} = \mathbf{G}_{c,k} \cdot \overrightarrow{\mathbf{r}}_{c,k}$ to the MSs. The receive signal of MS (c, k) is then

$$\overrightarrow{\mathbf{y}}_{c,k} = \sum_{d=1}^{C} \sum_{j=1}^{K} \overline{\mathbf{F}}_{k,j}^{(c,d)} \cdot \mathbf{G}_{d,j} \cdot \overrightarrow{\mathbf{r}}_{d,j} + \mathbf{w}_{c,k}^{(\mathrm{M})}, \tag{6.4}$$

with $\mathbf{w}_{c,k}^{(\mathrm{M})}$ being the noise in the MS.

In the UL, the MSs transmit their signals (6.2) and the (reverse) receive signal at the relays is written as

$$\overleftarrow{\mathbf{r}}_{c,k} = \sum_{d=1}^{C} \sum_{j=1}^{M} \mathbf{F}_{k,j}^{(c,d)} \cdot \mathbf{Q}_{d,j}^{(\mathrm{M})} \cdot \mathbf{s}_{d,j}^{(\mathrm{M})} + \overleftarrow{\mathbf{n}}_{c,k}. \tag{6.5}$$

After multiplication of $\overleftarrow{\mathbf{r}}_{c,k}$ with $\mathbf{G}_{c,k}$ and forwarding the resulting signal $\overleftarrow{\mathbf{t}}_{c,k} =$

$\mathbf{G}_{c,k} \cdot \overleftarrow{\mathbf{\Gamma}}_{c,k}$, BS c receives

$$\overleftarrow{\mathbf{y}}_c = \sum_{d=1}^{C} \sum_{j=1}^{K} \overline{\mathbf{H}}_j^{(c,d)} \cdot \mathbf{G}_{d,j} \cdot \overleftarrow{\mathbf{\Gamma}}_{d,j} + \mathbf{w}_c^{(B)}, \tag{6.6}$$

where $\mathbf{w}_c^{(B)}$ is the BS noise.

Once precoding and relay gain matrices are chosen, achievable rates can be formulated for both directions of communication. It is thereby assumed that the data symbols in the vectors $\mathbf{s}_{c,k}^{(B)}$ and $\mathbf{s}_{c,k}^{(M)}$ are i.i.d according to $\mathcal{CN}(0,1)$. The elements of the noise terms in the relays and terminals, $\mathbf{n}_{c,k}$, $\mathbf{w}_c^{(B)}$, and $\mathbf{w}_{c,k}^{(M)}$, are assumed to be i.i.d. $\mathcal{CN}(0,\sigma_n^2)$ and $\mathcal{CN}(0,\sigma_w^2)$, respectively.

In the DL, the achievable rate of the transmission from BS c to MS (c, k) is calculated by

$$\overrightarrow{R}_{c,k} = \log_2 \det \left(\mathbf{I}_{N_M} + \left(\overrightarrow{\mathbf{K}}_{c,k}^{(i+n)} \right)^{-1} \cdot \overrightarrow{\mathbf{K}}_{c,k}^{(sig)} \right), \tag{6.7}$$

where $\overrightarrow{\mathbf{K}}_{c,k}^{(sig)}$ and $\overrightarrow{\mathbf{K}}_{c,k}^{(i+n)}$ are covariance matrices of the desired signal and interference plus noise, summed over all relays. The signal covariance matrix is thus given by

$$\overrightarrow{\mathbf{K}}_{c,k}^{(sig)} = \sum_{d=1}^{C} \sum_{j=1}^{K} \sum_{d'=1}^{C} \sum_{j'=1}^{K} \overline{\mathbf{F}}_{k,j}^{(c,d)} \mathbf{G}_{d,j} \mathbf{H}_j^{(d,c)} \mathbf{Q}_{c,k}^{(B)} \cdot \mathbf{Q}_{c,k}^{(B)H} \mathbf{H}_{j'}^{(d',c)H} \mathbf{G}_{d',j'}^{H} \overline{\mathbf{F}}_{k,j'}^{(c,d')H} \tag{6.8}$$

and the one of the interference and noise is

$$\overrightarrow{\mathbf{K}}_{c,k}^{(i+n)} = \mathsf{E}\left[\overrightarrow{\mathbf{y}}_{c,k} \cdot \overrightarrow{\mathbf{y}}_{c,k}^{H} \right] - \overrightarrow{\mathbf{K}}_{c,k}^{(sig)}$$

$$= \sum_{\substack{d=1 \\ d \neq c}}^{C} \sum_{j=1}^{M} \sum_{b=1}^{C} \sum_{i=1}^{K} \sum_{b'=1}^{C} \sum_{i'=1}^{K} \overline{\mathbf{F}}_{k,i}^{(c,b)} \mathbf{G}_{b,i} \mathbf{H}_i^{(b,d)} \mathbf{Q}_{d,j}^{(B)} \cdot \mathbf{Q}_{d,j}^{(B)H} \mathbf{H}_{i'}^{(b',d)H} \mathbf{G}_{b',i'}^{H} \overline{\mathbf{F}}_{k,i'}^{(c,b')H}$$

$$+ \sum_{\substack{j=1 \\ j \neq k}}^{M} \sum_{b=1}^{C} \sum_{i=1}^{K} \sum_{b'=1}^{C} \sum_{i'=1}^{K} \overline{\mathbf{F}}_{k,i}^{(c,b)} \mathbf{G}_{b,i} \mathbf{H}_i^{(b,c)} \mathbf{Q}_{c,j}^{(B)} \cdot \mathbf{Q}_{c,j}^{(B)H} \mathbf{H}_{i'}^{(b',c)H} \mathbf{G}_{b',i'}^{H} \overline{\mathbf{F}}_{k,i'}^{(c,b')H}$$

$$+ \sigma_n^2 \sum_{b=1}^{C} \sum_{i=1}^{K} \overline{\mathbf{F}}_{k,i}^{(c,b)} \mathbf{G}_{b,i} \mathbf{G}_{b,i}^{H} \overline{\mathbf{F}}_{k,i}^{(c,b)H} + \sigma_w^2 \mathbf{I}_{N_M}. \tag{6.9}$$

In the UL, we assume that the BSs try to jointly decode all signals from the MSs within their corresponding cell. The achievable sum-rate of the UL at BS c is thus

$$\overleftarrow{R}_c = \log_2 \det \left(\mathbf{I}_{N_B} + \left(\overleftarrow{\mathbf{K}}_c^{(i+n)} \right)^{-1} \cdot \overleftarrow{\mathbf{K}}_c^{(sig)} \right), \tag{6.10}$$

where

$$\overleftarrow{\mathbf{K}}_c^{(\text{sig})} = \sum_{k=1}^{M}\sum_{d=1}^{C}\sum_{j=1}^{K}\sum_{d'=1}^{C}\sum_{j'=1}^{K} \overline{\mathbf{H}}_j^{(c,d)}\mathbf{G}_{d,j}\mathbf{F}_{j,k}^{(d,c)}\mathbf{Q}_{c,k}^{(\text{M})}\cdot\mathbf{Q}_{c,k}^{(\text{M})\mathsf{H}}\mathbf{F}_{j',k}^{(d',c)\mathsf{H}}\mathbf{G}_{d',j'}^{\mathsf{H}}\overline{\mathbf{H}}_{j'}^{(c,d')\mathsf{H}} \qquad (6.11)$$

is the covariance matrix of the desired signal at BS c that now contains the signals from all MSs in cell c. Accordingly, the covariance matrix of the interference and noise contains the noise as well as the signals originated from all other MSs:

$$\overleftarrow{\mathbf{K}}_c^{(\text{i+n})} = \sum_{\substack{b=1\\b\neq c}}^{C}\sum_{k=1}^{M}\sum_{d=1}^{C}\sum_{j=1}^{K}\sum_{d'=1}^{C}\sum_{j'=1}^{K} \overline{\mathbf{H}}_j^{(c,d)}\mathbf{G}_{d,j}\mathbf{F}_{j,k}^{(d,b)}\mathbf{Q}_{b,k}^{(\text{M})}\cdot\mathbf{Q}_{b,k}^{(\text{M})\mathsf{H}}\mathbf{F}_{j',k}^{(d',b)\mathsf{H}}\mathbf{G}_{d',j'}^{\mathsf{H}}\overline{\mathbf{H}}_{j'}^{(c,d')\mathsf{H}} +$$

$$+ \sigma_n^2\sum_{d=1}^{C}\sum_{j=1}^{K}\overline{\mathbf{H}}_j^{(c,d)}\mathbf{G}_{d,j}\mathbf{G}_{d,j}^{\mathsf{H}}\overline{\mathbf{H}}_j^{(c,d)\mathsf{H}} + \sigma_w^2\mathbf{I}_{N_\text{B}}. \qquad (6.12)$$

Note that with this formulation, no individual rates of the signals from the different users can be given. To this end, a specific decoder implementation at the BSs would have to be assumed which leads to a specific point in the achievable rate region of the UL signals. As we primarily focus on the DL, we refrain from that and limit ourselves to the achievable sum rate in the UL, which indicates how large the throughput from all MSs in a cell can be.

AF Two-Way Relaying

In two-way relaying, both directions of communication are combined and all BSs and MSs transmit their signals (6.1) and (6.2) simultaneously. Accordingly, the relays receive the superposition of all these signals

$$\mathbf{r}_{c,k} = \sum_{d=1}^{C}\sum_{j=1}^{M}\left(\mathbf{H}_k^{(c,d)}\mathbf{Q}_{d,j}^{(\text{B})}\mathbf{s}_{d,j}^{(\text{B})} + \mathbf{F}_{k,j}^{(c,d)}\mathbf{Q}_{d,j}^{(\text{M})}\mathbf{s}_{d,j}^{(\text{M})}\right) + \mathbf{n}_{c,k}. \qquad (6.13)$$

As before, the AF relays multiply their receive signal vector with a gain matrix $\mathbf{G}_{c,k}$ and broadcast then the resulting signal back to all terminal nodes. The resulting signals received by BS c and MS (c,k) are thus given by

$$\mathbf{y}_c^{(\text{B})} = \sum_{d=1}^{C}\sum_{j=1}^{K}\sum_{b=1}^{C}\sum_{i=1}^{M}\left(\overline{\mathbf{H}}_j^{(c,d)}\mathbf{G}_{d,j}\mathbf{H}_j^{(d,b)}\mathbf{Q}_{b,i}^{(\text{B})}\mathbf{s}_{b,i}^{(\text{B})} + \overline{\mathbf{H}}_j^{(c,d)}\mathbf{G}_{d,j}\mathbf{F}_{j,i}^{(d,b)}\mathbf{Q}_{b,i}^{(\text{M})}\mathbf{s}_{b,i}^{(\text{M})}\right) \qquad (6.14)$$

$$+ \sum_{d=1}^{C} \sum_{j=1}^{K} \overline{\mathbf{H}}_{j}^{(c,d)} \mathbf{G}_{d,j} \mathbf{n}_{d,j} + \mathbf{w}_{c}^{(\mathrm{B})}$$

and

$$y_{c,k}^{(\mathrm{M})} = \sum_{d=1}^{C} \sum_{j=1}^{K} \sum_{b=1}^{C} \sum_{i=1}^{M} \left(\overline{\mathbf{F}}_{k,j}^{(c,d)} \mathbf{G}_{d,j} \mathbf{H}_{j}^{(d,b)} \mathbf{Q}_{b,i}^{(\mathrm{B})} \mathbf{s}_{b,i}^{(\mathrm{B})} + \overline{\mathbf{F}}_{k,j}^{(c,d)} \mathbf{G}_{d,j} \mathbf{F}_{j,i}^{(d,b)} \mathbf{Q}_{b,i}^{(\mathrm{M})} \mathbf{s}_{b,i}^{(\mathrm{M})} \right) \qquad (6.15)$$

$$+ \sum_{d=1}^{C} \sum_{j=1}^{K} \overline{\mathbf{F}}_{k,j}^{(c,d)} \mathbf{G}_{d,j} \mathbf{n}_{d,j} + \mathbf{w}_{c,k}^{(\mathrm{M})}.$$

These signals not only include the desired signal and the interference terms from the one-way case, but also contain what the corresponding node has transmitted itself (self-interference) as well as additional interference from the other nodes of the same kind.

For the calculation of the achievable rates, this additional interference terms thus have to be taken into account. In the DL, we distinguish between the interference plus noise covariance matrix $\mathbf{K}_{\mathrm{M},c,k}^{(\mathrm{i+n})}$ from the self-interference covariance matrix $\mathbf{K}_{\mathrm{M},c,k}^{(\mathrm{self})}$. The achievable rate of the DL follows thus as

$$R_{c,k}^{(\mathrm{DL})} = \log_2 \det \left(\mathbf{I}_{N_{\mathrm{M}}} + \left(\mathbf{K}_{\mathrm{M},c,k}^{(\mathrm{i+n})} + \mathbf{K}_{\mathrm{M},c,k}^{(\mathrm{self})} \right)^{-1} \cdot \mathbf{K}_{\mathrm{M},c,k}^{(\mathrm{sig})} \right) \qquad (6.16)$$

with

$$\mathbf{K}_{\mathrm{M},c,k}^{(\mathrm{self})} = \sum_{d=1}^{C} \sum_{j=1}^{K} \sum_{d'=1}^{C} \sum_{j'=1}^{K} \overline{\mathbf{F}}_{k,j}^{(c,d)} \mathbf{G}_{d,j} \mathbf{F}_{j,k}^{(d,c)} \mathbf{Q}_{c,k}^{(\mathrm{M})} \cdot \mathbf{Q}_{c,k}^{(\mathrm{M})\mathrm{H}} \mathbf{F}_{j',k}^{(d',c)\mathrm{H}} \mathbf{G}_{d',j'}^{\mathrm{H}} \overline{\mathbf{F}}_{k,j'}^{(c,d')\mathrm{H}}, \qquad (6.17)$$

while $\mathbf{K}_{\mathrm{M},c,k}^{(\mathrm{sig})}$ and $\mathbf{K}_{\mathrm{M},c,k}^{(\mathrm{i+n})}$ are the same as in (6.8) and (6.9). When the MSs are able to estimate the effective channel that affects their own signals, they can subtract their self-interference. As the MSs already know what they have transmitted, the actual data symbols can also be used to estimate these channel coefficients [158]. When this can be accomplished perfectly, the term $\mathbf{K}_{\mathrm{M},c,k}^{(\mathrm{self})}$ can be removed from the achievable rate.

For the UL, we distinguish between interference that is caused by the BSs (including self-interference) and remaining interference from the MSs. The achievable sum rate at BS c follows as

$$R_c^{(\mathrm{UL})} = \log_2 \det\left(\mathbf{I}_{N_{\mathrm{B}}} + \left(\mathbf{K}_{\mathrm{B},c}^{(\mathrm{i+n})} + \mathbf{K}_{\mathrm{B},c}^{(\mathrm{BS\ int})}\right)^{-1} \cdot \mathbf{K}_{\mathrm{B},c}^{(\mathrm{sig})}\right), \qquad (6.18)$$

with the covariance matrices

$$\mathbf{K}_{\mathrm{B},c}^{(\mathrm{sig})} = \sum_{k=1}^{M}\sum_{d=1}^{C}\sum_{j=1}^{K}\sum_{d'=1}^{C}\sum_{j'=1}^{K} \overline{\mathbf{H}}_j^{(c,d)} \mathbf{G}_{d,j} \mathbf{F}_{j,k}^{(d,c)} \mathbf{Q}_{c,k}^{(\mathrm{M})} \cdot \mathbf{Q}_{c,k}^{(\mathrm{M})\mathrm{H}} \mathbf{F}_{j',k}^{(d',c)\mathrm{H}} \mathbf{G}_{d',j'}^{\mathrm{H}} \overline{\mathbf{H}}_{j'}^{(c,d')\mathrm{H}}, \quad (6.19)$$

$$\mathbf{K}_{\mathrm{B},c}^{(\mathrm{BS\ int})} = \sum_{\substack{b=1\\b\neq c}}^{C}\sum_{k=1}^{M}\sum_{d=1}^{C}\sum_{j=1}^{K}\sum_{d'=1}^{C}\sum_{j'=1}^{K} \overline{\mathbf{H}}_j^{(c,d)} \mathbf{G}_{d,j} \mathbf{H}_j^{(d,b)} \mathbf{Q}_{b,k}^{(\mathrm{B})} \cdot \mathbf{Q}_{b,k}^{(\mathrm{B})\mathrm{H}} \mathbf{H}_{j'}^{(d',b)\mathrm{H}} \mathbf{G}_{d',j'}^{\mathrm{H}} \overline{\mathbf{H}}_{j'}^{(c,d')\mathrm{H}},$$

$$(6.20)$$

and

$$\mathbf{K}_{\mathrm{B},c}^{(\mathrm{i+n})} = \mathrm{E}\left[\mathbf{y}_c^{(\mathrm{B})} \cdot \mathbf{y}_c^{(\mathrm{B})\mathrm{H}}\right] - \mathbf{K}_{\mathrm{B},c}^{(\mathrm{sig})} - \mathbf{K}_{\mathrm{B},c}^{(\mathrm{BS\ int})}. \qquad (6.21)$$

DF One-Way Relaying

In contrast to the AF case, DF relays completely decode the signals they receive before they forward them. The receive signal of relay (c,k) in the DL is the same as in (6.3). This signal can then be filtered by a receive combining matrix $\mathbf{G}_{c,k}^{(\mathrm{Rx})\mathrm{H}}$ which allows to adapt the BS transmit signals to lie in a subspace that has the least interference at the relays as described in Section 6.2. Applying such a receive filter leads to

$$\overrightarrow{\mathbf{r}}_{c,k} = \mathbf{G}_{c,k}^{(\mathrm{Rx})\mathrm{H}} \cdot \left(\mathbf{H}_k^{(c,c)} \mathbf{Q}_{c,k}^{(\mathrm{B})} \mathbf{s}_{c,k}^{(\mathrm{B})} + \overrightarrow{\mathbf{x}}_{c,k}^{(\mathrm{R,i+n})}\right), \qquad (6.22)$$

where $\overrightarrow{\mathbf{x}}_{c,k}^{(\mathrm{R,i+n})}$ contains all interference and noise terms. Relay (c,k) is thereby only interested in the signal intended for its associated MS. The symbol vector $\mathbf{s}_{c,k}^{(\mathrm{B})}$ is decoded, while $\overrightarrow{\mathbf{x}}_{c,k}^{(\mathrm{R,i+n})}$ is considered as noise. After that, the relays newly encode the data symbols, possibly with a different code book. Finally, the resulting symbols $\tilde{\mathbf{s}}_{c,k}^{(\mathrm{B})}$ are premultiplied by a transmit filter matrix $\mathbf{G}_{c,k}^{(\mathrm{Tx})}$ and forwarded to the MSs. The receive signal of MS (c,k) follows as

$$\overrightarrow{\mathbf{y}}_{c,k} = \sum_{d=1}^{C}\sum_{j=1}^{K} \overline{\mathbf{F}}_{k,j}^{(c,d)} \cdot \mathbf{G}_{d,j}^{(\mathrm{Tx})} \cdot \tilde{\mathbf{s}}_{d,j}^{(\mathrm{B})} + \mathbf{w}_{c,k}^{(\mathrm{M})}. \qquad (6.23)$$

In this case, we can derive an achievable rate as follows. When the BSs have transmitted their signals and relay (c,k) has applied its receive filter, it decodes the symbol

175

vector $\mathbf{s}_{c,k}^{(\mathrm{B})}$ that is contained in the receive signal (6.22). To this end, the interference in $\overrightarrow{\mathbf{x}}_{c,k}^{(\mathrm{R,i+n})}$ is treated as noise and the resulting rate on the BS-to-relay link can be given as

$$\overrightarrow{R}_{c,k}^{(\mathrm{BR})} = \log_2 \det \left(\mathbf{I}_{N_{\mathrm{R}}} + \left(\overrightarrow{\mathbf{K}}_{\mathrm{R},c,k}^{(\mathrm{i+n})} \right)^{-1} \cdot \overrightarrow{\mathbf{K}}_{\mathrm{R},c,k}^{(\mathrm{sig})} \right) \tag{6.24}$$

with the covariance matrices

$$\overrightarrow{\mathbf{K}}_{\mathrm{R},c,k}^{(\mathrm{sig})} = \mathbf{G}_{c,k}^{(\mathrm{Rx})\mathrm{H}} \mathbf{H}_k^{(c,c)} \mathbf{Q}_{c,k}^{(\mathrm{B})} \mathbf{Q}_{c,k}^{(\mathrm{B})\mathrm{H}} \mathbf{H}_k^{(c,c)\mathrm{H}} \mathbf{G}_{c,k}^{(\mathrm{Rx})} \tag{6.25}$$

and

$$\overrightarrow{\mathbf{K}}_{\mathrm{R},c,k}^{(\mathrm{i+n})} = \mathsf{E} \left[\mathbf{G}_{c,k}^{(\mathrm{Rx})\mathrm{H}} \overrightarrow{\mathbf{x}}_{c,k}^{(\mathrm{R,i+n})} \cdot \overrightarrow{\mathbf{x}}_{c,k}^{(\mathrm{R,i+n})\mathrm{H}} \mathbf{G}_{c,k}^{(\mathrm{Rx})} \right]. \tag{6.26}$$

The newly encoded data symbols $\tilde{\mathbf{s}}_{c,k}^{(\mathrm{B})}$ are multiplied with $\mathbf{G}_{c,k}^{(\mathrm{Tx})}$ and forwarded to the MSs. The achievable rate on the second hop can similarly be calculated and results in

$$\overrightarrow{R}_{c,k}^{(\mathrm{RM})} = \log_2 \det \left(\mathbf{I}_{N_{\mathrm{M}}} + \left(\overrightarrow{\mathbf{K}}_{\mathrm{M},c,k}^{(\mathrm{i+n})} \right)^{-1} \cdot \overrightarrow{\mathbf{K}}_{\mathrm{M},c,k}^{(\mathrm{sig})} \right), \tag{6.27}$$

with

$$\overrightarrow{\mathbf{K}}_{\mathrm{M},c,k}^{(\mathrm{sig})} = \overline{\mathbf{F}}_{k,k}^{(c,c)} \mathbf{G}_{c,k}^{(\mathrm{Tx})} \mathbf{G}_{c,k}^{(\mathrm{Tx})\mathrm{H}} \overline{\mathbf{F}}_{k,k}^{(c,c)\mathrm{H}} \tag{6.28}$$

$$\overrightarrow{\mathbf{K}}_{\mathrm{M},c,k}^{(\mathrm{i+n})} = \mathsf{E} \left[\overrightarrow{\mathbf{y}}_{c,k} \cdot \overrightarrow{\mathbf{y}}_{c,k}^{\mathrm{H}} \right] - \overrightarrow{\mathbf{K}}_{\mathrm{M},c,k}^{(\mathrm{sig})}. \tag{6.29}$$

Finally, an achievable rate of the two-hop link between BS and MS follows as [110]

$$\overrightarrow{R}_{c,k} = \min \left\{ \overrightarrow{R}_{c,k}^{(\mathrm{BR})}, \overrightarrow{R}_{c,k}^{(\mathrm{RM})} \right\}. \tag{6.30}$$

Thereby, we assume that equally long time slots or frequency bands are used for both hops. As we have seen in Chapter 4, the end-to-end rate could be improved by optimizing the time and frequency allocation of the two individual links. This, however, is unpractical in the cellular context as the different links of the UL and DL in adjacent cells would not necessarily be separated anymore. We will comment on that in more detail in Section 6.4.

In the UL, the relays receive the signals from the MSs. The receive signal at relay (c, k), after applying the receive filter, is

$$\overleftarrow{\tilde{\mathbf{r}}}_{c,k} = \mathbf{G}_{c,k}^{(\mathrm{Rx})\mathrm{H}} \cdot \left(\mathbf{F}_{k,k}^{(c,c)} \mathbf{Q}_{c,k}^{(\mathrm{M})} \mathbf{s}_{c,k}^{(\mathrm{M})} + \overleftarrow{\mathbf{x}}_{c,k}^{(\mathrm{R,i+n})} \right), \tag{6.31}$$

where $\overleftarrow{\mathbf{x}}_{c,k}^{(\mathrm{R,i+n})}$ contains the relay noise and all MS interference terms. The relay decodes the corresponding MS symbol vector $\mathbf{s}_{c,k}^{(\mathrm{M})}$, encodes it to the new symbol vector $\tilde{\mathbf{s}}_{c,k}^{(\mathrm{M})}$, and multiplies it with $\mathbf{G}_{c,k}^{(\mathrm{Tx})}$. Again, relay (c,k) serves only its associated MS, while it treats signals from other MSs as noise. After retransmission, BS c receives

$$\overleftarrow{\mathbf{y}}_c = \sum_{d=1}^{C} \sum_{j=1}^{K} \overline{\mathbf{H}}_j^{(c,d)} \mathbf{G}_{d,j}^{(\mathrm{Tx})} \tilde{\mathbf{s}}_{d,j}^{(\mathrm{M})} + \mathbf{w}_c^{(\mathrm{B})}. \tag{6.32}$$

The resulting end-to-end rate of the UL can be obtained in a similar way as in the DL. The rate of the first hop transmission from MS (c,k) to relay (c,k) is

$$\overleftarrow{R}_{c,k}^{(\mathrm{MR})} = \log_2 \det \left(\mathbf{I}_{N_{\mathrm{R}}} + \left(\overleftarrow{\mathbf{K}}_{\mathrm{R},c,k}^{(\mathrm{i+n})} \right)^{-1} \cdot \overleftarrow{\mathbf{K}}_{\mathrm{R},c,k}^{(\mathrm{sig})} \right) \tag{6.33}$$

with

$$\overleftarrow{\mathbf{K}}_{\mathrm{R},c,k}^{(\mathrm{sig})} = \mathbf{G}_{c,k}^{(\mathrm{Rx})\mathrm{H}} \mathbf{F}_{k,k}^{(c,c)} \mathbf{Q}_{c,k}^{(\mathrm{M})} \mathbf{Q}_{c,k}^{(\mathrm{M})\mathrm{H}} \mathbf{F}_{k,k}^{(c,c)\mathrm{H}} \mathbf{G}_{c,k}^{(\mathrm{Rx})} \tag{6.34}$$

$$\overleftarrow{\mathbf{K}}_{\mathrm{R},c,k}^{(\mathrm{i+n})} = \mathsf{E} \left[\mathbf{G}_{c,k}^{(\mathrm{Rx})\mathrm{H}} \overleftarrow{\mathbf{x}}_{c,k}^{(\mathrm{R,i+n})} \overleftarrow{\mathbf{x}}_{c,k}^{(\mathrm{R,i+n})\mathrm{H}} \mathbf{G}_{c,k}^{(\mathrm{Rx})} \right]. \tag{6.35}$$

On the second hop, the BSs decode all signals from the relays in their respective cell jointly. Depending on the specific decoding procedure applied in the BSs, different points in the achievable rate region can be achieved. In order to associate an achievable rate to each relay individually, we apply successive interference cancellation (SIC) at the BSs. With this, the end-to-end two-hop rates for each MS can be obtained. To this end, the receive signals at the BSs are sorted according to their signals strengths. While the individual user rates depend on the decoding order, they all lead to the same sum rate [143]. Here, we choose a fair approach and start to decode the strongest signal first. To this end, we evaluate signal strengths of the different signals according to the receive signal power of them, i.e. $\mathrm{Tr}\left\{ \overleftarrow{\mathbf{K}}_{\mathrm{R},c,k}^{(\mathrm{sig})} \right\}$. After the strongest signal is decoded, it is subtracted from (6.32). Then, the second strongest signal is decoded and also subtracted until the last signal can be decoded without any in-cell interference. With this ordering, the data rates of the different users are more balanced, and are thus fairer with respect to the users, than when e.g. the weakest signal would be decoded first.

With SIC, the rates of the links from the relays to the BS can be calculated. The data rate from relay (c,k) to BS c is thereby denoted by $\overleftarrow{R}_{c,k}^{(\mathrm{RB})}$. The resulting sum rate of the UL to BS c follows then as

$$\overleftarrow{R}_c = \sum_{k=1}^{M} \min \left\{ \overleftarrow{R}_{c,k}^{(\mathrm{MR})}, \overleftarrow{R}_{c,k}^{(\mathrm{RB})} \right\}. \tag{6.36}$$

Again, these achievable rates follow from the assumption that equally long time slots and the same bandwidth is used for both hops.

Note that we do not claim any optimality with this decoding strategy. So can certain rates that result from this decoding strategy potentially not be supported by the relays when their first hop rates are smaller. By choosing different decoding orders for the SIC that also take the first hop rates into account, the overall end-to-end rates might be improved. With this procedure, however, we can guarantee that the obtained end-to-end rates are achievable. But as we attach more importance to the DL rates, we refrain from complicated UL rate optimizations.

DF Two-Way Relaying

In the case of two-way relaying, the BSs and MSs transmit simultaneously and relay (c, k) receives

$$\tilde{\mathbf{r}}_{c,k} = \mathbf{G}_{c,k}^{(\mathrm{Rx})\mathsf{H}} \left(\mathbf{H}_k^{(c,c)} \mathbf{Q}_{c,k}^{(\mathrm{B})} \mathbf{s}_{c,k}^{(\mathrm{B})} + \mathbf{F}_{k,k}^{(c,c)} \mathbf{Q}_{c,k}^{(\mathrm{M})} \mathbf{s}_{c,k}^{(\mathrm{M})} + \mathbf{x}_{c,k}^{(\mathrm{R,i+n})} \right) \tag{6.37}$$

where the signal is again filtered with a receive filter $\mathbf{G}_{c,k}^{(\mathrm{Rx})\mathsf{H}}$. Now both data symbol vectors $\mathbf{s}_{c,k}^{(\mathrm{B})}$ and $\mathbf{s}_{c,k}^{(\mathrm{M})}$ are desired. These are also decoded by SIC by starting with the strongest signal first, as explained before. This leads to a pair of resulting first hop achievable rates $R_{c,k}^{(\mathrm{BR})}$ and $R_{c,k}^{(\mathrm{MR})}$, one for the signal from BS c intended for MS (c, k) and vice versa.

After successful decoding, the relays can combine the decoded data streams by an XOR operation with zero padding, i.e. the two codewords are made equally long and are combined to a single codeword [30]. The combined data symbol vector $\tilde{\mathbf{s}}_{c,k}^{(\mathrm{R})} \sim \mathcal{CN}(\mathbf{0}, \mathbf{I})$ in relay (c, k) is then precoded by $\mathbf{G}_{c,k}^{(\mathrm{Tx})}$ and the resulting signal is broadcasted to the terminals. BS c and MS (c, k) then receive this signal under interference from the other relays

$$\mathbf{y}_c^{(\mathrm{B})} = \sum_{d=1}^{C} \sum_{j=1}^{K} \overline{\overline{\mathbf{H}}}_j^{(c,d)} \cdot \mathbf{G}_{d,j}^{(\mathrm{Tx})} \cdot \tilde{\mathbf{s}}_{d,j}^{(\mathrm{R})} + \mathbf{w}_c^{(\mathrm{B})} \tag{6.38}$$

$$\mathbf{y}_{c,k}^{(\mathrm{M})} = \sum_{d=1}^{C} \sum_{j=1}^{K} \overline{\mathbf{F}}_{k,j}^{(c,d)} \cdot \mathbf{G}_{d,j}^{(\mathrm{Tx})} \cdot \tilde{\mathbf{s}}_{d,j}^{(\mathrm{R})} + \mathbf{w}_{c,k}^{(\mathrm{M})}. \tag{6.39}$$

After reception, MS (c,k) decodes the symbol vector $\tilde{\mathbf{s}}_{c,k}^{(\mathrm{R})}$. The achievable rate of this link is denoted by $R_{c,k}^{(\mathrm{RM})}$. When this signal is decoded, the MS can apply another XOR operation with the data bits it has previously transmitted. With this form of self-interference cancellation, the desired signal can be reconstructed at the terminal [30]. The BSs, in turn, receive signals from all relays within their cell. These signals are again decoded one after each other by SIC, starting with the strongest. The resulting achievable rates of the links between relays and BSs are denoted by $R_{c,k}^{(\mathrm{RB})}$. Finally, the resulting rates achievable on the two-hop UL and DL are then given by

$$R_{c,k}^{(\mathrm{DL})} = \min \left\{ R_{c,k}^{(\mathrm{BR})}, R_{c,k}^{(\mathrm{RM})} \right\} \tag{6.40}$$

$$R_{c}^{(\mathrm{UL})} = \sum_{k=1}^{M} \min \left\{ R_{c,k}^{(\mathrm{MR})}, R_{c,k}^{(\mathrm{RB})} \right\}. \tag{6.41}$$

Choosing the rates like this ensures that they lie inside the achievable rate region [30]. However, no optimality is claimed. Again, the SIC decoding order could be adapted to take also the rates on the other hop into account or the time or frequency slot assignment could be optimized in order to balance the different rates. Furthermore, the considered DF scheme requires the relays to decode the complete transmission blocks from both terminals before they can be newly encoded and retransmitted. This introduces additional delays as the retransmission can only start when the decoding and encoding process is completed. These delays could be reduced e.g. with block-Markov coding or other DF relaying strategies [110]. This is however not considered here.

6.1.2 Prelog Factor

In the derivations of the achievable rates, we have dropped the prelog factors which occur when multiple channel uses are required for the transmission of a single message. These factors (e.g. 1, $\frac{1}{2}$, or $\frac{1}{4}$) depend on the specific relaying protocol and the considered duplex mode. When in-band one-way relays are considered, a single transmission in one direction requires two resource blocks. As the UL and DL are separated in time or frequency, this would result in a total prelog factor $\frac{1}{4}$ in this case. With two-way relaying, two symbols (one in the DL and one in the UL) require two time slots. Also in a conventional network where no relays are used, the UL and DL are separated and the transmission of one symbol in either direction requires two resources. In this regard, the conventional network and two-way relaying can directly be compared with each other. When one way relays with out-band connection to the BSs are used, the same holds also for this case. Such relays can e.g. correspond to small cell nodes as discussed in Chapter 4. Moreover, the relays could convert their BS signals to frequency bands that are currently not used (cf. cognitive radio [137]) or lie in an ISM band. In this case, no additional costs have to be included into the spectral efficiency. The use of secondary links is especially motivated by the small transmit power of the relays that do not disturb other systems significantly.

In either case, and also in conventional networks, the resources have to be divided for the DL and UL. Thereby, the two directions of communication can use different fractions, which could be optimized as already indicated. In the following, we consider two scenarios:

- in-band relays with equal resources for both directions (DL and UL) when the two-way protocol is considered and

- a secondary link that is free for the second hop in the case of one-way relaying.

In this way, we can omit the prelog factors in order to compare the spectral efficiencies.

6.2   Transmission Schemes

In the following, we design specific transmission schemes for the different nodes that attempt to mitigate the interference in the entire network. In order to gain more understanding in what the limiting factors of the considered network are, we analyze the individual terms of the receive signals at the terminals. To this end, we apply spatially white signaling at all involved nodes and a scaled identity matrix at the relays. In this way, no interference is mitigated and the whole network is flooded with signals. This allows to measure the individual signal contributions for both the UL and DL and to identify the strongest interference sources. Based on this analysis, we can design precoding and relay gain matrices with the goal to mitigate the most severe interference terms.

We apply a per node transmit power of $P_\mathrm{B} = 40\,\mathrm{W}$ at the BSs and $P_\mathrm{R} = 6\,\mathrm{W}$ and $P_\mathrm{M} = 0.2\,\mathrm{W}$ at the relays and MSs. The precoding and relay gain matrices are accordingly

$$\mathbf{Q}_{c,k}^{(\mathrm{B})} = \sqrt{P_{\mathrm{B}}/(M \cdot N_{\mathrm{B}})} \cdot \mathbf{I}_{N_{\mathrm{B}}}, \tag{6.42}$$

$$\mathbf{Q}_{c,k}^{(\mathrm{M})} = \sqrt{P_{\mathrm{M}}/N_{\mathrm{M}}} \cdot \mathbf{I}_{N_{\mathrm{M}}}, \tag{6.43}$$

$$\mathbf{G}_{c,k} = \sqrt{P_{\mathrm{R}}/\mathrm{Tr}\left\{\mathsf{E}\left[\mathbf{r}_{c,k} \cdot \mathbf{r}_{c,k}^{\mathsf{H}}\right]\right\}} \cdot \mathbf{I}_{N_{\mathrm{R}}}. \tag{6.44}$$

The resulting (averaged) receive signal powers of the BSs and MSs, for one-way as well as two-way relaying are shown in Fig. 6.2. The network consists of $C = 19$ cells, each containing $M = K = 6$ MSs/relays. In the figure, we distinguish which relays have forwarded the different signal contributions (own relay, other in-cell relays, or relays from other cells in the DL and own relays and relays from other cells in the UL). More details on the simulation parameters are given in Section 6.3.

From the figure, we can conclude where the different interference contributions come from. In contrast to one-way relaying, additional interference terms appear in two-way relaying: the signals transmitted by the other terminals of the same kind, including self-interference. These signals are not present in one-way relaying because the different directions of communication are separated by orthogonal resources. The total signals received by the relays are thus of less power and one-way relays can apply a higher gain factor in order to meet the transmit power. Consequently, the (existing) signal contributions in one-way relaying are of higher power than in the two-way case. The dominant interference terms can be classified into:

1. BS signals intended for other MSs in the same cell ($\mathrm{I_A}$ in the figure)

and in the case of two-way relaying

2. self-interference ($\mathrm{I_S}$),

3. interference from other BSs in the UL ($\mathrm{I_B}$), and

4. remaining interference.

The terms $\mathrm{I_S}$ and $\mathrm{I_B}$ do not exist in one-way relaying.

In the following, we apply precoding and relay gain matrices that attempt to mitigate the interference seen by the terminals. A part of the interference can be cancelled with appropriate precoding or beamforming techniques. Other interference terms, however, cannot be reduced without global CSI or cooperation between all nodes. To reduce as much of the interference as possible, we combine different precoding schemes at the different nodes in a way that no sophisticated cooperation between them is required. Thereby, the individual schemes are chosen and combined such that a distributed signal

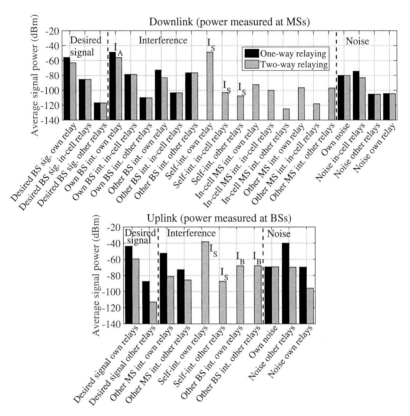

Figure 6.2.: Receive signal powers distinguished by their sources (one-way and two-way AF relaying protocol).

processing for interference mitigation is realized, i.e., the global task of improving the network performance is shared among the different nodes according to their complexity and abilities and each node computes its precoding or gain matrix based on *locally* available CSI. In order to apply schemes that are relevant for practical implementation, we focus on simple linear precoding techniques for which closed form solutions are known and can be computed in a non-iterative fashion. Consequently, we do not claim any optimality of the proposed schemes, but rather understand them as example implementations for the network that are, due to the low complexity, of high practical relevance. Moreover, we design the transmit signals on the BS-relay links such that they are independent of the ones on the relay-MS links. This has the advantage that the precoding at the BSs has not to be updated as often as the precoding on the relay-MS links. This is because the channels between BSs and fixed relays presumably have a much longer coherence time than the channels between the relays and the (possibly moving) MSs. The signaling of the MSs is spatially white such that they do not require any CSIT.

6.2.1 Block Zero-Forcing at the BSs

A strong interference source that degrades the performance in the DL is the BS signal intended for other MSs (I_A in Fig. 6.2). To cancel this interference, we apply block ZF as in the chapters before. In order to keep the precoding at the BSs of moderate complexity, we limit this scheme to be performed in each cell separately. With this, the strongest interference terms resulting at the relays are cancelled, but the BSs do not have to cooperate with others. The interference coming from BSs from adjacent cells are treated in the relays. The transmit signal of BS c is therefore

$$\mathbf{x}_c^{(\mathrm{B})} = \sum_{k=1}^{M} \mathbf{Q}_{c,k}^{(\mathrm{B})} \cdot \mathbf{s}_{c,k}^{(\mathrm{B})} = \sum_{k=1}^{M} \mathbf{Z}_{c,k} \cdot \tilde{\mathbf{Q}}_{c,k} \cdot \mathbf{s}_{c,k}^{(\mathrm{B})}, \qquad (6.45)$$

where

$$\mathbf{Z}_{c,k} = \mathrm{null}\left\{\left[\mathbf{H}_1^{(c,c)\mathsf{T}}, \ldots, \mathbf{H}_{k-1}^{(c,c)\mathsf{T}} \mathbf{H}_{k+1}^{(c,c)\mathsf{T}}, \ldots, \mathbf{H}_M^{(c,c)\mathsf{T}}\right]^{\mathsf{T}}\right\}$$

ensures that the signal intended for MS (c, k) is nulled at the other relays in this cell and $\tilde{\mathbf{Q}}_{c,k}$ is the power loading matrix as in Chapter 3.

As the BSs do not perform joint transmission, we impose a sum transmit power constraint for each BS

$$\mathrm{Tr}\left\{\sum_{k=1}^{M} \mathbf{Z}_{c,k}\tilde{\mathbf{Q}}_{c,k}\tilde{\mathbf{Q}}_{c,k}^{\mathsf{H}}\mathbf{Z}_{c,k}^{\mathsf{H}}\right\} \leq P_{\mathrm{B}}, \qquad \forall c. \qquad (6.46)$$

In order to avoid a computationally complex optimization algorithm for the power loading matrix $\tilde{\mathbf{Q}}_{c,k}$ as we have done it with the max-min optimization in previous chapters, we apply a standard waterfilling algorithm as in [133] or [143] that can be calculated in quasi closed-form. To this end, we decompose the power loading matrix to

$$\tilde{\mathbf{Q}}_{c,k} = \tilde{\mathbf{V}}_{c,k} \cdot \mathbf{P}_{c,k}, \qquad (6.47)$$

where $\tilde{\mathbf{V}}_{c,k}$ are the right hand singular vectors of the virtual channel $\tilde{\mathbf{H}}_k^{(c,c)} = \mathbf{H}_k^{(c,c)} \cdot$ $\mathbf{Z}_{c,k}$. The diagonal power loading matrix $\mathbf{P}_{c,k}$ weights each stream according to the waterfilling solution [133]

$$P_i^\star = \left(\frac{1}{\lambda} - \frac{\sigma_n^2}{\Lambda_i} \right)^+, \tag{6.48}$$

where $x^+ = \max(x, 0)$ and Λ_i is the i-th singular value of the SVD of $\tilde{\mathbf{H}}_k^{(c,c)}$, and the Lagrange multiplier λ is chosen such that the power constraint (6.46) is met. This maximizes the sum rate of the first hop links within a cell under the ZF constraints, when the interference from adjacent cells remains fixed. Note that only the first hop to the relays are considered in this precoding scheme. The BS signaling is thus independent of the link on the second hop. Consequently, the BS precoding does not have to be changed when the channel between relays and MSs changes.

6.2.2   AF relay gain matrices

In its simplest form, AF relaying is performed with a scaled identity matrix

$$\mathbf{G}_{c,k} = \sqrt{P_\mathrm{R}/\mathrm{Tr}\left\{ \mathsf{E}\left[\mathbf{r}_{c,k} \cdot \mathbf{r}_{c,k}^\mathsf{H} \right] \right\}} \cdot \mathbf{I}_{N_\mathrm{R}}. \tag{6.49}$$

These type A relays forward their receive signal scaled according to the power constraint, without modifying it. This form of AF relaying does not require any CSI at the relays.

More sophisticated type B relays that have access to local CSI can form linear combinations of all input streams to a beneficial output signal vector. The relay can e.g. design the relay gain matrix such that undesired signals are minimized while the desired signal components should remain at a good quality. To this end, the relay gain matrices are factorized to

$$\mathbf{G}_{c,k} = \sqrt{\alpha_{c,k}} \cdot \mathbf{G}_{c,k}^{(\mathrm{Tx})} \cdot \mathbf{G}_{c,k}^{(\mathrm{Rx})\mathsf{H}}, \tag{6.50}$$

where $\mathbf{G}_{c,k}^{(\mathrm{Rx})}$ is a receive filter, $\mathbf{G}_{c,k}^{(\mathrm{Tx})}$ a transmit filter, and $\alpha_{c,k}$ a scaling factor to adjust the transmit power.

For the design of the receive filter, we distinguish between one-way and two-way relaying. In the one-way case, the receive filter $\mathbf{G}_{c,k}^{(\mathrm{Rx})}$ is chosen to suppress the interference coming from the BSs of adjacent cells. Such a filter can be obtained by $\mathbf{G}_{c,k}^{(\mathrm{Rx})} = \left[\mathbf{v}_1^{(c,k)}, \ldots, \mathbf{v}_{d_\mathrm{s}}^{(c,k)} \right]$ [42]. Therein, $\mathbf{v}_i^{(c,k)}$ is the eigenvector corresponding to the

ith smallest eigenvalue of

$$\mathbf{\Gamma}_{c,k} = \sum_{\substack{d=1 \\ d \neq c}}^{C} \mathbf{H}_k^{(c,d)} \cdot \mathbf{H}_k^{(c,d)\mathrm{H}}. \tag{6.51}$$

With this, the receive signal is projected into the subspace that contains the least BS interference under the assumption of spatially white signaling. This has the advantage that $\mathbf{G}_{c,k}^{(\mathrm{Rx})}$ is independent of the actual BS signals and has thus not to be updated when a BS changes its precoding. Moreover, when the relay position is fixed, this covariance matrix is mainly static and simple to estimate.

In two-way relaying, we can additionally enhance the UL performance by choosing a receive filter that does not only reduce the interference from adjacent BSs but tries also to keep the signal from its MS at a good quality. To this end, $\mathbf{G}_{c,k}^{(\mathrm{Rx})}$ can be chosen as a filter that minimizes the BS interference (the terms indicated as other BS interference in Fig. 6.2) and noise under the constraint that the MS signal is kept constant. The resulting optimization problem

$$\mathbf{G}_{c,k}^{(\mathrm{Rx})} = \arg\min \mathrm{Tr} \left\{ \mathbf{G}_{c,k}^{(\mathrm{Rx})\mathrm{H}} \left(\mathbf{\Gamma}_{c,k} + \sigma_n^2 \mathbf{I}_{N_\mathrm{R}} \right) \mathbf{G}_{c,k}^{(\mathrm{Rx})} \right\}$$
$$\text{s.t. } \mathbf{G}_{c,k}^{(\mathrm{Rx})\mathrm{H}} \mathbf{F}_{k,k}^{(c,c)} = \mathbf{I}_{N_\mathrm{M}} \tag{6.52}$$

can be solved in closed form and its solution is given by

$$\mathbf{G}_{c,k}^{(\mathrm{Rx})} = \left(\mathbf{\Gamma}_{c,k} + \sigma_n^2 \mathbf{I}_{N_\mathrm{R}} \right)^{-1} \cdot \mathbf{F}_{k,k}^{(c,c)} \cdot \left(\mathbf{F}_{k,k}^{(c,c)\mathrm{H}} \left(\mathbf{\Gamma}_{c,k} + \sigma_n^2 \mathbf{I}_{N_\mathrm{R}} \right)^{-1} \mathbf{F}_{k,k}^{(c,c)} \right)^{-1}. \tag{6.53}$$

This approach is a MIMO extension of the minimum variance distortionless response (MVDR) filter [51]. With this, the covariance matrix $\mathbf{\Gamma}_{c,k}$ has to be known as in the one-way case and, additionally, the channel to the "own" MS (c,k).

The transmit filter of the relay (in the downlink direction) is chosen as a transmit matched filter (MF) matched to the channel between relay and its associated MS:

$$\mathbf{G}_{c,k}^{(\mathrm{Tx})} = \overline{\mathbf{F}}_{k,k}^{(c,c)\mathrm{H}}. \tag{6.54}$$

In the uplink direction (also for two-way relaying), this filter is a receive matched filter. The combined relay gain matrix is then scaled with

$$\alpha_{c,k} = \frac{P_\mathrm{R}}{\mathrm{Tr} \left\{ \mathbf{G}_{c,k}^{(\mathrm{Tx})} \mathbf{G}_{c,k}^{(\mathrm{Rx})\mathrm{H}} \mathrm{E} \left[\mathbf{r}_{c,k} \mathbf{r}_{c,k}^\mathrm{H} \right] \mathbf{G}_{c,k}^{(\mathrm{Rx})} \mathbf{G}_{c,k}^{(\mathrm{Tx})\mathrm{H}} \right\}}. \tag{6.55}$$

Note that the gain matrices of these type B relays are chosen such that the relays mainly improve the links to the MSs, because the BS-relay links are presumably already strong due to the high transmit power and the ZF at the BSs. Also note that the receive filters at the relays depend only on the covariance matrix of the BS-relay interference. The individual channel coefficients need not to be known. Moreover, the relay receive filters do not have to be updated very often, since these channels change only slowly when the relays are at fixed positions. Additionally, the precoding at the BSs can be done with respect to the effective channel that includes the specific relay receive filters, i.e. the block ZF and waterfilling is given as a function of the effective channel $\mathbf{G}_{c,k}^{(\mathrm{Rx})\mathsf{H}} \cdot \mathbf{H}_k^{(c,c)}$ instead of $\mathbf{H}_k^{(c,c)}$ only. This further improves the overall performance. This form of BS precoding has no additional requirements on CSI as the effective channels (including the relay receive filters) can be estimated by the same pilot symbols as before. In the following, we thus apply this precoding scheme.

6.2.3   DF Relay Filter Design

The same filter techniques can also be applied to DF relays. When type A DF relays are considered, the relay filter matrices are $\mathbf{G}_{c,k}^{(\mathrm{Rx})\mathsf{H}} = \mathbf{I}_{N_\mathrm{R}}$ and $\mathbf{G}_{c,k}^{(\mathrm{Tx})} = \sqrt{P_\mathrm{R}/N_\mathrm{R}} \cdot \mathbf{I}_{N_\mathrm{R}}$.

For the more sophisticated type B relays, the filters from the AF case can be adopted. In this case, the receive filter in the one-way protocol contains, as for AF relaying, the eigenvectors corresponding to the d_s smallest eigenvalues of $\boldsymbol{\Gamma}_{c,k}$, i.e. $\mathbf{G}_{c,k}^{(\mathrm{Rx})} = \left[\mathbf{v}_1^{(c,k)}, \dots, \mathbf{v}_{d_\mathrm{s}}^{(c,k)}\right]$. This projection not only reduces the BS interference, but also results in a smaller dimension of the (effective) signal space seen by the BSs. As a result, the BSs need to zero-force fewer dimensions and thus have additional antennas to improve their beamforming. As a downside, however, less data streams can be transmitted to the relay as without the dimension reducing receive filter. The combination of this filter and the BS precoding with respect to this filter however improves the overall performance. The transmit filter is a scaled transmit MF

$$\mathbf{G}_{c,k}^{(\mathrm{Tx})} = \sqrt{\dfrac{P_\mathrm{R}}{\mathrm{Tr}\left\{\overline{\mathbf{F}}_{k,k}^{(c,c)\mathsf{H}} \cdot \overline{\mathbf{F}}_{k,k}^{(c,c)}\right\}}} \, \overline{\mathbf{F}}_{k,k}^{(c,c)\mathsf{H}}, \tag{6.56}$$

such that it meets the relay transmit power constraint.

For two-way relaying, the receive filter can be replaced by the MVDR solution as in (6.53).

6.2.4   Self- and BS-Interference Cancellation

In two-way relaying, both directions of communication are combined into the same physical channel. A strong contribution of interference is thus the self-interference that propagates back from the relays (I_S in Fig. 6.2). In DF relaying, this self-interference is cancelled at the terminals by the XOR operation with the previously transmitted data. Also in AF relaying, this interference can be cancelled in each node. Thereby, the effective channel from itself via the relays back to it has to be known. At the MSs, this effective channel is described by an $N_M \times N_M$ matrix which can be estimated with pilot symbols that are included in the MS signal. Alternatively or in addition, the self-interference itself can be used to obtain CSI estimates [158]. When the self-interference is completely cancelled, the covariance matrix $\mathbf{K}_{M,c,k}^{(\text{self})}$ in (6.16) disappears and the resulting rate is significantly improved.

Self-interference can also be canceled at the BSs in the same way. However, this might not be sufficient to achieve high UL rates, as the sum of signals from all other BSs $d \neq c$ is a strong contribution of the interference at BS c (I_B in Fig. 6.2). Therefore, we propose that (at least close) BSs cooperate with each other in a way that they share their transmit symbols, as already introduced in Chapter 5. In this way, the BSs can not only cancel their self-interference, but can also reconstruct and cancel the interference caused by neighboring BSs. The known data symbols or pilot/training sequences included in the signals can be used to estimate the effective channels via the relays and no CSI needs to be shared. As a result, the covariance matrix $\mathbf{K}_{B,c}^{(\text{BS int})}$ disappears in (6.18) completely. This form of BS cooperation improves the UL rates of two-way relaying drastically. The BSs need however to exchange their user data with each other. For this, a backhaul connection of sufficient capacity is required between them. In practice, however, it might be sufficient when only neighboring BSs share their data such that the interference from them can be cancelled as only close BSs have a strong impact. Interference from BSs further away can be assumed to be weak compared to other signal contributions. Thereby, the data exchange between BSs can be kept moderate. If this data exchange is realized, it can however also be exploited for the downlink transmission. In the following, we consider this form of BS cooperation to show the potential of AF two-way relaying. In Chapter 8, we extend BS cooperation to the case where this knowledge is also used for the design of transmit signals in the downlink.

6.3 Simulation Results

In the following, we compare the different relaying schemes by means of computer simulations in a realistic setup. We focus on the sum rate that is achievable in a cell of interest and compare the performance to a non-cooperative reference scenario, that is a cellular network without relays in which the BSs serve multiple MSs by block ZF and waterfilling on the direct BS-MS channels. The reference scheme thus applies the same transmission scheme as described previously, just to the MSs directly instead of via relays. With this, however, an important property of the approach with relays is lost in the reference. As the transmission to the MSs requires accurate CSIT at the BSs for the block ZF, the channels to them are required to have a sufficiently long coherence time. In the case with relays, this has not to be the case, as the block ZF is there applied to the relays which are static. This difference in the schemes is however not reflected in the simulations.

6.3.1 Simulation setup

The network consists of $C = 19$ hexagonal cells, where 18 cells are arranged in two circles around a middle cell that is the cell of interest (cf. Fig. 6.3). The distance between adjacent BSs is 1000 m. Each cell contains $M = K$ MSs/relays with $N_M = 2$ and $N_R = 4$ antennas. The BS antenna arrays have $N_B = M \cdot N_R$ antennas. All antennas are omnidirectional and we apply the WINNER II channel model as before to get a realistic network model. The channels are drawn according to the urban WINNER II scenario C2 with LOS condition for all channels between a BS and its associated relays. As we consider the relays as dedicated infrastructure nodes, we assume that the relays are intentionally placed at locations with good propagation conditions to the BS of their cell. For all other channels, specifically the ones to MSs but also the ones between BSs and other-cell relays, we impose a NLOS condition.

If not stated otherwise, the chosen transmit powers at the BSs, relays, and MSs are $P_B = 40\,\mathrm{W}$, $P_R = 6\,\mathrm{W}$, and $P_M = 0.2\,\mathrm{W}$. Assuming a total bandwidth of 100 MHz and a noise figure of 5 dB at all nodes, the noise variances are $\sigma_n^2 = \sigma_w^2 = 5 \cdot 10^{-12}\,\mathrm{W}$. Based on the insights from the previous chapters, we place the BSs further apart from each other and reduce their transmit power as compared to the network topology discussed in Chapter 3. With this, the interference the BSs cause to adjacent cells is reduced as compared to the previous urban micro-cell setup and the increased pathloss due to the

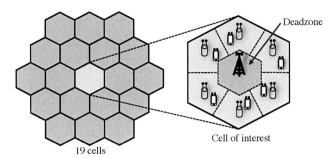

Figure 6.3.: The network model considered in the simulations consists of 19 cells, 18 arranged in two circles around a cell of interest.

larger cell size can be compensated by the installation of the relays. Moreover, when we consider six relays in each cell, the total transmit power of the BS and the relays in each cell ($P_B + 6 \cdot P_R = 76\,\text{W}$) is comparable to the one assumed previously.

In the basic network setup that we consider in the following, $K = 6$ relays are placed in each cell at a distance of $d_{BR} = 350\,\text{m}$ in a circle around the BS, as depicted in Fig. 6.3. The MSs are randomly placed with a uniform distribution, one in each small relay cell such that each user is served by one relay. In a zone of $\frac{2}{3}d_{BR}$ around the BS, no MSs are considered. Users in this area can be served by other relays operating in other frequency bands. Alternatively, static MSs located close to the BS can also be served by the BS directly. By applying such a "deadzone", we only consider MSs that are located towards the cell edge. Such cell-edge users are particularly challenging in the context of interference-limited cellular networks. The radius of the deadzone is chosen such that the strengths of the signals from the BS and from the relays are equal on average. MSs that are located within the deadzone thus receive stronger signals from the BS and as we have seen in the previous chapter, the direct signal from the BSs can offer already quite large data rates.

6.3.2 Comparison of different Relaying Schemes

In the following, we compare the achievable rates of all aforementioned transmission schemes with the rates of a reference case in which no relays are used. In this reference, the BSs serve the MSs directly by block ZF and apply sum-rate optimal power loading according to waterfilling [143]. The BSs thereby have to track the channels to the MSs which are not necessarily static. In the simulations, the BSs of all $C = 19$ cells

Scheme	Protocol	Type	Duplex	Requirements
Reference	Block ZF with waterfilling			CSIT to MSs (not static)
AF relaying	1-way	A	TDD	Static CSI at BS, reciprocal
			FDD	Static CSI at BS, pilots by RS
		B	TDD	CSI at RS, reciprocal
			FDD	CSI at RS, pilots by RS
	2-way	A	TDD	exchange BS data
			FDD	exchange BS data, pilots by RS
		B	TDD	exchange BS data, CSI at RS
			FDD	exchange BS data, CSI at RS
DF relaying	1-way	A	TDD	CSI at RS (MS link not static)
			FDD	CSI at RS (MS link not static)
		B	TDD	CSI at RS (MS link not static)
			FDD	CSI at RS (MS link not static)
	2-way	A	TDD	CSI at RS (MS link not static)
			FDD	CSI at RS (MS link not static)
		B	TDD	CSI at RS (MS link not static)
			FDD	CSI at RS (MS link not static)

Table 6.1.: Transmission schemes applied in this chapter and their requirements.

apply the same beamforming strategy. The different transmission strategies compared in this chapter are summarized in Table 6.1, the achievable sum rates of the different schemes are shown in Fig. 6.4. As discussed in Section 6.1.2, the prelog factors of the relaying schemes are intentionally omitted for TDD and FDD relaying. For comparison, however, we also include the rates of TDD relaying when this factor that arises from the multiple channel uses for one-way relaying is considered (in-band relays); this factor has no impact on two-way relaying.

It can be seen that significant gains can be achieved with the relays, even with the simple type A relays. When the prelog factor is taken into account, two-way relaying leads to the best results, as the UL and DL are combined in a spectrally efficient way. For one-way relaying, the performance is somewhat diminished. Nevertheless, the use of the relays has still its advantages. The acquisition of CSI is drastically simplified which can enable massive MIMO at the BSs. The achievable rates of the reference are thus rather optimistic, as the overhead to obtain the required CSI, especially from moving MSs, is not considered. The performance of TDD and FDD relaying is comparable. The reciprocal channels in the TDD case do not have a significant impact. For FDD systems, very similar results can be expected. The following simulations are therefore limited to the TDD case and the prelog factor is no longer considered.

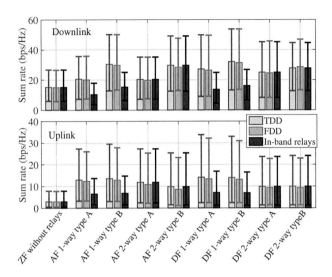

Figure 6.4.: Achievable sum rates of the different relaying schemes compared to a conventional network without relays in which the BSs serve the MSs directly with block ZF and waterfilling. Shown are the mean sum rates (bars) as well as 5% and 95% percentiles.

In order to gain more understanding in the performance of the different schemes, we show the empirical CDFs of average user rates (sum rate per cell divided by $M = 6$) for the DL in Fig. 6.5. As stated in Section 6.1.2, prelog factors are not considered in the presented achievable rates. Hence, in case all resources have to be counted and the relays are in-band half-duplex relays, the rates of all considered variants of one-way relaying must be scaled with $1/2$ before comparing them to the reference scenario or the two-way relaying schemes (cf. Fig. 6.4). Then, two-way relaying outperforms one-way relaying in all investigated schemes. If the resources of the second hop do not have to be accounted for or full-duplex relays can be used, the one-way curves show their potential gains compared to the two-way approach due to higher gain factors and less interference.

In the DL, we can see that all type B relays (solid curves) achieve high performance gains as compared to the reference. With AF relaying, both one- and two-way relaying achieve a very similar performance, one-way is only slightly better due to the higher amplification gains as less signals are involved. In the DF counterpart, the rates with

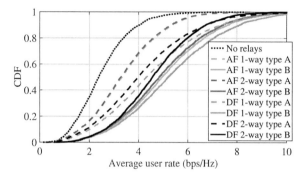

Figure 6.5.: CDF of average user rates for the DL. The transmit powers are $P_B = 40\,\text{W}$, $P_R = 6\,\text{W}$, $P_M = 0.2\,\text{W}$, and noise variances $\sigma_n^2 = \sigma_w^2 = 5 \cdot 10^{-12}\,\text{W}$ per $100\,\text{MHz}$ bandwidth.

two-way relaying are not that large in the high rate regime. When a MS is close to such a relay, i.e. when this user could profit significantly from the relay, the UL signal from the MS received by this relay is strong. This affects also the DL rate for this MS as the DF two-way relay has to decode both signals and they cause interference to each other. In the low rate regime, however, the two-way approach is similarly good as one-way relaying. In this case, the decoding of the DL signal is not strongly affected by the MS signals. Note that this behavior can be reversed when the decoding order strategy in the SIC is changed. We have however chosen the SIC order such that the weaker signal can profit more to achieve more fairness, which is reflected in the better low rate behavior. In the case of AF relaying, the relays do not have to distinguish the different signals. Hence, these rates outperform the DF two-way case. When the DF type B relays are applied in the one-way case, its potential is the highest. In this case, no MS interference is present and due to the ZF at the BS and the filtering of the residual interference, the relay can decode its intended signal with a high rate.

The type A relays, which do not filter any out-of-cell signals, are less affective. Again, AF one-way and two-way relaying are very similar; the advantage of one-way due to the smaller relay receive power is only small. When no relay filters are applied, DF relaying clearly outperforms AF relaying. Due to the decoding process, interference (and also relay noise) is removed while in the AF case these signals are also amplified and forwarded and affect the rates seen by the MSs more. As the DF relays need to decode two signals if the two-way protocol is applied, the performance of one-way relaying is better as only one signal is desired and less interference is present.

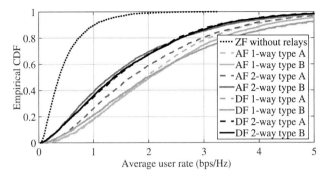

Figure 6.6.: CDF of average user rates for the UL. The transmit powers are $P_B = 40\,\text{W}$, $P_R = 6\,\text{W}$, $P_M = 0.2\,\text{W}$, and noise variances $\sigma_n^2 = \sigma_w^2 = 5\cdot10^{-12}\,\text{W}$ per $100\,\text{MHz}$ bandwidth.

When we consider the UL rates, shown in Fig. 6.6, we observe that the type B relays are not necessarily better than the simpler type A relays. The relay filters are specifically chosen to improve the DL. Thereby, the focus is set on reducing the interference from other BSs and to keep the interference that the relays cause to other MSs low. In the uplink, however, the BS signals affect only the two-way relaying schemes. This is reflected in the UL rates of DF two-way relaying that are lower than most other schemes. Only AF two-way relaying with type B relays is similar. The other schemes, particularly the type A AF relays show better performance. In two-way AF relaying, the relay filters decrease the performance. While the receive filter in these relays tries to reduce the other-BS interference, these signals are canceled in the BSs anyway. Filtering these signals has no impact on the UL rates in this case. In order to improve them, the relay filters could be designed such that they reduce the interference coming from MSs in other cells. For this, however, the relays would have to estimate the channels to them, which might be difficult if the mobile users are not static. Moreover, the DL rates would then suffer as the BS interference would not be mitigated as with the filters applied here. Nevertheless, the performance in the UL is still significantly improved as compared to the reference without relays.

To summarize the insights, we can identify the following schemes as the ones of the highest practical interest: AF 1-way type A or B, DF 1-way type B, and DF 2-way type B. For DF relays, the extra effort for applying transmit and receive filters (type B) do not increase their complexity as the relays need accurate CSIR for successful decoding anyway. DF type A relays do therefore not exploit their full potential. If the costs for the second hop (relay-to-MSs) do not have to be taken into account, AF 1-way relays are of much lower complexity and achieve a similarly good performance. If the costs of the second hop are taken into account by the prelog factor, AF 2-way relaying requires the additional complexity of exchanging BS signals for interference cancellation. In this case, DF 2-way relaying seems more practical as this is not required to achieve good performance.

In the following, we look at the impact of the relay positions within the cell. To this end, we plot the mean DL rates (user rates averaged over 1000 channel realizations) versus the distance between the BS and the relays in Fig. 6.7. In all cases there are still $K = 6$ relays located in a circle around the BS in each cell. Only the curves for the type B relays are shown. The results represented by solid lines (with a deadzone of radius $\frac{2}{3}d_{\mathrm{BR}}$ around the BSs) confirm the results in the CDFs of Fig. 6.5, which were found for $d_{\mathrm{BR}} = 350\,\mathrm{m}$. With increasing d_{BR}, the performance of the considered relaying schemes improves up to 400 m; only for $d_{\mathrm{BR}} < 200\,\mathrm{m}$ the reference scenario performs better. If the relays are close to the BS and serve users located towards the cell edge, the relays cannot contribute to high rates. As the relays transmit with lower power than the BS, the receive signal strength at the MSs is thus comparable or lower as the one from the BS. The direct transmission is therefore better in this case, also because there is less interference in the network when no relays are present. When the relays are located further away from the BS, the power gain from the relays helps in improving the performance while the precoding and relay filters keep the additional interference low.

If we compare the rates also for the case without deadzones, shown as dashed lines, the gains look less impressive. In the reference scenario, MSs that are very close to a BS achieve very high rates by the direct BS transmission with waterfilling, which favors strong users. The rates, which do not vary with the BS-to-RS distance, are thus dominated by the users close to the cell center. In the case with the deadzone, the users are forced to be further away from the BSs, which is reflected in the rates that drop with increasing distance. When relaying is applied but the users are not restricted to lie further away from the BS, only those users close to a relay can profit from them. The average user rates are thus lower without a deadzone as the probability is high that some MSs are closer to the BS than to their relay. Nevertheless, the users that are close to a relay can strongly benefit from the additional nodes. In the considered network setup, the relays all have the same distance to its BS. The power allocation by

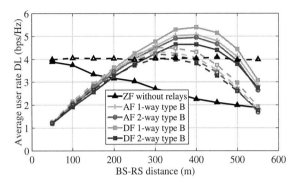

Figure 6.7.: Mean DL user rates for different d_{BR}. The solid lines are with a deadzone of $2d_{BR}/3$ around the BSs, the dashed lines are without deadzone.

waterfilling that is applied to the BS beamforming thus allocates a similar amount of power to all users as only the channels to their relays are considered for this. All relays are affected by the same pathloss (but possibly with different shadowing). The data rates are thus more balanced than when the BS precodes its signals with respect to the channels to the MSs directly which can differ in their strengths by orders of magnitude due to the different distances. Accordingly, the relay schemes achieve much higher rates on the cell edge whereas in the case of direct transmission, the high rates that contribute most to the average are for MSs located very close to the BS. The relaying schemes thus lead to a more balanced and fairer rate distribution. Additionally, the direct transmission seems to be an aggressive reference because the BSs would have to track the channels to mobile users with many antennas. With the relays, the CSI estimation at the BSs is simplified as fast fading can be eliminated from the point of view of the BSs, since the relays are, in contrast to the MSs, not moving.

The behavior of the UL rates for the same settings are shown in Fig. 6.8. Again, the curves show that the performance increases when the users are forced to be further away from the BS. This is even more pronounced as in the DL: the peak rates are achieved at a BS-to-RS distance of 450 m. When no deadzone is applied, the performance drop is larger when the relays are further away from the BS. Again, the area in which the MSs can be located is larger and the probability that some users are much closer to the BS than to a relay increases. The direct signal from the MSs to the BS is thus stronger and the relays less effective. Interestingly however, the AF one-way relays outperform all other schemes when they are closer to the BS ($d_{BR} \leq 200$ m). In this case, the

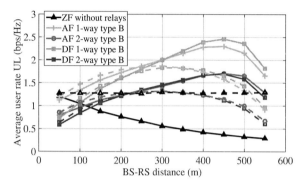

Figure 6.8.: Mean UL user rates for different d_{BR}. The solid lines are with a deadzone of $2d_{BR}/3$ around the BSs, the dashed lines are without deadzone.

relays can collect all signals within their cell, while the interference from MSs in other cells is weak and there is no other BS interference as in the case of two-way relaying. The relays then act as signal collectors and forward all desired signals to the BS. As the BS jointly decodes all signals within its cell and the AF relays do not distinguish between different signal contributions as the DF relays, every forwarded signal helps in improving the UL rates.

Achievable mean user rates of the DL for varying transmit powers and selected schemes are shown in Fig. 6.9. Here, the distance between BSs and relays is again $d_{BR} = 350\,\text{m}$ and a deadzone of $\frac{2}{3}d_{BR}$ is applied. In contrast to the cases discussed before, all nodes, i.e. BSs, relays, and MSs, transmit here with the same power. The interference coming from other MSs has thus a higher impact than before. With this, we can see which schemes are more affected by interference limitedness than others and which schemes can exploit more degrees of freedom. In the plot, we can observe that all curves saturate when the transmit power is larger than 50 dBm, the network is thus interference limited. In this high power regime, the AF type A two-way relaying scheme performs worse than the reference. As these relays are not able to filter out interference and the unwanted signals from MSs from other cells are particularly strong, there is not much room for improvement on the DL rates. The AF and DF one-way type B relays are significantly better in this regime. As only BS signals are included in these schemes, the stronger MS signals have no effect and the type B filters can reduce the BS interference. When two-way relaying is applied, these filters still reduce the BS interference, but the strong MS interference lets the rates decrease somewhat.

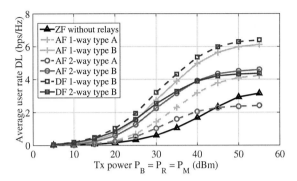

Figure 6.9.: Mean user rates in the DL for varying transmit powers. The noise variances are fixed to $\sigma_n^2 = \sigma_w^2 = 5 \cdot 10^{-12}\,\mathrm{W}$.

AF one-way relays without interference reducing filters behave thereby similarly. The strong MS interference is not present in this case, but the signals are affected more by unwanted signals from other BSs.

In the intermediate transmit power regime (between 20 and 40 dBm), the MS interference is not that high and all type B relays perform better than the simpler type A relays. The one-way relays are thereby better as less interference is included by the protocol. They achieve not only higher rates but also a steeper slope of the curve. This indicates that more spatial degrees of freedom can be exploited in this regime. Interestingly, AF relaying performs very good, even though this relaying strategy also amplifies noise and interference. Similar performance can only be achieved by DF relaying in the one-way protocol. This type of relaying, however, requires four orthogonal resources for one transmission in each direction. It is therefore less efficient than two-way relaying which requires only two resources. Even though the simple type A AF relays in the one-way protocol achieve smaller rates, their slope in the curve is similar in the intermediate power regime as the better type B relays. The additional interference that cannot be reduced by the relay filters affects the rates, but from the curve we can see that this scheme is less affected by interference limitedness than the two-way protocol.

In Fig. 6.10, we show the UL rates for the same settings and transmit powers. There, the schemes behave differently than in the DL. All AF relaying schemes perform similarly good and achieve the highest rates in the high power regime. One- and two-way relaying does thereby not differ notably. This is because the BS interference has

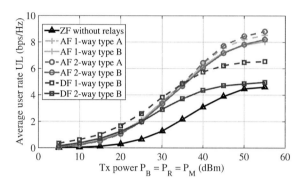

Figure 6.10.: Mean user rates in the UL for varying transmit powers. The noise variances are fixed to $\sigma_n^2 = \sigma_w^2 = 5 \cdot 10^{-12}\,\mathrm{W}$.

no impact in either case: in one-way relaying these signals are not present and in two-way relaying they are subtracted and cancelled at the BSs. The type A relays are somewhat better than the relays with filters. The explanation of this is again that the relay filters are specifically designed to improve the DL rates and the interference from other MSs is not considered in them. This has the consequence that the signals from other MSs in the same cell are not treated as interference and thus filtered, but are also amplified and forwarded to the BS which jointly decodes all these signals. In either case, however, AF relays performs better than DF relaying in this regime. In the latter case, all signals except the associated ones are treated as interference and limit the decoding process. One-way relaying is thereby better than two-way relaying as no BS signals are present.

In the intermediate transmit power regime, all relaying schemes behave similar. As the interference is not that strong, DF one-way relaying performs best in this case. The AF relays have however a steeper slope and overtake the performance of DF one-way relaying at about 35 dBm. In all cases, the schemes with relays outperform the reference. DF two-way relaying saturates however to a similar rate as the reference when the transmit power is high. With more interference, especially the one from other MSs, these relays cannot decode their intended signals with a high rate anymore. This effect is less pronounced with AF whose performance gain increases as compared to the reference when the transmit powers are higher. The UL can thus benefit from these AF relays, no matter if they apply filtering or not, as they collect all signals in the cell which are useful for the BS to decode the signals from its users.

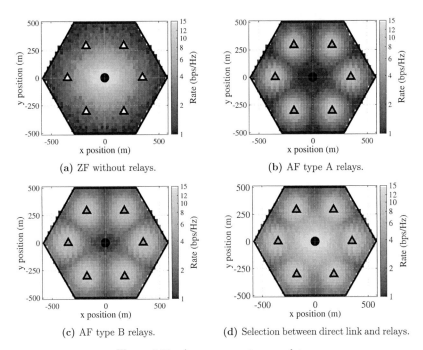

(a) ZF without relays.

(b) AF type A relays.

(c) AF type B relays.

(d) Selection between direct link and relays.

Figure 6.11.: Average user rates area plots.

The rate distribution in the area of a cell is shown in Fig. 6.11. In order to keep the simulations manageable, we focus on the average user rates in each grid point. We compare the reference with block ZF but without relays to the schemes with AF type A and type B relays. As already indicated before, one-way and two-way relaying show only a small difference and the average user rates for the DF counterparts are also similar in each grid point. We see that while the reference without relays achieves very high rats in the center of the cell, the rates drop below 4 bps/Hz after a distance of more than 250 m from the BS. In the case of relays, the rates close to the BS are small. Communicating via relays does not make much sense in this area as the pathloss to a relay that is further away from the BS than the MS itself is much larger than of the direct link. In this area, the users are better off when they are served directly by the BS or by other relays that are closer to the cell center. In the vicinity of a relay, the data rates are however drastically boosted. In the case of the simple type A relays that do not apply interference reducing filtering, the rates between two relays drop to low values due to the higher interference in these areas. The relay cell borders where strong interference is present are clearly visible. When type B relays with filters are applied, this interference is mitigated and the rates are more balanced. A larger area can be served with higher rates. The coverage of a relay is thus larger. But also here, the rates are low close to the BS. Also shown in Fig. 6.11d is the potential of a combination of the direct link from the BS and relaying. In this case, each user can select if it is served by the BS directly or via its associated relay. The two transmissions (direct or via a relay) are thereby separated, e.g. by a FDMA scheme which assigns different resources for these links. A prelog factor has however not been taken into account in the plot. In this case, the data rates are more balanced throughout the entire cell, similar to the cooperation schemes in the chapters before. There are however still spots of poor rates between the relays, especially towards the cell edge. In order to boost the data rates also in these locations, additional relays could be located there. Different relays, and also the direct links, could be separated by allocating different resource blocks. With this, a very homogenous rate distribution could be achieved without the rather complex cooperation scheme introduced in Chapter 3. The transmission schemes presented in this chapter are much simpler in the sense that the nodes do not have to apply computationally complex iterative optimization schemes that have to be applied jointly for all involved nodes together.

In Fig. 6.12, we compare the relaying schemes described in this chapter with the cooperation schemes from Chapter 3. While the transmission schemes and the network settings cannot be compared with each other directly (due to different network setups, different numbers of users in each cell, and different transmit powers at the BSs), we can still compare the outage rates each user can achieve in the network. The rather poor 5% outage rate from the initial reference scheme where each sectorized BS transmits spatially white to a single user is clearly outperformed by all other schemes. Regarding the rates around the 5% outage line, the block ZF without relays achieves already a much higher performance. Note here however, that more antennas are involved (24 instead of 12 BS antennas for each cell) and that six instead of previously three users are served in each cell. When the relaying schemes with type B relays are considered, they all achieve a similar performance which is close to the six BS super-cell cooperation scheme that also serves six users. The 5% outage rates with the type B relays are almost the same even though the transmission schemes are simpler and the performance is better than transmit cooperation between three BSs. With type A relays, a performance degradation has to be taken into account but the 5% outage

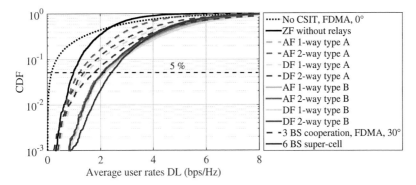

Figure 6.12.: CDF of instantaneous user rates for the DL. The relaying schemes from this chapter are compared with the reference without CSIT, 3 BS cooperation, and 6 BS super-cells from Chapter 3.

rates are still significantly larger than with the reference schemes. In this case, the relays can be implemented with very low complexity, e.g. as AF FDD relays that do not require any CSI.

6.3.3 Denser Networks

In the previous simulations, all cells contain $K = M = 6$ relays and MSs. The gains achievable with relays can however further be increased when more nodes are present. Fig. 6.13 shows average user rates for different numbers of users, where $M = K$ and $N_B = M \cdot N_R$ grow accordingly. The rates for both the DL (solid lines) as well as the UL (dashed lines) are shown, normalized by the number of users M. The relays are randomly placed with a uniform distribution in the cell with a deadzone of 300 m around the BSs. In each relay cell there is one MS also randomly placed. The transmit powers are again $P_B = 40\,\text{W}$, $P_R = 6\,\text{W}$, $P_M = 0.2\,\text{W}$. While adding more and more relays into the system, the total transmit power of each cell also increases, as the transmit power of each node is fixed. To this end, a curve of the reference scheme in which the BSs apply ZF directly to the MSs and transmit with a power that corresponds to the total power of all nodes in the cell, i.e. $\tilde{P}_B = P_B + K \cdot P_R$, is also included (black dotted line). With this comparison, the performance gains that are due to the higher power can be differentiated from the improvements that come from the relaying schemes. The higher transmit power has only a visible impact on the DL. If two-way relaying is applied,

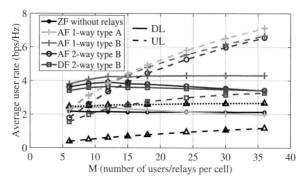

Figure 6.13.: Average user rates for increasing number of relays/users.

the higher BS power has also an impact on the UL due to the stronger signals the relays receive. Due to the imposed power constraint at each relay, the signals cannot be amplified as much as when the BS signals are weaker. The difference in the UL rates is however marginal.

In the DL, it can be seen that all schemes stay constant or drop slightly with the number of users/relays. In the reference case when the power stays constant, the interference power does not increase with the number of users while the increasing number of BS antennas can still perform block ZF. The interference situation for each user does therefore not change and the average user rate stays constant. When the BS sum transmit power increases with the number of nodes, the users experience a slight increase in their rates due to this additional power. The impact of the additional power is however small and the performance gain saturates as also more interference is injected into the network. When relays are applied, the type A relays as well as the two-way relaying schemes lead to a small reduction in the average user rates with more nodes. As the number of nodes increases, the interference the relays cause to other users gets higher. The performance drop is however small due to the distributed interference reduction, especially with type B relays. The performance with relays can thus be maintained when more nodes are present and the increased interference does not have a high impact. The type B one-way relays show thereby a particularly good behavior. With two-way relaying, there is a performance drop with more nodes. Due to the additional MSs that are close to the relays when the network gets denser, they also cause stronger interference which affects the DL rates as this interference cannot be cancelled at the MSs.

In the UL, the relay schemes can profit more from a larger number of nodes. The performance steadily increases with M and the performance gain as compared to the reference without relays gets bigger. Because all relays within a cell receive and forward all signals from the MSs therein, each additional relay helps in collecting signal power which can be exploited by the BS in the joint decoding of all these signals. The gain in the received signal strengths at the BSs grows thereby faster than the increased interference from the other cells. The spatial separation and the interference reducing signal processing help thereby. If two-way relaying with DF relays is applied, the performance gain is however small as the MSs get closer to the relays which increases the interference that limits the decoding ability at the DF relays. The performance of the AF relays scales however very beneficially with the number of involved nodes. In this case, multiple relays forward the signals of multiple MSs which can be resolved by the joint decoding at the BS. Particularly for this case, a large number of relays can help in increasing the performance in cellular networks drastically.

6.4 Critical Discussion

By the use of ubiquitous relaying, interference can be reduced and coverage can be made more homogeneous. Through the distributed form of interference management, the spatial degrees of freedom can be better exploited and the frequency reuse factor can be improved towards one. Turning the cellular network into a two-hop network also simplifies CSI estimation at the terminals. Because each node only needs locally available CSI from the nodes it directly communicates with, no exchange of channel knowledge is required. Furthermore, the BSs, which are potentially equipped with large antenna arrays, have only to estimate the channels to their relays which we assume to be static over a longer time period than those to the possibly moving MS. This can enable massive MIMO at the BSs as the performance is no longer limited by the estimation of rapidly changing channels. The approach of ubiquitous relaying is not only scalable in terms of the number of involved nodes/antennas, but it is also transparent to the implemented communication technology and can be applied on top of other approaches such as CoMP, heterogeneous networks, or others.

AF relays show already a good behavior when number of nodes is small. With increasing node density, their potential can further be exploited. DF relays have more problems to decode their signals due to the stronger interference when the network becomes dense. AF relays on the other hand do not distinguish between different signals. Especially on the UL, they can profit from more nodes as all signals within a cell is useful for decoding at the BS. When more and more AF relays are deployed, the throughput thus grows beneficially.

With BSs that can cooperate with each other to cancel other BS interference, two-way AF relaying shows large performance gains and proves to be very efficient for cellular networks. This is in contrast to previous results shown in [104] in which two-way relaying achieves rather poor rates as compared to one-way relaying. Therein, however, no BS interference cancellation is performed. As a result, the achievable rates are rather weak as compared to one-way relaying. In the schemes presented here, the BSs exchange their user data and can cancel this additional interference. This is an important mechanism in achieving high user rates with two-way relaying. The BS interference cancellation, however, adds additional complexity to the BSs which is similar to BS cooperation. Nevertheless, computationally expensive iterative optimization procedures do not have to by applied for the proposed schemes and also no CSI has to be exchanged as the transmission schemes all work with local CSI. As the channels to the relays presumably have a long coherence time, the limited form of BS cooperation thus introduces only a small overhead. Moreover, no clustering of BSs is required; any channel information that helps to reconstruct and cancel interference or to apply relay filters is beneficial. But when user data is readily available at the BSs, they could also perform joint beamforming to further reduce or cancel interference to other cells in the DL. In Chapter 8, we apply such a scheme that can also be supported by a variable or high number of relays.

If the prelog loss due to the use of multiple channel uses for one transmission is considered, two-way relaying clearly outperforms one-way relaying. On the other hand, the one-way schemes with simple AF relays already show a very beneficial performance scaling with the number of nodes. The relays can thereby be of very low complexity; especially in FDD, they can be implemented as simple frequency converters. By deploying a large number of them, the throughput of cellular networks can still be enhanced significantly with comparably low costs. If full-duplex relays can be used or when the second hop is for free (e.g. as a secondary link), one-way AF relays can lead to a better performance than two-way relaying.

The sample transmission schemes applied in this chapter already show a significant gain as compared to a conventional multi-user MIMO approach. We thus consider the approach to include a large number of relays into the network to be a promising option for future networks. In the following chapter, we focus on very simple AF relays and study in more detail how the concept of ubiquitous relaying can further be developed to improve the performance of future mobile communication by the required factors. The schemes proposed in this chapter are however not optimal in any way. The performance could be further increased, e.g. when the schemes are combined with power control and/or transmit cooperation at the BSs, not least as some user data and CSI is already available at these nodes. Also the direct link between BSs and MSs and the relay links could be combined for a further enhancement of the performance.

Also the DF relaying schemes can further be improved. In this case, a specific proto-col with a SIC order is applied. Thereby, no time slot or frequency band optimization is performed for the two links. This optimization has in Chapter 4 been identified as an important ingredient for DF relaying. If each link is scaled individually, however, this poses difficulties to the signaling protocol. Some relays might be in transmitting mode while others are still receiving. In two-way relaying, the UL would interfere with the DL and vice versa as the two directions are not separated anymore. In one-way relaying, a system wide time slot optimization between the two hops could help. So can a smaller fraction of the resources be allocated to the BS-to-relay transmission and a larger to the one between relays and MSs as the BS-to-relay links are presum-ably stronger than the relay-to-MS links. Especially in type A relays, where the BSs can transmit four spatial data streams to each relay, the rates of the first hop can be expected to be higher than in the second hop. With type B relays, the rates of the two links might be more balanced as the relay filters reduce the number of effective antennas seen by the BS from four to two. In this case, only two spatial streams are sent on the BS-to-relay link, which reduces the data rates on this link.

In Fig. 6.14, we show by how much the DL data rates can be increased by such a time sharing approach for DF relaying. For the simulation, the basic network configuration with $M = K = 6$ relays and MSs is considered. Therein, the first hop (BS-to-relay link) uses a fraction of $t \in [0, 1]$ of the total transmission time while $1 - t$ is allocated for the second hop. Note that all transmissions in the entire network use the same fraction t. With this, the different transmissions are still separated. Compared to equal time sharing (at the black dashed line at $t = 0.5$), this can lead to a performance gain for DF relaying. The average user rates can be increased by about 0.3 bps/Hz for both the type A and type B DF relays. The type A DF relays can thereby achieve the same performance as the more sophisticated type B AF relays and the performance gain of type B DF relays over the type B AF relays is somewhat enlarged. For a fair comparison, a prelog factor of $1/2$ is applied to the AF relaying schemes. When

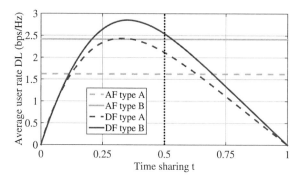

Figure 6.14.: Network wide time sharing between the two hops in one-way DF relaying. Also shown are the average user rates for AF one-way relaying. A prelog factor of 1/2 is taken into account.

more relays are applied, the AF relays scale however better than the DF counterparts as previously discussed. The performance gain due to optimized time sharing in DF relaying is thus not particularly pronounced.

In all schemes and simulations discussed in this chapter, we have made idealized assumptions regarding CSI, interference cancellation and precoding as well as relay filtering. In all cases, the CSI at the different nodes was perfect and all coefficients of the locally available channels known. In the following, we discuss the impairments of the applied schemes when this CSI is not perfect any more and what implications on the system design this has. Particularly as we mentioned that ubiquitous relaying can simplify the channel estimation process due to the static relays, this aspects have to be considered such that we can further evolve the idea of massive relay deployment in the upcoming chapters.

6.5 Aspects of Channel Estimation

For the transmission schemes introduced in the previous sections, CSI is necessary at the BSs, relays, and MSs in different forms. We distinguish between CSIR and CSIT. Usually, acquiring CSIR (e.g. based on a training sequence) is not considered as difficult as obtaining CSIT. In TDD systems assuming channel reciprocity, CSIT can be determined from the CSIR which has been obtained as part of the decoding process in a previous transmission. In case of FDD this is not possible due to the different frequencies. One way of acquiring CSIT nevertheless is using a feedback channel: the receiver is feeding its CSIR (possibly quantized and compressed) back to the sender. However, CSIT may then be outdated or noisy (e.g. due to quantization). Another way to acquire CSIT in a FDD system would be that the receiver transmits a training sequence on the transmit frequency of the transmitter in a separate time slot. The transmitter then estimates the CSIR and determines the CSIT by assuming channel reciprocity. In the following, we discuss which nodes need which form of CSI, how they can acquire it, and what impact this has on the node complexity. Furthermore, we introduce error models for CSI imperfections in order to determine and discuss the robustness of the transmission schemes.

6.5.1 Acquisition of CSI

At the BSs, channel estimation is necessary for different tasks: CSIR to decode the UL signals, CSIR to cancel self- and BS-interference and CSIT for the calculation of the beamforming matrices. Whereas CSIR can be acquired at the BS based on training sequences as described above, CSIT needs to be estimated at the relay and fed back to the BS, or the relay can transmit a training sequence on the transmit frequency of the BS on demand.

At the MSs, no CSIT is required by the schemes presented here, only CSIR for decoding the DL signal and, in case of two-way relaying, for canceling the self-interference.

At the relays, the necessity of CSIR and CSIT depends on the type of the relay and the signal processing. Whereas a type A AF relay does not need any CSI at all, a type B AF relay needs to know the relay-MS channel $\mathbf{F}_{k,k}^{(c,c)}$ as well as the BS interference covariance matrix $\mathbf{\Gamma}_{c,k}$ for the computation of the transmit and receive filter. For TDD relays, the CSIT can be acquired via the CSIR. For FDD, either a feedback from the BS/MS is necessary or the transmission of training sequences by the BS and MS on the transmit frequency of the relay. As only the channel covariance matrix from the BSs is required, the estimation is much simpler than for the full channel. Its dimensions are only $N_{\mathrm{R}} \times N_{\mathrm{R}}$ and a sample covariance matrix can be obtained by observing the received signal over time. For DF relays, CSIR is always necessary for the decoding, also for type A relays. When type B DF relays are used, the CSIT can be obtained from CSIR when the relays operate in TDD mode. In the FDD case, this is not possible and the acquisition of CSIT by feedback or pilot transmission comes on top.

6.5.2   Node Functionality

The simplest form of relays considered in this work are type A AF relays in FDD mode. In this case, the relays can be seen as simple frequency converters that amplify their input signal without the requirement of any CSI. In order to allow its BS to estimate the BS-relay channel, these relays have to be able to transmit a training sequence on demand. This kind of relay can be referred to as a "drilled" relay, as it only responds to requests of the BS. Apart from some synchronization mechanisms, such relays do not need any additional functionalities. If the relays operate in TDD mode, an additional buffer to store the received signal before it can be retransmitted is required.

The more sophisticated type B AF relays additionally need to acquire CSI such that they can calculate their receive and transmit filters. To this end, the relays need either to be able to estimate the required channels themselves or to receive the CSI that is delivered from their BS and/or MS. As a result, such relays require a decoding functionality that does not differ much from the one in DF relays.

DF relays are the most complex relays considered in this chapter. Additional to the CSIR necessary for the decoding, the signals need to be re-encoded. For type B DF relays, also CSIT is required that can be obtained as in the AF case.

While the relaying protocol (whether one-way or two-way) does not matter for the relay complexity in AF relays, it influences the tasks of the terminal nodes. For one-way relaying, the terminal nodes just need to evaluate the training sequences and decode the signal. For two-way relaying instead, they additionally need to estimate and subtract the self interference (and the interference of the other BSs). Especially for the BSs, that cancel the other BS interference, two-way relaying thus adds some complexity to the terminals. However, when the relays are static, the CSI for interference cancellation needs to be tracked with a comparably low frequency. If DF relays are used, the task of interference cancellation is simpler. Only self-interference has to be compensated, which can be done in the digital domain by an XOR operation.

6.5.3   Estimation Error Models

As the positions of BSs and relays are fixed, we consider the channel between a BS and a relay as quasi-static. Acquiring CSIT of a certain quality for this link seems possible and less difficult than for the link between a relay and a possibly moving MS. These considerations motivate the chosen transmission schemes. Block ZF, suffering stronger from imperfect CSIT, is only used on the channel between BS and relay, while the more robust relay filters are used on the link between relay and MS. In the following, we investigate the robustness of the considered schemes regarding imperfect CSI. These imperfections can arise from channel estimation errors, quantization of the channel estimates in the feedback channel, outdated CSI, etc. In order to capture these effects, we apply simple models that are based on additive Gaussian errors as e.g. in [150].

Complete Channel Matrix: For the BS beamforming and the relay filters, the actual channel matrices $\mathbf{H}_k^{(c,c)}$ and $\mathbf{F}_{k,k}^{(c,c)}$ need to be known at the respective nodes. Imperfections on this type of CSI are modeled by

$$\hat{\mathbf{H}}_k^{(c,c)} = \sqrt{\frac{1}{L_{c,k}}} \left(\sqrt{1 - \vartheta_{\mathrm{H}}^2} \mathbf{H}_k^{(c,c)} + \vartheta_{\mathrm{H}} \mathbf{W}_k^{(c,c)} \right), \qquad (6.57)$$

where $\vartheta_{\mathrm{H}}^2 \in [0,1]$ is the CSI noise scaling factor and the pathloss $L_{c,k}$ is assumed to be known perfectly (averaged over time). Only the small scale fading is affected by the estimation error $\mathbf{W}_k^{(c,c)}$ with elements i.i.d. $\mathcal{CN}(0,1)$. We define the estimation SNR as $\mathrm{SNR}_{\mathrm{H}} = \frac{1 - \vartheta_{\mathrm{H}}^2}{\vartheta_{\mathrm{H}}^2}$ as a measure for the quality of the CSI. As the channels between the BSs and the relays are considered quasi-static, high SNRs can be expected.

For the estimation of $\mathbf{F}_{k,k}^{(c,c)}$, the same model is used. Thereby, the estimation SNR given by $\mathrm{SNR}_{\mathrm{F}} = \frac{1 - \vartheta_{\mathrm{F}}^2}{\vartheta_{\mathrm{F}}^2}$ can differ from the one at the BS, as this channel cannot be assumed to be quasi-static.

Channel Covariance Matrix: For the error of the estimation of the covariance matrix $\mathbf{\Gamma}_{c,k}$, required for the calculation of $\mathbf{G}_{c,k}^{(\mathrm{Rx})}$, we use the model

$$\hat{\mathbf{\Gamma}}_{c,k} = \mathbf{\Gamma}_{c,k} + \sigma_{\Gamma}^2 \mathbf{W}_{c,k} \mathbf{W}_{c,k}^{\mathsf{H}}, \qquad (6.58)$$

where $\mathbf{W}_{c,k}$ is again an estimation error matrix as above and $\sigma_{\Gamma}^2 \in [0, \infty)$ the noise scaling factor. The instantaneous estimation SNR of this model is defined as $\mathrm{SNR}_{\Gamma} = \frac{\mathrm{Tr}\{\mathbf{\Gamma}_{c,k}\}}{N_{\mathrm{R}} \sigma_{\Gamma}^2}$. The estimation error is assumed to be small, as the sample covariance matrices can be averaged over time.

CSI for Interference Cancellation: For the cancellation of the self-interference at the BSs and the MSs, we consider the compound channels (from the BS/MS to the relays and back) denoted by $\mathbf{H}_{c,k}^{(\mathrm{comp})}$ and $\mathbf{F}_{c,k}^{(\mathrm{comp})}$. These can be estimated e.g. with training sequences contained in the transmit signals. The estimation error of the compound channels is modeled by

$$\hat{\mathbf{H}}_{c,k}^{(\text{comp})} = \mathbf{H}_{c,k}^{(\text{comp})} + \sigma_{\text{s}}\mathbf{W}_{c,k}, \tag{6.59}$$

with $\mathbf{W}_{c,k}$ the estimation error matrix as above and $\sigma_{\text{s}} \in [0, \infty)$ the CSI noise scaling factor. For the cancellation of the self-interference, the BS/MS subtracts the estimated self-interference. For the achievable rate, only the remainder of the self-interference covariance matrix is of importance

$$\hat{\mathbf{K}}_{\text{M},c,k}^{(\text{self})} = \sigma_{\text{s}}^2 \mathbf{W}_{c,k}\mathbf{W}_{c,k}^{\text{H}}. \tag{6.60}$$

To relate the estimation noise power to the actual self-interference power, we define the instantaneous estimation SNR of this error model as $\text{SNR}_{\text{self}} = \frac{\text{Tr}\left\{\hat{\mathbf{K}}_{c,k}^{(\text{self})}\right\}}{N_i \sigma_{\text{s}}^2}$, for $i \in \{\text{B}, \text{M}\}$.

The same model is used for the cancellation of the interference from other BSs. The remainder of the other BS signal covariance matrix is modeled as

$$\hat{\mathbf{K}}_{\text{B},c}^{(\text{BS int})} = \sigma_{\text{B}}^2 \mathbf{W}_{c,k}\mathbf{W}_{c,k}^{\text{H}}, \tag{6.61}$$

with all parameters as above. As these channels are assumed to be quasi-static and all data is expected to be known at the receiver, high SNRs can be expected.

6.5.4 Performance Evaluation with Imperfect CSI

So far, perfect CSI was assumed for all simulations, i.e. the beamforming and relay gain matrices are all computed based on the correct channels. In the following, we study the influence of CSI imperfections as discussed above. The influence of the CSI noise on the UL and DL performance for the standard setup with $K = M = 6$ relays/MSs per cell is shown in Fig. 6.15. In the first three rows, only one type of CSI imperfections is considered at one time: i) only at the BSs for the calculation of the beamforming, ii) only at the relays, and iii) only for interference cancellation at the terminals. With this, we can study which types of CSI imperfections have which impact on the performance. In the lowermost row, all nodes are affected by CSI imperfections in the same way, i.e. all estimation SNRs are equal. The following can be observed:

i) It can be seen that the BS beamforming requires good CSI. Otherwise, the performance degrades rapidly. When the BS applies block ZF, wrong CSI prevents

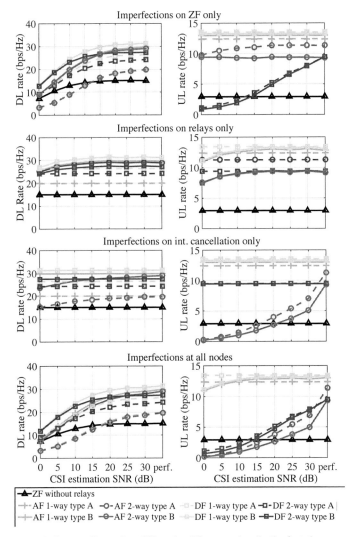

Figure 6.15.: Influence of imperfect CSI at the different nodes. In the first three rows, only
one type of CSI estimation is affected by imperfections, the others are assumed
to be perfect. In the lowermost row, the estimation SNR is the same at all
involved nodes.

the cancellation of interference within its cell. With a lower CSI estimation SNR, more interference is thus injected and the user rates drop with noisy CSI. Nonetheless, as we consider the channels between BSs and relays as quasi-static, a high CSI estimation SNR can be expected in our setup. In the UL, only two-way relaying depends on the BS beamforming. The one-way schemes are thus independent of this type of imperfections. With two-way relaying, the higher interference received by the relays affects both directions of communication. Especially with DF relays, this effect is clearly visible, as these relays, when applied in the two-way mode, cannot decode their desired signals with high data rates anymore. AF two-way relaying is less affected as the relays do not distinguish between different signals and the strong BS interference can be subtracted again at the BS.

ii) At the relays, CSI imperfections only have an influence on type B relays. The chosen relaying schemes are however quite robust; the interference mitigating receive filter and the transmit matched filter do not degrade the performance significantly at low SNRs. This can be observed for both the DL as well as the UL.

iii) In the case of AF two-way relaying, the cancellation of self- and BS-interference is crucial, especially in the UL. This type of interference is very strong at all nodes and has thus to be known accurately in order to get good end-to-end performance. This form of relaying is thus only beneficial if the terminals can estimate the corresponding channels appropriately. Especially at the BSs where the interference from other BSs has also to be cancelled, wrong CSI degrades the performance drastically. In the DL, the effect of noisy CSI for the self-interference cancellation is less severe, as the MSs only cancel their own interference while the interference from other MSs is still present. Self-interference cancellation thus reduces the overall interference only to a moderate extent. The schemes other than two-way AF relaying are not affected by these imperfections.

iv) When all nodes and interference mitigating measures are affected by the same level of CSI estimation noise, the DL performance is dominated by the block ZF that does not work properly anymore. In the UL, only two-way relaying is strongly affected by the imperfections, with AF relays mostly due to the self- and BS-interference cancellation, while DF two-way relaying suffers mostly from noisy block ZF at the BSs.

In the following, we apply imperfections at all nodes and look at the behavior of selected relaying schemes when the network density is increased. To this end, the same setup as discussed in Fig. 6.13 with an increasing number $M = K$ of relays/users is applied. The curves in Fig. 6.16 show the sum rates (UL plus DL) normalized by the number of users in the network for the case of perfect CSI at all nodes (solid lines) as well as for the case in which the different nodes are affected by CSI estimation errors (dashed lines). In the latter case, the BS beamforming is based on CSI with an estimation SNR of 20 dB, the CSI at the relays has an SNR of 10 dB, and the one for self- and BS-interference cancellation has an SNR of 30 dB. With this choice of CSI estimation SNRs, we can reflect that the CSI estimation at the BSs is simpler due to the static relays than at the relays which also have to take the channels of possibly moving MSs into account. Furthermore, the high SNR for BS- and self-interference cancellation can be justified as the signals these nodes have transmitted themselves as well as the user data that is available at the BSs can be used for channel estimation on top of training sequences included in the signals.

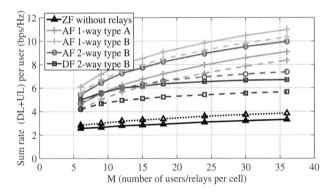

Figure 6.16.: Increasing number of relays/users with (dashed) and without (solid) imperfections. The black dotted line corresponds to the reference where the BS transmits with the sum power that the BS and the relays would have together.

It can be seen that the performance of AF relaying improves with the number of relays/MSs, while the one with DF relays and the reference scheme without relays (once with $P_B = 40\,\mathrm{W}$ (black solid) and once with $\tilde{P}_B = P_B + K \cdot P_R$ (black dotted)) tend to saturate with the number of users. For high nodes densities, this is mainly due to the UL that gains more from a larger number of AF relays.

When imperfections are included, a performance drop can be observed in all schemes. With AF one-way relaying however, the scaling behavior stays the same as with perfect CSI. The additional interference affects these relays only moderately and their rate increase can be maintained. In the case of two-way relaying, especially with AF type B relays, the performance also saturates when the CSI is affected by noise. In this case, the network also becomes more flooded by interference which cannot be handled by the interference mitigating schemes. The performance behaves in this case similar to DF relaying. Especially the simple type A AF relays achieve a good sum performance and the degradation with CSI imperfections is small. As these relays are of very low complexity, more of these relays can be deployed with little costs. The lower rates as compared to the more complex type B relays can thus be recovered by deploying more of them. Also the use of idle MSs as relays can further improve the performance, as a growing network increases the throughput. However, the two-way gain is not as pronounced in networks with high density when the CSI is imperfect. Hence, the possible performance loss of one-way relaying due the multiple channel uses (when a prelog factor has to be taken into account) can be recovered by a larger amount of such relays.

7

The Cellular Relay Carpet

In the last chapter, we have seen that it is beneficial when MSs are served via relays instead of the BS directly. Thereby, simple AF relays are particularly interesting. They can be implemented with low complexity, show a beneficial performance scaling when the node density increases, and they are robust with respect to imperfections. AF type A relays do not require any CSI and, if they operate in FDD mode, need to apply only a simple frequency conversion and amplification of their receive signals. This can be realized with a low-complexity implementation and the relays do not introduce any further delays as they would occur e.g. in the TDD case when entire signal blocks would have to be stored before retransmission or in DF relays where complete codewords need to be decoded first. As such delays at least double the round trip time, networks relying on such techniques would not be feasible for time critical applications. For practical networks on the other hand, the delays should be kept as small as possible, while the QoS should be improved. Due to the beneficial scaling behavior shown in Section 6.5, even with corrupted CSI, the simple AF FDD relays thus prove to be a valid option for an ubiquitous usage in cellular networks.

In this chapter, we go one step further and propose a network concept in which the mobile users are no longer served by BSs, but by a large amount of relays that are spread over the entire area of the network, similar to a carpet. With this "relay carpet", we combine and enable different proposals for 5G networks that include massive MIMO, network densification, distributed cooperation, and sophisticated multi-user beamforming within a single network architecture. With a massive deployment of relays, the entire network is turned into a two-hop network (see Fig. 7.1). Accordingly, all BSs and MSs see only the relays as the nodes they communicate with. If the relays are dedicated infrastructure nodes and mounted at fixed positions with good propagation conditions to their BS, fast fading between them is virtually eliminated. This allows to equip the BSs with a large number of antennas and to apply sophisticated multi-user MIMO transmission with accurate CSI. The relays on the other hand are in

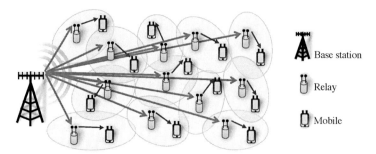

Figure 7.1.: The relay carpet: a sophisticated BS serves a large amount of MSs in the same physical channel by the help of many distributed relays.

close vicinity of the MSs which experience less pathloss and better coverage. Moreover, when many relays are installed, multiple relays can be in coverage range of one MS, which allows to benefit from additional diversity. The relay carpet thus has numerous advantages: the BSs only have to track quasi static channels which drastically simplifies the estimation of CSI, while for the MSs the network appears like a much simpler network consisting only of nodes with few (effective) antennas. The distributed relays also lead to a more equally distributed signal quality in the area of service. Moreover, the signal processing at the BSs and relays can jointly contribute to an effective distributed interference management. The relay carpet does thereby not only simplify the signal processing, but can also offer significant performance gains throughout the network.

In this chapter, we describe the concept of the relay carpet and argue how simple relay implementations can enable sophisticated signal processing with massive MIMO BSs. A disadvantage of simple AF relays is however that they inject additional noise into the network and also amplify and forward interference. The relays thus need to be accessed in an appropriate manner and only the relays that are beneficial to the QoS experienced by the users should be active. To this end, we first study the behavior of these relays when multiple of them are assigned for certain users and identify required measures to be able to benefit from massive deployment of relays.

One important aspect thereby is power control to manage the residual interference and the energy consumption. By allocating the transmit power in an optimized way, the relays can also lead to considerable power savings as compared to conventional networks. To this end, existing power control schemes designed for conventional cellular networks (see [22] for an overview) or for pure relay channels (e.g. [70]) have to be adjusted such that they are feasible for the relay carpet network with its inherent combination of large antenna arrays, interfering two-hop links across different cells, and distributed beamforming. The proposed schemes are all of low complexity and thus relevant for practical implementation. As shown by the results in a realistic setup, the relay carpet thereby proves to be an enabler for the required performance in future cellular systems.

7.1 The Relay Carpet Concept

In order to meet the demands of an expected thousand-fold increase in data traffic and number of devices in the next decade, several new technologies are under discussion for next generation (5G) networks. Among others that we have summarized in Chapter 2, the most important and promising approaches are the extreme densification of the networks, up to very small cells that are served with millimeter wave technologies, very large antenna arrays (massive MIMO) that allow to serve many users at the same time, and cooperation to allow sophisticated signal processing with multi-user beamforming over wide areas to mitigate or cancel interference over the border of individual cells or sectors [5,45].

By the deployment of very small cells, the total throughput of the network can drastically be increased. If, in the extreme case, each user is served by its own small cell, many more MSs can transmit or receive data at the same time. The network can thus deal with high user densities [98]. The adaptation to the user position can be achieved by handovers between such small smells, which is easy to realize and requires little overhead. In order to limit the interference seen by the users in a dense environment, the classical reuse partitioning can be applied which allocates different resource blocks for neighboring cells. When realized with millimeter wave signals around a carrier frequency of 60 GHz, the high pathloss at these frequencies allows to keep the reuse factor moderate [114]. For this approach, it is however difficult to find and install sufficiently many new BSs sites, e.g. due to social acceptance or due to the costs of deployment and maintenance.

Installing very large antenna arrays and letting different nodes cooperate with each other allows to separate different users by beamforming. With this, many users can be served in parallel in the same resource blocks. Massive MIMO allows thereby not only to increase the spatial multiplexing gain and diversity in the network to a large extent, but also helps in keeping the signal processing simple. When the number of antennas grows large, the channels between different users become more and more orthogonal to each other [94]. By beamforming, many users can thus be served simultaneously with high data rates. This, however, requires to track the instantaneous channels to each MS. An increasing number of antennas leads therefore to a rapidly increasing overhead and/or the acquired CSI is corrupted by pilot contamination [89]. Moreover, BSs that cooperate to perform joint beamforming also require very high backhaul rates, not only to support the data rates of their users, but also to exchange user data and CSI with their cooperation partners.

An attempt to combine the advantages of the aforementioned approaches, while avoiding their disadvantages, is to support the BSs by a large amount of relays. If the relays are of low cost and low power, they can be installed in massive numbers across the entire network such that the area appears like a carpet of small relay cells. In this "relay carpet", few BSs that are equipped with very large antenna arrays and high computational power as well as a backhaul connection of high capacity are organized similarly as in conventional micro- or macro-cell networks. With their massive antenna arrays, they can cover wide areas and perform spatial multiplexing to separate many users in the spatial domain. By the large amount of antennas, also a high diversity gain can be achieved. The BSs do however not serve the MSs directly, but a vast amount of simple relays without a fixed connection to the backhaul supports the communication. These relays serve the MSs in their close vicinity within their small cells. Thereby, many more relays can be deployed than MSs have to be served. As a result, different subsets of relays can be activated or deactivated, e.g. by BSs that transmit only to selected relays through beamforming or by more sophisticated scheduling functionalities. In this way, static users are served by the relay cells they are located in or multiple relays can follow the movements of mobile users. If GPS information is available, direction and speed of the users can be predicted to assign the relays appropriately. The potential disadvantage that moving users might require many handovers can thus be mitigated. A sketch of such a network is depicted in Fig. 7.2.

As a result, the entire network is turned into a two-hop network in which the BSs as well as the MSs communicate via relays. The BSs see only the relays as their communication partners. If dedicated relays are mounted at fixed positions, fast fading between them is eliminated. For the transmit CSI, the BSs thus only have to track quasi static channels. This simplifies the channel estimation and allows to equip the BSs with

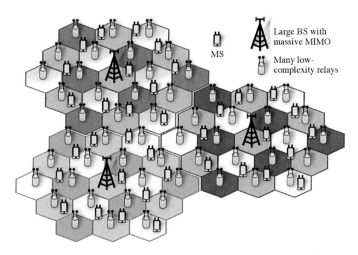

Figure 7.2.: BSs with large antenna arrays serve a wide area while the MSs receive their signals from relays that are spread throughout the area.

(very) large antenna arrays and to apply sophisticated multi-user MIMO transmission. In order to serve mobile users, only the static relays have to be addressed, which is much simpler than to follow possibly fast moving MSs. Accordingly, the static relays enable massive MIMO at the BSs. The MSs on the other hand see a much simpler network of relays with only few antennas in their close vicinity. The relays thus lead to a more equally distributed signal quality and the users experience less pathloss and better coverage.

Additionally, the relays can shape the (effective) channel between BSs and MSs in a beneficial way. Accordingly, network operators do not have to rely on random properties of the propagation channel, which can result in deep fades or shadowed users, but can achieve much more homogeneous coverage. To this end, the relays can perform simple signal processing tasks that allow for signal amplification or even distributed interference management. As a side-benefit, the angular spread of the effective channel can be increased by allocating multiple relays to one user (active scattering [151]) and MSs can be equipped with more antennas in a compact space. As a large amount of relays is deployed, the complexity of the relay nodes is crucial. Different relay architectures can assist the communication between BSs and MSs in different ways, depending on their available CSI and computational power. The node density and the relay complexity thus lead to a tradeoff in which the performance and the infrastructure costs can be balanced.

7.1.1   Low-Complexity Relay Implementation

Depending on their functionalities, the relays can fulfill different signal processing tasks. For a massive deployment, however, the relay nodes should be of very low complexity such that they can be implemented in an inexpensive way. As seen in the previous chapter, type A AF relays offer thereby promising gains and scale in a beneficial way. In their simplest form, these relays apply a frequency conversion from the input frequency band around f_1 to a band around f_2 and amplify the input signals with a scaled identity matrix

$$\mathbf{G}_{c,k} = \sqrt{\alpha_{c,k}} \cdot \mathbf{I}_{N_\mathrm{R}}, \tag{7.1}$$

where $\alpha_{c,k}$ is the amplification factor of relay k in cell c and N_R is the number of antennas at the relay. As a result, the transmit and receive phase of the relays are orthogonalized but no additional delays are introduced. If the relays convert their BS signals to frequency bands that are currently not used (cf. cognitive radio [137]) or lie in an ISM band, the spectrum of the second hop does not have to be included into the spectral efficiency as additional costs. The use of secondary links is especially motivated by the small transmit power of the relays that do not disturb other systems significantly.

A conceptual schematic of such a type A relay is sketched in Fig. 7.3. Apart from the frequency conversion and amplification, it contains an input and an output filter as well as a simple control unit that can adjust the relay gains or the local oscillator. A control channel from the corresponding BS is also included. This channel can be of very low rate and can be used for synchronization, to control the timing of the relays, or to transmit wake up patterns to activate or deactivate the relays appropriately. It can also be used to initiate a training phase with which the BSs can estimate the channels to the relays. To this end, the relays can transmit predefined signals a the receive frequency f_1 that are modulated by specific variations of the amplifications gains. If the channel between BSs and relays can be assumed to be reciprocal, this allows then to estimate the channel coefficients between the BSs and the relay nodes for sophisticated beamforming or precoding. Note in this regard that if the relays are mounted at fixed

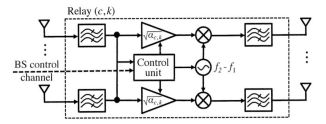

Figure 7.3.: Conceptual schematic of an FDD AF relay.

positions, the channel between BSs and relays can be assumed to be slowly fading or quasi static. This further simplifies channel estimation, as the estimation process can take place over longer time intervals.

As an example of the frequency conversion, the relays could convert licensed frequency bands around 2.6 GHz of the long range communication from the BSs to the relays up to a frequency around 60 GHz for the short range communication between the relays and the mobile users. With such millimeter wave links, high bandwidths and hence high data rates could be allocated to users that are close to access points [114]. When applied to the relay carpet concept, sophisticated BS transmission on low(er) frequencies could provide signals over larger distances to the relays as in conventional systems, and the relays then serve the MSs with LOS connections over short distances. To this end, however, a massive number of relays needs to be deployed, especially in indoor environments. But when the relays are of low cost, multiple of these relays could be placed in each room or office in a building. Each mobile user would then have multiple relays in its close vicinity and can be served by them even though the signals suffer from higher pathloss on frequencies as high as 60 GHz [108, 115]. In the following, however, we will focus on relays that convert their signals within the same frequency range around 2.6 GHz to maintain the comparability to the other schemes, as for frequencies much higher than 5 or 6 GHz, the applied WINNER II channel model would not be applicable anymore [76].

More sophisticated type B relays that have access to local CSI can additionally apply a linear processing to reduce the interference present in the network. To this end, the gain matrix of relay (c, k) can be factorized to

$$\mathbf{G}_{c,k} = \sqrt{\alpha_{c,k}} \cdot \mathbf{G}_{c,k}^{(\mathrm{Tx})} \cdot \mathbf{G}_{c,k}^{(\mathrm{Rx})\mathsf{H}}. \tag{7.2}$$

As an example, the receive filter $\mathbf{G}_{c,k}^{(\mathrm{Rx})}$ can be chosen as an MVDR filter or, as a simpler alternative, to only suppress the interference coming from the BSs of adjacent cells as described in Chapter 6. Assuming that N_R is larger than the number of transmitted data streams d_s, this filter can be obtained by $\mathbf{G}_{c,k}^{(\mathrm{Rx})} = [\mathbf{v}_1^{(c,k)}, \ldots, \mathbf{v}_{d_\mathrm{s}}^{(c,k)}]$. Therein, $\mathbf{v}_i^{(c,k)}$ is the eigenvector corresponding to the ith smallest eigenvalue of

$$\mathbf{\Gamma}_{c,k} = \sum_{b \neq c} \mathbf{H}_k^{(c,b)} \cdot \mathbf{H}_k^{(c,b)\mathrm{H}}. \tag{7.3}$$

With this, the receive signal is projected into the subspace that contains the least BS interference under the assumption of spatially white signaling. Accordingly, $\mathbf{G}_{c,k}^{(\mathrm{Rx})}$ is independent of the actual BS signals and has thus not to be updated when a BS changes its beamforming. Moreover, when the relay position is fixed, the covariance matrix (7.3) is mainly static and simple to estimate.

The transmit filter of the relay can be chosen as a transmit matched filter $\mathbf{G}_{c,k}^{(\mathrm{Tx})} = \mathbf{F}_{k,k}^{(c,c)\mathrm{H}}$ with respect to the channel to the corresponding MS, which is also simple to estimate, as the dimensions are small. In this case, the functionality of the relays shown in Fig. 7.3 needs to be extended such that the gain matrix can be applied and allows the relays to obtain the required CSI. Nevertheless, the relay gain matrices can be calculated based on local CSI only and no cooperation with other nodes is required. Alternatively, the relay filters can be calculated at the BSs and the gain coefficients sent via the control channel to the relays.

7.1.2 Beam Switching

In order to serve the MSs in the small relay cells, one or multiple relays that are closest to an MS can be activated. The BS then transmits the signal intended for this user to its associated relays. Thereby, relays belonging to different users can be separated by multi-user beamforming such as block ZF as discussed in the last chapter. To this end, the BS has to know only the channel to the static relays. If a user moves and comes out of range of a relay and into the coverage of a new one, the BS does not have to track the channel to the MS but it can deactivate the old relay and switch its beamforming to the relay that is now in coverage range of the new position of the MS. The activation/deactivation can thereby either be handled via the control channel between the BS and the relays or with the beamforming itself. If the relays only forward signals if their input signal strength is above a certain threshold, the beamforming at

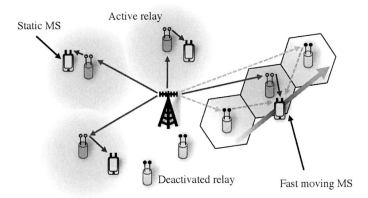

Static MS

Active relay

Deactivated relay

Fast moving MS

Figure 7.4.: With beam switching, the BS can follow the movements of mobile users. Thereby, the BS transmits signals to the relays which offer the best service to the MSs.

the BS can take care that the relays only receive signals above this threshold when they are selected for serving the user associated with this transmit signal. Relays that are not provided with signals remain silent so as they do not amplify and forward noise and interference.

For the BS transmission itself, only the location of the MSs and the channels to the relays have to be known. The former can be acquired e.g. via GPS that is included in all modern smartphones or with other means. The required CSI for the beamforming on the other hand is simpler to obtain than when the instantaneous channels to the MSs were required. When the relays are static, the channels to them are constant for a longer period of time and less overhead is required for accurate channel estimates.

In order to avoid the problem of fast handovers between different relay cells if some mobile users move with a high velocity, multiple relays can be assigned to the same MS. To this end, the direction and speed of the MS can be predicted with location information of the MS. Multiple relays can then be assigned to this MS in a way that a bigger cell is formed and that additional relays are activated according to the movement of the user as depicted in Fig. 7.4. Relays that are no longer in the service range of the MS can then be deactivated again. In this way, the task of tracking the MSs lies with the BS that adjusts the beamforming to the assigned relays, while the relays can be implemented in a simple way. If the relays do not have CSI, the channels to the

MSs are unknown. In this case, the signals from different relays add up in power as no coherent combination is possible. With CSI, i.e. when type B relays are used, the performance can be improved by coherent signaling between different relays.

7.1.3 Channel Estimation & Pilot Contamination

A big issue for DL beamforming at BSs with many antennas is channel estimation. Accurate CSIT is an essential component in order to maximize the network performance by precoding. In order to obtain such CSI, the coefficients of all channel components of a transmitting BS to all its receivers need to be known. To this end, either the BS can send training sequences or pilots and the receivers can estimate the channels and feed the CSI back to the BS. Alternatively, when the channels can be assumed to be reciprocal, the receivers can transmit pilots which allow to estimate the channel coefficients at the BS directly. The former method is usually applied in FDD systems where the UL and DL are separated in frequency, which is the case in most current cellular systems. While this approach works well with a small number of antennas, the burden of training and channel estimation grows rapidly with larger antenna arrays. As each BS antenna has to transmit its own pilots and these pilots have to be separated by orthogonal resources, the training overhead is proportional to the number of BS antennas. With large antenna arrays at the BSs, this would therefore always impose severe limitations on the size of the BS antenna arrays [67].

If channel training and DL transmission is organized in TDD, i.e. when the receivers transmit pilots on the DL frequency, channel reciprocity can be utilized and the BS can estimate the channels. Thereby, the differences in the transfer characteristics of the transceiver chains, amplifiers, and filters in the two directions have to be accounted for. This can however be handled by measuring the different gains in each direction [67]. With this reverse channel estimation, the overhead is independent of the number of BS antennas but in turn limited by the number of users or receivers. If the BSs are equipped with large antenna arrays, with more antennas than users, TDD channel training is thus the only way to acquire timely CSI for rapidly changing channels. UL training eliminates the need for feedback and together with channel reciprocity it is sufficient to provide the BSs with the desired CSI.

When multi-cell scenarios are considered, the training sequences transmitted by the users have to be separated. The use of orthogonal pilots across many cells is thereby infeasible when the number of users is large. Orthogonal pilots would need a number of symbols that is at least as large as the total number of users. Particularly short coherence times due to mobility do not allow for such long training phases. Accordingly, different users need to reuse the same pilot symbols. This, however, causes pilot contamination as the BSs receive the same pilots from different users. In a simple scenario with two users that use the same pilot symbol, the estimate of the channel $\mathbf{h}_1$ of the first user would be [67]

$$\hat{\mathbf{h}}_1 = c_1\mathbf{h}_1 + c_2\mathbf{h}_2 + c\mathbf{w}, \tag{7.4}$$

where $\mathbf{h}_2$ is the channel of the second user, c_1, c_2, and c are constants that depend on the propagation conditions and transmit powers, and $\mathbf{w}$ is additive noise. As a consequence, the channels between two users cannot be separated anymore and the CSI is corrupted by $\mathbf{h}_2$. In order to keep the effect of pilot contamination small, one can introduce a pilot reuse factor $\beta \leq 1$ similar to the frequency reuse factor of classical cellular networks. This reuse factor describes the fraction of the number of cells that can use the same pilots. By applying a small reuse factor, users with the same pilots are further apart from each other and the CSI gets more accurate. This in turn increases the overhead, as many orthogonal resources have to be allocated to separate the channel estimation from users that are close to each other. If B pilot symbols are available, the channels to $K = \beta \cdot B$ MS antennas can be estimated in one cell. Thereby, B has to be significantly smaller than the block length of the transmission such that the overhead over the actual user data does not reduce the data rates significantly.

For a given system with specific B and β, the number of users that can be served simultaneously in each cell is thus limited. Depending on the coherence time of the channels, a certain update rate of the CSI is necessary. In the following, we illustrate the number of required channel updates for a moving user. A mobile user that is served directly by the BS moves with a velocity v. By the rule of thumb given by [113], we can specify the coherence time T_c of the channel to this user by

$$T_c = \sqrt{\frac{9}{16\pi}} \cdot \frac{1}{f_d}, \tag{7.5}$$

where

$$f_d = f_c\frac{v}{c} \tag{7.6}$$

is the maximum doppler shift that can result from the velocity v (if the MS is moving directly towards the BS) with the carrier frequency f_c and the speed of light c. When we assume that the CSI and the precoding at the BS is updated once within each time frame of length T_c, the update frequency is as shown by the dashed blue line in Fig. 7.5. For the evaluation, the carrier frequency was set to $f_c = 2.6\,\mathrm{GHz}$. When the user moves with 20 km/h, at least 100 updates per second are required to track the channel, with speeds above 120 km/h, this number increases towards 1000. The maximum number of users that can be served by this BS is thus limited by T_c, the number of orthogonal pilots B and the reuse factor β.

When we now consider the relay carpet where the users are not served by the BS directly but via relays, the required number of CSI and precoding updates can be lowered significantly. If the BS precoding is performed only with respect to the channels to the relays and these channels remain constant, a precoding update is only required when a user switches relay cells. When a user traverses a relay cell of diameter d_{rc}, the time in which the user is in this cell is given by

$$T_{\mathrm{rc}} = \frac{d_{\mathrm{rc}}}{v}. \tag{7.7}$$

An update of the BS precoding is thus only required once in a time frame of length T_{rc} when the relays are at fixed positions and the channels between the BS and the relays remain constant. The resulting update frequencies are shown as solid lines for different relay cell diameters d_{rc}. When the relays are 5 m apart from each other, which leads to rather small relay cells, and the user moves with 20 km/h, the number of updates is decreased from 100 to 1 per second as compared to the case where the channel between the BS and the MSs is tracked directly. With larger relays cells (50 m) this number drops further by one decade to one update in 10 seconds.

When we assume that a system is designed for a certain amount of channel and precoding updates, we can argue that the number of supported users within a cell can be increased by a factor of 100 or 1000 when the users are served via relays instead of by the BS directly. The number depends thereby on the size of the relay cells. Placing relays between 5 and 50 m apart from each other leads to a reasonable realization of the relay carpet with a massive deployment of relays. The analysis presented here is however only valid if the channels between the BS and its relays remain constant between two update time instances. If the moving users themselves lower the coherence time of the BS-to-relay channel, e.g. when they appear as moving scattering objects, more updates might be required. When the relays are mounted at positions, e.g. a couple of meters above street level, with good connection to the BS, we can hope

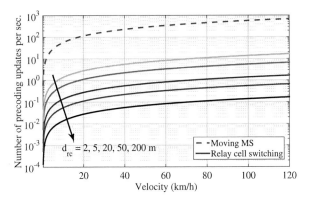

Figure 7.5.: Number of required channel updates when a user moves with a certain velocity.

that the moving users do not affect the BS-relay channels significantly. The coherence time in the relay carpet can also be reduced when mobile users do not traverse the relay cells but follow a path that leads to a more frequent relay cell switching. A user can e.g. zig-zag such that it often switches back an forth between two cells. In this case, however, the BS does not necessarily need to re-estimate the channel each time. When the channels to the relays remain unchanged over a longer time, the same pre-calculated BF matrices can be reused. Additionally, when the relay cells are small and a user moves with a high velocity, the coherence time for relay cell switching can also be prolonged by assigning multiple relays to the same moving user as already indicated above. The relay carpet thus allows for a drastic increase in the coherence time as compared to the case where massive MIMO BSs serve their users directly. The number of supported users can increase by orders of magnitude. The ubiquitous deployment of relays in the relay carpet concept is therefore an enabler for massive MIMO with accurate and timely channel estimation.

When more sophisticated type B relays are applied, they also need CSI to calculate their receive and transmit filters. The relays are however equipped with only a small number of antennas. The channel estimation does therefore not change as compared to conventional systems where no massive MIMO is applied. For the receive filters, only the covariance matrix of the channels from the BSs $\Gamma_{c,k}$ needs to be estimated. The size of this matrix is however independent of the number of BS antennas. For the transmit MF applied to the relays, the channel between a single relay and MS is sufficient. Applying massive antenna arrays at the BSs does therefore also not affect the estimation of these channels.

In the following, we study the scaling behavior of the relay carpet with respect to achievable rates when the number of relays increases. To this end, we first define the system model in the next section. Thereby, we assume that the CSI can be obtained accurately and in time and consider type A relays with the implementation shown in this chapter that transmit with a fixed amplification gain.

7.2 Network Model

The organization of the network and the system model is similar to the one introduced in Chapter 6. The area is divided into C cells, each with one BS that serves multiple MSs. For notational simplicity, we assume that all cells have M active MSs and that all nodes of the same kind have the same number of antennas, although a generalization is straightforward. The number of antennas at the BSs is denoted by N_B, the one of the MSs by N_M. In the DL, BS c, with $c \in \{1, \ldots, C\}$, wants to transmit $d_s \leq N_M$ data streams to MS (c, j) (the jth MS in cell c). The communication is assisted by $K \geq M$ relays, such that each relay serves one MS but a MS can be served by multiple relays, e.g. to avoid many handovers when users are moving. The relays are equipped with N_R antennas, where $N_M \leq N_R \leq N_B$. The channel between BS b and relay (c, k) is denoted by $\mathbf{H}_k^{(c,b)} \in \mathbb{C}^{N_R \times N_B}$, the one between relay (c, k) and MS (b, j), possibly in a different frequency band, by $\mathbf{F}_{j,k}^{(b,c)} \in \mathbb{C}^{N_M \times N_R}$. Direct links between BSs and MSs are not considered.

The transmit symbol vector from BS c intended for MS (c, j), denoted by $\mathbf{s}_{c,j} \in \mathbb{C}^{d_s}$, is premultiplied by the corresponding beamforming matrix $\mathbf{Q}_{c,j} \in \mathbb{C}^{N_B \times d_s}$. The receive signal of relay (c, k) can thus be written as

$$\mathbf{r}_{c,k} = \sum_{b=1}^{C} \mathbf{H}_k^{(c,b)} \sum_{j=1}^{M} \mathbf{Q}_{b,j} \mathbf{s}_{b,j} + \mathbf{n}_{c,k}, \qquad (7.8)$$

where $\mathbf{n}_{c,k}$ is the relay noise. The relays multiply their receive signal (7.8) with a gain matrix $\mathbf{G}_{c,k} \in \mathbb{C}^{N_R \times N_R}$ and, after a frequency conversion, retransmit $\mathbf{t}_{c,k} = \mathbf{G}_{c,k} \cdot \mathbf{r}_{c,k}$.

With $\mathbf{w}_{c,k}$ being the noise induced in MS (c,k), its receive signal is

$$\mathbf{y}_{c,k} = \sum_{b=1}^{C} \sum_{j=1}^{K} \mathbf{F}_{k,j}^{(c,b)} \cdot \mathbf{G}_{b,j} \cdot \mathbf{r}_{b,j} + \mathbf{w}_{c,k}. \tag{7.9}$$

The BSs are able to accurately estimate the channels to the relays within their cell. As before, we apply ZF with waterfilling at the BSs, but also other techniques are possible. With this specific choice, the BSs only have to track the quasi-static channels to their relays and can cancel the interference at all relays within their cell. For the acquisition of the required transmit CSI, the relays have to enable channel estimation at their BS. To this end, the relays can transmit training sequences triggered by a request on the control channel. When the channel to the BS is quasi-static, this is required only on a slow time scale. Note that with block ZF, only the interference between the BSs and their in-cell relays is cancelled. The remaining interference is further reduced by the relay filters if the type B architecture is used and/or when power control is applied.

For given BS precoding and relay gain matrices, the achievable rate for MS (c,k) can be calculated by

$$R_{c,k} = \frac{1}{2} \log_2 \det \left(\mathbf{I}_{N_\mathrm{M}} + \left(\mathbf{K}_{c,k}^{(\mathrm{i+n})} \right)^{-1} \cdot \mathbf{K}_{c,k}^{(\mathrm{sig})} \right), \tag{7.10}$$

with the covariance matrix of the desired signal

$$\mathbf{K}_{c,k}^{(\mathrm{sig})} = \sum_{b,j} \sum_{b',j'} \mathbf{F}_{k,j}^{(c,b)} \mathbf{G}_{b,j} \mathbf{H}_{j}^{(b,c)} \mathbf{Q}_{c,k} \mathbf{Q}_{c,k}^{\mathsf{H}} \mathbf{H}_{j'}^{(b',c)\mathsf{H}} \mathbf{G}_{b',j'}^{\mathsf{H}} \mathbf{F}_{k,j'}^{(c,b')\mathsf{H}} \tag{7.11}$$

and of the interference and noise

$$\mathbf{K}_{c,k}^{(\mathrm{i+n})} = \mathsf{E}\left[\mathbf{y}_{c,k} \cdot \mathbf{y}_{c,k}^{\mathsf{H}} \right] - \mathbf{K}_{c,k}^{(\mathrm{sig})}. \tag{7.12}$$

Note that the prelog factor $1/2$ in (7.10) could be excluded if we considered a second hop in a frequency band that is unlicensed or unused at the moment.

In the UL, the achievable sum rate with joint decoding of each cell is given by

$$\overleftarrow{R}_{c,k} = \frac{1}{2} \log_2 \det \left(\mathbf{I}_{N_\mathrm{M}} + \left(\overleftarrow{\mathbf{K}}_c^{(\mathrm{i+n})} \right)^{-1} \cdot \overleftarrow{\mathbf{K}}_c^{(\mathrm{sig})} \right), \tag{7.13}$$

with

$$\overleftarrow{\mathbf{K}}_c^{(\text{sig})} = \sum_{k=1}^{M} \sum_{b,j} \sum_{b',j'} \mathbf{F}_{k,j}^{(c,b)} \mathbf{G}_{b,j} \mathbf{H}_j^{(b,c)} \mathbf{Q}_{c,k} \mathbf{Q}_{c,k}^{\mathsf{H}} \mathbf{H}_{j'}^{(b',c)\mathsf{H}} \mathbf{G}_{b',j'}^{\mathsf{H}} \mathbf{F}_{k,j'}^{(c,b')\mathsf{H}} \qquad (7.14)$$

$$\overleftarrow{\mathbf{K}}_c^{(\text{i+n})} = \sum_{\substack{d=1 \\ d \neq c}}^{C} \sum_{k=1}^{M} \sum_{b,j} \sum_{b',j'} \mathbf{F}_{k,j}^{(c,b)} \mathbf{G}_{b,j} \mathbf{H}_j^{(b,c)} \mathbf{Q}_{d,k} \mathbf{Q}_{d,k}^{\mathsf{H}} \mathbf{H}_{j'}^{(b',c)\mathsf{H}} \mathbf{G}_{b',j'}^{\mathsf{H}} \mathbf{F}_{k,j'}^{(c,b')\mathsf{H}} \qquad (7.15)$$

$$+ \sigma_n^2 \sum_{b,j} \mathbf{F}_{k,j}^{(c,b)} \mathbf{G}_{b,j} \mathbf{G}_{b,j}^{\mathsf{H}} \mathbf{F}_{k,j}^{(c,b)\mathsf{H}} + \sigma_w^2 \mathbf{I}_{N_\mathrm{B}} \qquad (7.16)$$

the corresponding covariance matrices for the desired signal and interference plus noise.

7.3 Scaling Behavior

In the following, we study the performance of the relay carpet when multiple relays are assigned to the MSs. To this end, we apply the same simulation environment as described in Chapter 6, where each cell contains six relay cells. Here, however, there are a single MS and K_m relays placed randomly in each relay sector. A deadzone of radius 200 m is applied around the BS. We apply the simple type A relays with fixed amplification gains $\alpha_{c,k}$ which are obtained in the following way: the pathloss of the channel between the relay and its corresponding BS is compensated and multiplied by a factor, i.e. the relay gains are

$$\alpha_{c,k} = L_{c,k} \cdot \gamma_{c,k}, \qquad (7.17)$$

where $L_{c,k}$ is the pathloss between the BS of cell c and the k-th relay in this cell and $\gamma_{c,k}$ is the scaling factor that we choose to be in the range of -20 to $+20$ dB.

In Fig. 7.6, we study the performance in a seven cell setting where six cells are arranged in a circle around the cell of interest whose average sum rate is shown. Each cell has a diameter of 1000 m and consists of $M = 6$ relay cells, each with one MS. In each of these relay cells there are $K_m = 3$ relays randomly placed and the number of antennas at the BS, relays, and MSs are $N_\mathrm{B} = 24$, $N_\mathrm{R} = 2$, and $N_\mathrm{M} = 2$, i.e. the number of BSs antennas suffices to apply ZF to at most $K_1 + \cdots + K_6 = 12$ relays when $d_\mathrm{s} = 2$ data streams have to be transmitted to each MS. For the simulations, 1, 2, or 3 relays are chosen for each MS, first the closest one, then the two closest and

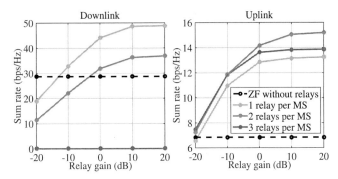

Figure 7.6.: Sum rates for increasing number of active relays per MS. Six cells are arranged in a circle around the cell of interest.

all three of the relay cell. The relays that are not activated remain silent and are not considered in the BS beamforming.

On the left hand side of Fig. 7.6, where the DL rates are shown, we can observe that using only a single relay for each MS leads to the best performance, i.e., without an appropriate scheme to handle multiple relays, the achievable rates drop when additional relays are activated. This has two reasons: firstly, the relays inject additional noise and interference into the system which degrades the performance and secondly, the number of BS antennas is fixed to $N_B = 24$, which leaves less room for optimization with waterfilling when more relays are active. With $K_m = 3$ relays per MS, the rates even drop to zero, as the ZF beamforming at the BSs does not have enough spatial degrees of freedom. In this case, a different precoding scheme should be applied, such as e.g. a leakage based approach as described in Chapter 8. In the DL, only the relay that is closest to a user should be activated and forward signals to it. With $K_m = 1$ or 2 relays, the performance improves with higher amplification gains until it saturates with $\gamma_{c,k} = +10\,\text{dB}$ due to the interference that is also increased with the signal powers.

In the UL shown on the right in Fig. 7.6, however, the behavior with more relays looks different. If only a single cell is considered, all signals involved therein contribute to the desired signal when the BS can jointly decode all UL signals. In this case, all active relays collect and forward signals that are all useful for the BSs, except the noise, and no interference is present. The relays are thus rather energy collectors than noise and interference generators as in the DL and the sum rate of the UL strictly increases. In Fig. 7.6, the performance with two relays per MS is thus better than with only a single relay per MS. But as there are seven cells considered in this simulation setup, the performance saturates and even drops again when more relays are used due to the interference the relays of adjacent cells cause. Hence, dense networks with many cells (or BSs) also require appropriate techniques to mitigate the inter-cell interference. With type B relays that can reduce this interference, a better behavior can be expected.

In the cases discussed above, the BS antenna arrays consist of a fixed number of $N_B = 24$ antennas, regardless of how many relays are active. This implies that not enough spatial degrees of freedom are available to apply the ZF precoding when more than 12 relays are active in a cell. In the following, we look at the scaling behavior when the number of BS antennas grows proportionally with the number of relays and focus on the DL only. In Fig. 7.7, we consider only a single BS cell and vary the number of relays per MS from $K_m = 1$ to $K_m = 10$ for different amplification gains. All relays are placed randomly in their respective relay cell. Two relay selection schemes are considered. In the first case (relay selection) only one out of the K_m relays is chosen, the one that is closest to the MS, and the others are deactivated. In the second case (relay combination) all relays are active and are jointly served by the BS. The number of BS antennas grows with the number of relays according to $N_B = K_m \cdot M \cdot N_R$.

We can observe that the performance continuously improves with the number of relays when the closest relay is selected. With more than two relays to choose for each MS, the rates are higher than the reference case in which the BS serves the MSs directly with ZF beamforming, even with the very small relay gain of -20 dBm. As already seen before, relay gains that are higher than the compensated pathloss do not

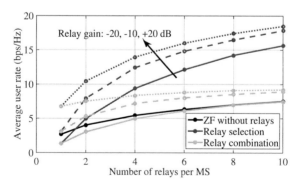

Figure 7.7.: Average user rates for increasing number of relays per MS. Only a single cell is considered.

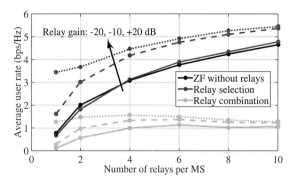

Figure 7.8.: Average user rates for increasing number of relays per MS. Six cells are arranged in a circle around the cell of interest.

lead to a further improvement as it saturates due to the interference limitedness that arises from the relay interference to other MSs. When all relays are used jointly, the improvement with the number of relays saturates much quicker. Moreover, relatively high relay gains are required to achieve rates that are better than the reference case without relays.

This effect is even more pronounced in Fig. 7.8, where an additional ring of seven cells around the cell of interest is considered. In this case, increasing the number of relays leads to lower rates when more than 4 relays are active for each MS. The overall rates are also reduced considerably due to the additional interference caused by the surrounding cells. With relay selection however, a similar scaling behavior can be observed as in the case before.

7.3.1 Improvements for the Relay Carpet

It is thus important to apply adequate measures against the additional interference and the amplified noise the relays inject into the system. Otherwise, the benefits of higher signal power, better coverage and simplified channel estimation are mitigated or even destroyed when many low-complexity relays are deployed. On one hand, the beamforming can be improved. As shown by the results in Chapter 6, the approach of block ZF at the BSs can already offer good results when the same number of relays is active as MSs are served and when no transmit cooperation at the BSs is applied. If mobile users are served by multiple relays and if the number of relays within a cell can grow, e.g. when additional relays are installed, the ZF approach seems however to be too inflexible and can lead to less performance gains as compared to conventional cellular networks. A potentially more promising and flexible approach can be provided be a leakage based scheme such as SLNR precoding. This approach has the advantage that no requirements on the number of antennas have to be fulfilled and that it can easily be applied to transmit cooperation between multiple BSs. For this, no fixed cooperation clusters need to be applied as in the block ZF approach, but dynamic cooperation clusters can be obtained, such that each MS is served by its own, possibly overlapping, set of BSs. This approach is developed and studied in Chapter 8.

A second problem of the approach discussed above is the choice of the relay amplification gains. As the relays inject additional and amplified noise into the system and cause additional interference to other users, the relay gains have to be chosen appropriately that the network does not become interference limited. It is thus important to apply power control to the active nodes in order to control the interference in the network. Moreover, with power control the network performance can be further improved as it provides an efficient means for optimization of different objective functions. We will thus develop power control algorithms that optimize different aspects of the relay carpet network. For the time being, the BS transmission schemes remain as defined in this chapter but we extend the schemes to include also power control at the transmitting nodes. An alternative beamforming scheme and an extension to transmit cooperation is discussed in the next chapter.

7.4   Power Control

Without power control, all nodes are enforced to transmit with full power. Even with the interference reduction of the BS/relay processing, the relays forward residual interference to other users. Optimization of the power allocation can thus offer further improvements as the remaining interference can be controlled. Additionally, it can lead to savings regarding energy consumption and gains in terms of QoS or outage probability. Such power control schemes are studied e.g. in [70] for pure relay channels or in [22,95] for traditional cellular networks without relays. However, these schemes cannot directly be applied to the network considered here, as the BS and relay powers of multiple links are coupled across different cells. This schemes would result in situations in which the BS and relay power optimization block each other and do not converge to a solution. In the following, we outline low-complexity power control algorithms for different objectives that guarantee convergence and show that the low-complexity relays offer large gains regarding power savings and coverage. Thereby, we focus on two objectives that we have already introduced and applied before: minimize the required transmit power to achieve a certain target rate and to maximize the minimum rate within a certain area to achieve the maximum fairness among all users therein. Furthermore, we also apply an adaptation of the latter that attempts to maximize the coverage. Thereby, not the performance of the worst user is maximized, but rather the number of users that can fulfill a target rate. With this, we can avoid cases in which a bad data rate of one user can worsen the rates of all others such that none of them can achieve the target anymore. To evaluate the power control schemes, we apply the same simulation settings as in Chapter 6.

7.4.1 Minimize Power

The first goal is to minimize the required transmit power to achieve a target rate $R^\star$ at each of the MSs. To this end, the scaling factor $\alpha_{c,k}$ from (7.1) at relay (c,k) can be adjusted. On the BS side, we assign a scaling factor $\beta_{c,k} > 0$ for the signal to each MS in the corresponding cell, i.e. the beamforming matrix of the signal from BS c to MS (c,k) is $\mathbf{M}_{c,k} = \sqrt{\beta_{c,k}} \cdot \mathbf{Q}_{c,k}$. The objective is now to minimize both the relay and BS transmit powers by adjusting $\alpha_{c,k}$ and $\beta_{c,k}$ such that the MSs achieve the target rate $R^\star$.

As in other power minimization problems, there are situations in which no feasible power allocation exists due to the stringent rate constraints [95]. Additionally, feasible scenarios can lead to transmit powers that are too high for practical systems with regulatory restrictions. We thus introduce a maximal power at each node that must not be exceeded: a maximal power $P_{\text{B,max}}$ and $P_{\text{R,max}}$ is assigned to the BSs and relays. These powers are then minimized in an alternating fashion.

Relay Power Minimization

Assuming that the beamforming matrices of the BSs as well as the relay gain matrices of all surrounding cells are fixed, the relays of a cell of interest c can iteratively optimize their transmit powers. The relays initialize their scaling factors according to $P_{\text{R,max}}$ by setting

$$\alpha_{c,k}^{(0)} = P_{\text{R,max}} \big/ \text{Tr} \left\{ \mathsf{E} \left[\mathbf{G}_{c,k} \mathbf{r}_{c,k} \cdot \mathbf{r}_{c,k}^{\mathsf{H}} \mathbf{G}_{c,k}^{\mathsf{H}} \right] \right\}, \quad \forall k. \tag{7.18}$$

Then, in iteration step $n = 0, 1, \ldots$, the relay with the highest rate $R_{c,l}^{(n)}$ at the corresponding MS is identified and the power of this relay is updated according to

$$P_{\mathrm{R},c,l}^{(n+1)} = P_{\mathrm{R},c,l}^{(n)} - \mu P_{\mathrm{R},c,l}^{(n)} \left(1 - \frac{R^\star}{R_{c,l}^{(n)}} \right), \tag{7.19}$$

where μ is a step size parameter that has to be small enough that the resulting rate cannot fall below $R^\star$. This can be realized by a backtracking line search. The update equation (7.19) reduces the power based on the ratio of the desired and the actual rate and guarantees that there is no change as soon as the target rate is achieved. The relay gain matrix is then scaled by

$$\alpha_{c,l}^{(n+1)} = P_{\mathrm{R},c,l}^{(n+1)} / \mathrm{Tr} \left\{ \mathsf{E} \left[\mathbf{G}_{c,l} \mathbf{r}_{c,l} \cdot \mathbf{r}_{c,l}^{\mathsf{H}} \mathbf{G}_{c,l}^{\mathsf{H}} \right] \right\}. \tag{7.20}$$

These steps are repeated as long as there are rates that exceed $R^\star$ by more than some tolerance ε. In each step, the relay transmit power is reduced and the interference for all other MSs is strictly decreased and their rates improved. Therefore, the algorithm converges and any further change in the scaling factors $\alpha_{c,k}$ cannot reduce the transmit power without letting a rate fall below $R^\star$. The solution is thus a local optimum. The algorithm, however, does not necessarily lead to a solution in which all rates are higher than $R^\star$ (e.g. if $R^\star$ is set too high for the interference scenario). After the optimization, some MSs with poor initial rates might stay in outage, i.e. their rates remain smaller than $R^\star$.

BS Power Minimization

Similar to the optimization of the relay powers, the transmit power of BS c can also be minimized. To this end, the power allocated to each beamforming matrix $\mathbf{Q}_{c,k}$ is controlled while the relay gain matrices are fixed. Starting with equally allocated power, i.e.

$$\beta_{c,k}^{(0)} = \frac{P_{\mathrm{B,max}}}{M \cdot \mathrm{Tr} \left\{ \sum_{i=1}^{M} \mathbf{Q}_{c,i} \cdot \mathbf{Q}_{c,i}^{\mathsf{H}} \right\}}, \quad \forall k, \tag{7.21}$$

the highest rate $R_{c,l}$ in the selected cell c is identified and the corresponding power is reduced according to

$$\beta_{c,l}^{(n+1)} = \beta_{c,l}^{(n)} - \frac{R_{c,l}^{(n)} - R^\star}{m}. \tag{7.22}$$

Here, the step size can be chosen as

$$m = \frac{1}{\ln(2)} \cdot \mathrm{Tr} \left\{ \left(\mathbf{K}_{c,l}^{(i+n)} \right)^{-1} \mathbf{K}_{c,l}^{(sig)} \right\}, \qquad (7.23)$$

which corresponds to the derivative of $R_{c,l}$ evaluated at $\alpha_{c,l} = 0$. This step size is thus an upper bound on the slope of $R_{c,l}$ with respect to $\alpha_{c,l} > 0$ and guarantees that the resulting rate after the update cannot fall below $R^\star$.

Alternating Optimization

We can now combine both schemes such that the BS and the relay powers are minimized and the algorithm can be extended to the whole network. When the BS power is optimized in one cell, lowering the relay powers within this same cell cannot lead to further improvements, as MSs that already achieve $R^\star$ could fall below that value. In order to guarantee convergence, the algorithm is extended to all (or a cluster of) cells. Running the BS optimization once in each cell offers a potential to optimize the relays in a second turn, as the rates are further increased by the lower interference of the neighboring cells. The relay powers can thus be further reduced and we can again iterate over the relay and BS optimization until all MSs achieve $R^\star$ within the tolerance ε or are in outage as no further improvement is possible. The alternating procedure is summarized in Algorithm 1. The order of the BS and relay power optimization can also be swapped.

Algorithm 1 Minimize power

1: Initialization: $P_{B,c,i} = P_{B,max}$, $P_{R,c,i} = P_{R,max}$, $\forall c, i$
2: **while** some $R_{c,i} > R^\star + \epsilon$ **do**
3:    **for** $c = 1 : C$ **do**
4:       **while** $\exists l : R_{c,l} > R^\star + \epsilon$ **do**
5:          update $\beta_{c,l}$ according to (7.22), calculate $R_{c,i}$, $\forall i$
6:       **end while**
7:    **end for**
8:    **for** $c = 1 : C$ **do**
9:       **while** $\exists l : R_{c,l} > R^\star + \epsilon$ **do**
10:         update $\alpha_{c,l}$ according to (7.20), calculate $R_{c,i}$, $\forall i$
11:       **end while**
12:    **end for**
13: **end while**

The optimization can be realized in a distributed way, where each node updates its scaling factor itself, or centralized at the BSs. In the latter case, only signal covariance matrices need to be fed back to the BSs. After computation, the relays can then be informed about their scaling factors via their control channel. This does not increase the signaling overhead significantly, as similar feedback and control signals are already included in current systems. If the optimization is distributed among different nodes, communication to exchange the necessary information between them would be required, which might introduce additional overhead.

7.4.2   Maximize Minimum Rate

In order to achieve fairness across the users, power control can also be applied to maximize the minimum rate under a sum transmit power constraint. To this end, Algorithm 1 is adapted such that the transmit power for the strongest MS is not only reduced, but transferred to the weakest user in the cell. For the derivation of this scheme, we focus on BS power control.

As before, the transmit power at BS c is equally distributed among all users of this cell. Then, in iteration step n, the MS (c, j) that achieves the lowest rate $R_{c,j}^{(n)} = R_{\min}^{(n)}$ and the one that achieves the highest rate $R_{c,l}^{(n)}$ are identified. The power allocated to MS (c, l) is then reduced by updating

$$\beta_{c,l}^{(n+1)} = \beta_{c,l}^{(n)} - \frac{R_{c,l}^{(n)} - R_{\min}^{(n)}}{m}, \qquad (7.24)$$

in which m as in (7.23) guarantees $R_{\min}^{(n+1)} \geq R_{\min}^{(n)}$. The updated BS power is

$$P_{B,c,l}^{(n+1)} = \beta_{c,l}^{(n+1)} \cdot \frac{P_{B,\max}}{M} \qquad (7.25)$$

and the saved power $\Delta P = P_{B,c,l}^{(n)} - P_{B,c,l}^{(n+1)}$ can be allocated to the weakest user (c, j) according to

$$\beta_{c,j}^{(n+1)} = \left(P_{B,c,j}^{(n)} + \Delta P \right) \cdot \frac{M}{P_{B,\max}} \qquad (7.26)$$

in order to scale the corresponding beamforming matrix. These steps can be repeated until all rates in the cell are equal within a tolerance ε. The scheme summarized in Algorithm 2 can also be applied to the relays when a sum transmit power constraint among all relays of the same cell is imposed.

Algorithm 2 Maximize minimum rate

1: Given: desired cell c
2: Initialization: $P_{\mathrm{B},c,i} = P_{\mathrm{B,max}}/M$, $\forall i$
3: **while** $\left(\max_{l}(R_{c,l}) - \min_{j}(R_{c,j}) \right) > \epsilon$ **do**
4:     identify strongest and weakest user with $R_{c,l}$ and $R_{c,j}$
5:     update $\beta_{c,l}$ according to (7.24)
6:     update $\beta_{c,j}$ according to (7.26)
7:     calculate new rates $R_{c,i}$, $\forall i$
8: **end while**

7.4.3   Outage Reduction

While the max-min algorithm attempts to make the rates equal, resulting in maximal fairness, it can happen that a single user with very poor conditions can lower all other (possibly much higher) rates to a value that is not useful anymore for acceptable communication. To avoid this, we can slightly modify Algorithm 2 such that the probability that an MS is in outage is reduced.

To this end, we can proceed similar as in Algorithm 2 but with a different power allocation update. Initially, the power $P_{\mathrm{B,max}}$ is equally distributed for each MS. Then, the power for the MS with the highest rate $R_{c,l}$ is reduced as in (7.22). The step size m guarantees $R_{c,l}^{(n+1)} \geq R^{\star}$, which is important to not get an additional outage. The saved power of the strongest user,

$$\Delta P = P_{\mathrm{B},c,l}^{(n)} - P_{\mathrm{B},c,l}^{(n+1)}, \tag{7.27}$$

is allocated to MS (c, j) with

$$j = \arg\min_{j} \left(R^{\star} - R_{c,j} \right)^{+}, \tag{7.28}$$

where $(\cdot)^{+} = \max\{0, \cdot\}$, i.e. the MS whose rate is *closest below* $R^{\star}$. The power for this MS is updated as in (7.26) and the steps are repeated as long as there exists an l such that $R_{c,l} > R^{\star} + \varepsilon$. This scheme is summarized in Algorithm 3.

7.4.4   Evaluation of Power Control

For the evaluation of the power control algorithms, we consider a network of the same form as described in Chapter 6. The network consists of $C = 7$ regularly arranged

---

Algorithm 3 Minimize outage

1: Given: desired cell c
2: Initialization: $P_{B,c,i} = P_{B,max}/M$, $\forall i$
3: **while** $\exists l : R_{c,l} > R^\star + \epsilon$ **do**
4:    Identify strongest user with $R_{c,l}$ and user with $R_{c,j}$ closest below $R^\star$
5:    update $\beta_{c,l}$ and $\beta_{c,j}$ according to (7.22) and (7.26)
6:    calculate new rates $R_{c,i}$, $\forall i$
7: **end while**

---

hexagonal cells. The BSs are located in the center of each cell and the distance between adjacent BSs is 1000 m. The number of MSs and relays in each cell is $M = K = 6$, where the relays are regularly placed and the MSs randomly. Each MS, relay, and BS is equipped with $N_M = 2$, $N_R = 4$, and $N_B = 24$ antennas. All antennas are omnidirectional and we apply the usual channel model with Rayleigh fading, pathloss, and shadowing according to the urban scenario of the WINNER II model. Assuming a system bandwidth of 100 MHz and a noise figure of 5 dB, the noise variances are $\sigma_n^2 = \sigma_w^2 = 5 \cdot 10^{-12}$ W and, if not stated otherwise, the maximal allowed transmit powers are $P_{B,max} = 40$ W and $P_{R,max} = 6$ W. The target rate is $R^\star = 1$ bps/Hz. In the following simulations, we exclude the prelog factor $1/2$ in (7.10), as it would occur with half-duplex in-band relays, as we are mainly interested in the performance of power allocation. Moreover, as discussed before, we can consider a second hop in a frequency band that is unlicensed or unused at the moment. Otherwise, the achievable rates with relaying have to be divided by two to obtain the two-hop in-band rates.

Figure 7.9 shows the empirical CDFs of achievable user rates in the center cell and the required sum transmit power (BS plus relay power) allocated for one user when Algorithm 1 is applied. The performance of the relay carpet is also compared to the case of a conventional network without relays where the same power control scheme is applied. The BSs perform block ZF on the direct channels to the mobiles within their cell. The conventional network is shown once with the BS transmit power $P_B = 40$ W and once with the power $\tilde{P}_B = P_B + K \cdot P_R = 76$ W, which corresponds to the same sum power of a BS and the relays together.

In the upper plot (CDF of the user rates), we can observe that not all users achieve the target rate $R^\star = 1$ bps/Hz. With a certain probability, some users are in outage. This outage probability can however significantly be lowered with the help of the relays. While the reference without relays has an outage probability of 43 % and 35 % with increased power, the use of simple type A relays already leads to an outage probability of 33 % when no power control is applied. With power control, this value is reduced

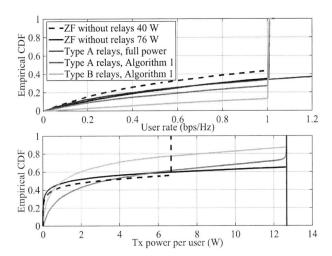

Figure 7.9.: Empirical CDFs of instantaneous user rate and the required transmit power (BS plus relay power) allocated to the transmission for one user.

to about 25 % or to 12 % with type B relays. Also the required transmit power is reduced to a large extent. When no power control is applied, each user is served with a fixed power of 12.67 W. With power control, this value is only required when the corresponding user is in outage. In all other cases, the transmit power can be reduced. With the help of relays, the power required to achieve the target can be lowered down to 1 W with type B relays. On average, these relays achieve an user rate of 0.94 bit/channel use and the average power per user is 3.51 W, whereas the conventional network with $P_{B,max} = 76$ W (same sum power as BS and relays together) achieves an average user rate of 0.80 bit/channel use and require a mean power of 5.38 W per user. By the help of the relays, the power consumption as well as the outage probabilities can thus be improved.

Algorithm 2, that attempts to maximize the minimum rate, is considered in Fig. 7.10 which shows the empirical CDF of the minimal rate within the cell of interest. Also here, the relay carpet shows a significantly better behavior than the conventional network without relays, even for the relays that do not have CSI (type A). With type B relays, the minimal rates are even more increased.

Also shown in the plot is a comparison of the minimal rates when the network with type A relays is optimized by a gradient search method, within a single cell as well

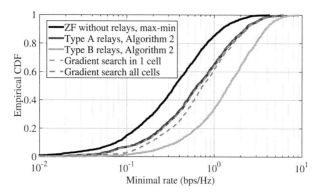

Figure 7.10.: Empirical CDF of the minimal rates after max-min optimization with Algorithm 2. For comparison, also the rates with optimization with a gradient search for type A relays are shown.

as jointly over all cells. The gradient based optimization, providing an upper bound, is similar to the one described in Chapter 3 and attempts to solve the optimization problem

$$\max_{\alpha_{c,k}} \min_{l} \{R_{c,l}\} \qquad \text{s.t.} \quad \sum_{k=1}^{K} P_{\mathrm{R},c,l} = P_{\mathrm{R,tot}}, \quad \forall c, \tag{7.29}$$

where the gradient of the achievable rate $R_{c,l}$ is calculated with respect to $\alpha_{c,i}$. Alternatively, the algorithm can also be applied to optimize the BS power allocation by optimizing the factors $\beta_{c,i}$. Note that the optimization can either be performed in each cell separately, or jointly over multiple cells. The performance of the gradient optimization is identical to the scheme proposed in Algorithm 2 and the extension to the whole network does not lead to large improvements. When it is applied in each cell separately, the interference between the users is already decreased significantly.

In Fig. 7.11, we study the performance of Algorithm 3 that attempts to minimize the outage probability of the users. We can also observe here that the relay carpet concept leads to a much lower outage probability and thus offers a significant improvement compared to conventional cellular networks. The algorithms can be further improved, when MSs achieving very low rates are discarded and the power for them is assigned to the remaining MSs.

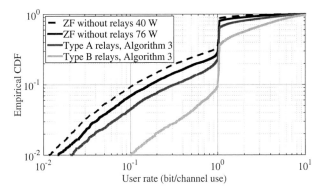

Figure 7.11.: Empirical CDF of the user rates after outage reduction with Algorithm 3.

7.5 Conclusions

The relay carpet with its simple transmission and power control schemes presented in this an the previous chapter offers a significant performance gain as compared to a conventional multi-user MIMO approach, not only in terms of achievable rates but also with regard to power savings and outage behavior. By considering Fig. 6.16, we see that the performance of relaying scales beneficially with the node density, particularly with the simple type A AF relays. With these relays, many users can be served by each BS and high data rates can be achieved. When power control is applied, further gains can be achieved and less power is required for a good QoS in the entire network. By using simple frequency conversion relays, the two-hop concept for cellular networks does not lead to additional delays and the relays can be implemented in a low-complexity and inexpensive fashion. A massive deployment of relays can thus act as an enabler of massive MIMO at the BSs and combines sophisticated multi-user beamforming with small cells. We thus consider the relay carpet to be a promising concept for cellular networks that can improve their performance by the factors required for future generations.

Nevertheless, the transmission schemes applied so far do not exploit the available resources to its full extent. The BS precoding that zero-forces the unintended data streams to other relays within the same cell is not flexible enough for future networks. Especially when relaying is combined with transmit cooperation at the BSs, other beamforming schemes such as SLNR precoding offer more appropriate and dynamic possibilities. In the next chapter, we thus expand the relay carpet concept to a further step and introduce post-cellular networks in which the cellular structure is completely abandoned. The MSs are then served in a highly flexible way in which we combine BS cooperation and low-complexity relaying with optimized link selection, BS cluster optimization, and power control. In this way, the mobile users can benefit from all advantages of the aforementioned schemes discussed in this thesis so far. Each MS is then served by its own subset of BSs and multiple relays and all infrastructure nodes cooperate together in a distributed way that takes the different node complexities into account.

8

Post-Cellular Networks

In order to overcome the high expectations of future cellular networks, sophisticated interference management is inevitable as we have seen in the previous chapters. High data rates, good coverage, and reliability can only be achieved when all resources of the network can be allocated in an efficient way to the users. The classical approach thereby is to introduce a spatial reuse that ensures a certain separation between BSs that use the same physical channel [33]. By exploiting the pathloss, each BS has to be closer to the users it serves than to other users it interferes with, which leads to the usual *cellular* network topology. To increase the network capacity, multiple BSs should cooperate with each other and additional BSs can be installed in the areas where it is most needed. With such a densification, the cell sizes are reduced down to pico- or femto-cells [41, 98]. In practice, however, finding sufficiently many new BS sites is in many cases difficult or even impossible, e.g. due to the costs, availability of backbone access, or public acceptance.

To this end, it might be more convenient to abandon the cellular network layout but rather let backhaul access points operate in places where they can most easily be installed. With this, the typical cells of traditional networks vanish and the backhaul access points have to serve MSs that are possibly far away. Such a "post-cellular" network topology, however, requires an aggressive spatial multiplexing that separates interfering users. This can be achieved by large antenna arrays at the BSs (massive MIMO) [94] or by forming large virtual antenna arrays with cooperative transmission across different BS sites [39, 80]. Dedicated stationary nodes not belonging to the wired infrastructure, such as relays, can thereby assist the communication between BSs and MSs. As shown by the relay carpet concept with ubiquitous relaying in the chapters before, even relays of very low complexity that are spread in large numbers can substantially reduce system complexity and enable massive MIMO while a favorable performance scaling is maintained.

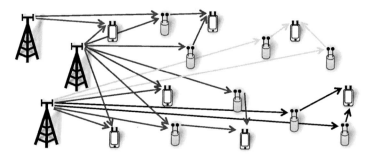

Figure 8.1.: A "post-cellular" network in which a cluster of BSs serves a wide area by the assistance of many distributed low-complexity relays.

Example scenarios of such post-cellular networks could include cities where sophisticated backhaul access points/BSs are installed *around* the city where more space is available and the public acceptance is higher or the roll-out of a new technology or a developing country where a high density of BSs is available in a confined region (e.g. the capital city) but no wired infrastructure in the backcountry. The MSs in the city center or backcountry could then be served by massive relay enabled spatial multiplexing or the help of possibly available existing BSs or residential internet access points. But also in dense scenarios, e.g. in urban hot spots, the joint use of sophisticated BS beamforming and ubiquitous relaying can enable the required performance gains for future mobile communication networks. The network becomes thereby closer to an ad-hoc like network in which new nodes can seamlessly be integrated into the existing infrastructure.

In this chapter, we consider such post-cellular networks where sparsely located BSs serve MSs that are possibly far away by cooperation and the help of low-complexity relays (see Fig. 8.1). To this end, we combine the insights and approaches from all previous chapters and bring them together in one framework in which we can profit from the advantages from the previously studied concepts and schemes:

- BS cooperation with large or massive MIMO arrays,

- network densification by the deployment of many relays,

- link selection with dynamic BS clustering and relay routing,

- power control to manage the residual interference and to optimize the network performance.

With this, the most important approaches that are under discussion for 5G network are combined. The massive deployment of relays acts thereby as an enabler for massive MIMO antenna arrays and builds a bridge between sophisticated large scale beamforming and densification to ultra small cells with small and low-power nodes.

By BS cooperation studied in Chapter 3, we have already seen that promising gains can be achieved when multiple BSs transmit their signals with joint precoding. The applied block ZF approach has thereby been chosen as a technique that allows precoding with relatively low complexity. Combined with optimized power loading, different objectives such as maximizing the minimum rate, the sum rate, or others can efficiently be solved. When the network geometry is adapted to BS cooperation, this form of precoding can lead to homogeneous rate distributions and good QoS in large parts of the network. On the other hand, the ZF approach imposes severe restrictions on the number of antennas, number of supported users, and requires fixed and non-overlapping BS clusters. Moreover, in order to formulate optimization problems that can efficiently be solved, the OCI the BSs cause to other cooperation areas is ignored. Better performance can be achieved if this OCI was also controlled and mitigated. With block ZF, this would however require a vast amount of excess antennas that can then no longer be used for optimizing the desired signals but are needed to null (also very weak) interference links.

In order to obtain a more flexible solution, we apply a leakage based precoding scheme that is based on SLNR precoding [121, 123]. Interference is thereby not zero-forced but the signal and the interference caused by the BSs is balanced in a way that allows to improve the performance in the entire network. The precoding can be calculated in closed form, which does not require any complex iterative procedures, no restrictions regarding number of antennas are imposed, and the cooperation clusters can be built flexibly and dynamically for each user separately. With such dynamic cooperation sets, the precoding can quickly be adapted to changing user distributions and rate requirements. As the SLNR-based precoding approach decouples each transmission from each other, we can also easily incorporate power control to manage the residual interference between different users. The dynamic BS clustering goes thereby one step further as e.g. in [37], where BS clusters are built dynamically and optimized, but have to remain fixed during relatively long time periods to allow to cancel intra-cluster interference. An alternative is proposed in [125] where the BS beamforming and the cooperation clusters are jointly optimized. The solution requires however a complex iterative optimization procedure. Nevertheless, the dynamic clustering can achieve large improvements over static approaches which can be expected to carry over also to the SLNR approach used here.

In most cases considered so far, the networks were very regularly set up. When the infrastructure nodes are located more randomly, as it is the case in current real world networks and even more so in future networks where new BSs or internet access points are installed wherever they can be, the interference scenario and the pathloss distributions are different [138]. While we have seen in the macro diversity UB scheme that link selection can already lead to large improvements of the network performance, we can expect that a dynamic BS-MS assignment in non-regular network topologies can lead to even higher gains. The higher differences in pathloss and interference at different user locations can be exploited and we can profit from selection diversity while the gains from cooperation can be maintained or even extended.

However, when different types of nodes coexist or cooperate with each other, they might have different backhaul connections and different computational capabilities which have to be taken into account by the transmission schemes. From a practical perspective, it is not feasible that all BSs cooperate with each other in a global fashion, but each MS is desirably served only by a subset of BSs. Such a dynamic BS clustering approach has already been studied in [28], where a sparse beamforming design is developed. This approach, however, requires very complex iterative optimization procedures that are computationally very costly, accurate CSI of the entire network and cannot be extended to relay assisted networks.

When BS cooperation is assisted by additional relays, the performance can further be boosted as we have seen in Chapter 4. The densification of the network allows thereby to profit from smaller pathloss and the additional signal power of the supporting nodes contributes to a better coverage. Based on the analysis in Chapter 6 and the arguments provided in Chapter 7, we limit ourselves to simple AF relays in the following. Such AF relays can forward multiple signals and can thereby assist the communication to multiple MSs simultaneously which is not directly possible with DF relays. Simple relay implementations further allow to deploy them in massive numbers which simplifies channel estimation at the BSs when the users are served via such static relays and enable massive MIMO. When considering the downlink, the relays inject and amplify however additional noise and interference which needs to be controlled. When many relays are deployed, only those that contribute to a better performance should be used. To this end, relays with weak signals have to be turned off and the residual interference and relay noise has to be controlled. Relay routing and power control are thus important to exploit the available resources and keep the interference on a low level. Power control is thereby an efficient means as shown by the results in the previous chapter.

Relay routing and link selection, on the other hand, allows to profit from selection diversity and the interference seen by the different users can further be minimized by a clever signal path through the relays. As shown in [20], joint relay scheduling and resource allocation can indeed lead to large improvements in relay assisted networks. The studies therein are however limited to DF relays that can only forward signals of a single user and no BS cooperation is considered. By the combination of dynamic BS clustering, joint beamforming, relay selection, and power control, we can expect to gain from all the advantages of these schemes jointly.

To this end, we develop a framework that allows to optimize the network performance by dynamic BS clustering with joint coherent transmission and extend the scheme such that it can also be applied to relay assisted two-hop networks. The extension is thereby not straight forward, as we take the limited CSI of the relays also into account and consider practical conditions such as backhaul rate constraints and power control. Thereby, we apply a leakage-based beamforming scheme [123] that can be calculated in closed-form and develop an extension thereof for two-hop communication with simple AF relays where the fading coefficients of the second hop channel are unknown. The BS cooperation clustering, relay routing, and power allocation are then optimized by an evolutionary algorithm [83] that is very efficient and flexible with respect to changing requirements and/or constraints.

By simulation results, we show that such a relay enabled post-cellular network can efficiently deliver high data rates to areas where no wired backhaul access is present. The advantages of the different concepts studied in the previous chapters as cooperative precoding with large antenna arrays, relaying, and massive deployment of small cells are thereby beneficially combined such that mobile users can be served with high data rates in a very dynamic and flexible way. We conclude that such post-cellular networks topologies offer many advantages over the more traditional network setups.

8.1 System Model

In the following, we focus on DL transmission. Regarding the UL, each BS can perform joint decoding of all desired signals. As we have seen in the last chapter, additional relays thereby improve the rates when they forward signals with a sufficiently high SNR. When multiple BSs can cooperate and jointly decode the signals of their users together, further improvements can be achieved. In the DL, however, each additional node also imposes additional interference and, in the case of AF relays, also injects additional noise into the system which can potentially lower the data rates again. The DL is thus the more challenging direction of communication.

In this section, we introduce the network and signal models that are applied in the rest of this chapter. Thereby, we distinguish two different network topologies. Firstly, we describe a post-cellular network in which the BSs transmit directly to the MSs, i.e. no relays are present in this case. In this way, we can develop the precoding with dynamic BS clustering based on the simpler network description. This basic network is then extended to the case where also relays are deployed. The signaling and optimization procedures introduced for the direct transmission can then be adapted such that also two-hop transmissions are handled. Thereby, we assume that the MSs are either all served directly by their associated BSs or they are served via relays in a two-hop fashion where the direct channels from the BSs are not used. A combination of direct and two-hop communication could also be included in the framework, but we will comment on that later.

8.1.1 Direct Transmission

The basic network with direct BS-to-MS transmission under consideration consists of B BSs that transmit signals to M MSs. For the sake of notational simplicity, we assume that all BSs are equipped with N_B antennas and all MSs with N_M antennas. With BS cooperation, each MS $j \in \{1, \ldots, M\}$ can be served jointly by a subset of BSs. In contrast to the BS cooperation schemes applied so far, each user can here be served by its own BS cluster that possibly overlaps with clusters of other MSs. In order to represent which BSs serve which MSs, we define the *routing* or *BS clustering* matrix $\mathbf{C} \in \{0,1\}^{M \times B}$ as the matrix whose element $\mathbf{C}[j, b]$ in the j-th row and b-th column is 1 if MS j is served by BS b and 0 otherwise. Each MS can thus be served by an arbitrary number of BSs and the BS cooperation clusters for different MSs can overlap or contain different nodes. Furthermore, we use $\mathcal{I}_j$ to denote the set of indices of the non-zero elements of the j-th row $\mathbf{C}[j, :]$ (the BSs that serve MS j) and $\mathcal{J}_b$ to denote the index set of the non-zero elements of the b-th column $\mathbf{C}[:, b]$ (the MSs that are served by BS b).

For a practical implementation, we limit ourselves to linear precoding and describe the transmit signal of BS b by $\mathbf{x}_b = \sum_{j \in \mathcal{J}_b} \mathbf{Q}_{j,b} \cdot \mathbf{s}_j$, where $\mathbf{s}_j \in \mathbb{C}^{d_s}$ is the data symbol vector with d_s elements i.i.d. $\mathcal{CN}(0,1)$ intended for MS j and $\mathbf{Q}_{j,b} \in \mathbb{C}^{N_B \times d_s}$ the corresponding precoding matrix from BS b. The transmit signals are constrained to a per-BS sum transmit power constraint

$$\mathrm{Tr}\left\{ \sum_{j \in \mathcal{J}_b} \mathbf{Q}_{j,b} \cdot \mathbf{Q}_{j,b}^{\mathsf{H}} \right\} \leq P_{\mathrm{B}}. \tag{8.1}$$

For the case of direct BS-to-MS transmission, we use $\mathbf{H}_{j,b} \in \mathbb{C}^{N_M \times N_B}$ to describe the block fading channel between BS b to MS j and $\mathbf{H}_{j,\mathcal{I}_j} = \left[\{\mathbf{H}_{j,b}\}_{b \in \mathcal{I}_j} \right]$ for the concatenated channel matrix from all BSs $b \in \mathcal{I}_j$ to MS j. With $\mathbf{w}_j$ being the additive noise with elements i.i.d. $\mathcal{CN}(0, \sigma_w^2)$, the receive signal at MS j, split into desired signal, interference, and noise, is given by

$$\mathbf{y}_j = \mathbf{H}_{j,\mathcal{I}_j} \cdot \mathbf{Q}_j \cdot \mathbf{s}_j + \sum_{\substack{i=1 \\ i \neq j}}^{M} \mathbf{H}_{j,\mathcal{I}_i} \cdot \mathbf{Q}_i \cdot \mathbf{s}_i + \mathbf{w}_j, \tag{8.2}$$

where $\mathbf{Q}_j = \left[\{\mathbf{Q}_{j,b}^{\mathsf{T}}\}_{b \in \mathcal{I}_j} \right]^{\mathsf{T}}$ is the concatenated precoding matrix for the signal to MS j from all its serving BSs.

For given clustering and precoding matrices $\mathbf{C}$ and $\mathbf{Q}_j$, the achievable rate for MS j can be computed by

$$R_j = \log_2 \det \left\{ \mathbf{I} + \left(\mathbf{K}_j^{(\mathrm{i+n})} \right)^{-1} \cdot \mathbf{K}_j^{(\mathrm{sig})} \right\}, \tag{8.3}$$

with the covariance matrices of the desired signal and interference plus noise which are given by

$$\mathbf{K}_j^{(\mathrm{sig})} = \mathbf{H}_{j,\mathcal{I}_j} \mathbf{Q}_j \mathbf{Q}_j^{\mathsf{H}} \mathbf{H}_{j,\mathcal{I}_j}^{\mathsf{H}} \tag{8.4}$$

$$\mathbf{K}_j^{(\mathrm{i+n})} = \sum_{\substack{i=1 \\ i \neq j}}^{M} \mathbf{H}_{j,\mathcal{I}_i} \mathbf{Q}_i \mathbf{Q}_i^{\mathsf{H}} \mathbf{H}_{j,\mathcal{I}_i}^{\mathsf{H}} + \sigma_w^2 \cdot \mathbf{I}_{N_M}. \tag{8.5}$$

The signal model and the resulting achievable rates thus differ only in the form of the BS clusters from the other cooperative networks considered in the previous chapters.

8.1.2 Two-Hop Transmission

When the transmission is assisted by additional relays, we let the MSs be served in a two-hop fashion and do not include the direct link between BSs and MSs. With this, the BSs see only the channel to the relays and we can apply and profit from the advantages discussed in the relay carpet concept. The channels to the possibly moving MSs, which are hard to estimate when many BS antennas are involved, are thereby not required by the precoding scheme as we will see. For the network description, we assume that K relays, each with N_R antennas, are present in the network. The descriptions of the BS clusters by the matrix $\mathbf{C}$ and the sets $\mathcal{I}_j$ and $\mathcal{J}_b$ as well as the model of the BS transmit signals are adopted from the direct case.

For the two-hop communication, we use $\mathbf{H}_{k,\mathcal{I}_j}^{(1)}$ to denote the first hop channel from all BSs in the set $\mathcal{I}_j$ to relay k. The receive signal of relay k is then

$$\mathbf{r}_k = \sum_{j=1}^{M} \mathbf{H}_{k,\mathcal{I}_j}^{(1)} \cdot \mathbf{Q}_j \cdot \mathbf{s}_j + \mathbf{n}_k, \tag{8.6}$$

with $\mathbf{n}_k$ being the relay noise with elements i.i.d. $\mathcal{CN}(0, \sigma_n^2)$. In order to use relay nodes of very low complexity and low cost, such that they can be implemented in massive numbers throughout the network, we consider the simple type A AF relays described in Chapter 7. These relays perform a frequency conversion and amplify the received signal with a fixed gain matrix $\mathbf{G}_k = \alpha_k \cdot \mathbf{I}_{N_R}$, where α_k is the amplification factor. Such relays can be implemented in a very inexpensive way (analog) and introduce no additional delays. With $\mathbf{H}_{j,k}^{(2)}$ describing the second hop channel between relay k and MS j, the receive signal at MS j follows as

$$\mathbf{y}_j = \sum_{k=1}^{K} \sum_{i=1}^{M} \mathbf{H}_{j,k}^{(2)} \mathbf{G}_k \mathbf{H}_{k,\mathcal{I}_i}^{(1)} \mathbf{Q}_i \mathbf{s}_i + \sum_{k=1}^{K} \mathbf{H}_{j,k}^{(2)} \mathbf{G}_k \mathbf{n}_k + \mathbf{w}_j. \tag{8.7}$$

For given clustering, precoding, and relay gain matrices $\mathbf{C}$, $\mathbf{Q}_j$, and $\mathbf{G}_k$, the achievable rate for MS j over two hops follows as

$$R_j = \frac{1}{2} \log_2 \det \left\{ \mathbf{I} + \left(\mathbf{K}_j^{(\text{i+n})} \right)^{-1} \cdot \mathbf{K}_j^{(\text{sig})} \right\}. \tag{8.8}$$

Here we assume in-band relays, i.e. the separation of the two-hop with different frequency bands with FDD relays is reflected in the prelog factor $\frac{1}{2}$. When out-band relays are used or when the second hop is for free as discussed in Chapter 6, this prelog factor can be omitted. The covariance matrices in (8.8) are given by

$$\mathbf{K}_j^{(\text{sig})} = \sum_{k=1}^{K} \sum_{l=1}^{K} \mathbf{H}_{j,k}^{(2)} \mathbf{G}_k \mathbf{H}_{k,\mathcal{I}_j}^{(1)} \mathbf{Q}_j \mathbf{Q}_j^{\mathsf{H}} \mathbf{H}_{l,\mathcal{I}_j}^{(1)\mathsf{H}} \mathbf{G}_l^{\mathsf{H}} \mathbf{H}_{j,l}^{(2)\mathsf{H}} \tag{8.9}$$

$$\mathbf{K}_j^{(\text{i+n})} = \mathsf{E}\left[\mathbf{y}_j \mathbf{y}_j^{\mathsf{H}}\right] - \mathbf{K}_j^{(\text{sig})}. \tag{8.10}$$

In the following, we develop how the precoding and the cooperation clustering can be designed to maximize the network performance. To this end, we first consider the case of direct transmission without relays and extend the framework afterwards to the two-hop case.

8.2 Cooperative Precoding

In order to achieve a good network performance, the cooperation sets described by $\mathbf{C}$ need to be chosen appropriately and the precoding matrices $\mathbf{Q}_j$ need to be optimized. In practical systems, it is thereby hardly possible that all BSs of the entire network can cooperate with each other. Not only would the computational complexity be prohibitive but also the required data traffic between BSs would exceed any feasible backhaul capacity. It is therefore desirable that each MS is only served by a small or moderate number of BSs, i.e. that the clustering matrix $\mathbf{C}$ is *sparse*, and that practical conditions as limited backhaul rates are considered in the optimization. In this way, the complexity of sharing data between different BSs and joint signal processing can be reduced and kept feasible for practical implementation.

This aspect has already been studied in [28], where the authors attempt to maximize the system performance under per-BS transmit power as well as backhaul rate constraints. By considering the weighted sum rate as the performance measure, the resulting optimization problem can be formulated as [28]

$$\max_{\mathbf{Q}_{j,b}} \sum_{j=1}^{M} \alpha_j \cdot R_j \tag{8.11}$$

$$\text{s.t.} \quad \text{Tr}\left\{\sum_{j \in \mathcal{J}_b} \mathbf{Q}_{j,b} \mathbf{Q}_{j,b}^{\mathsf{H}}\right\} \leq P_{\text{B}}, \quad \forall b \tag{8.12}$$

$$\sum_{j \in \mathcal{J}_b} R_j \leq C_{\text{B}}, \quad \forall b. \tag{8.13}$$

The difficult backhaul rate constraint (8.13) is then rewritten as an ℓ_0-norm constraint

$$\sum_{j=1}^{M} \left\| \|\tilde{\mathbf{Q}}_{j,b}\|_{\mathrm{F}}^2 \right\|_0 \cdot R_j \leq C_{\mathrm{B}}, \tag{8.14}$$

where $\tilde{\mathbf{Q}}_{j,b} = \mathbf{Q}_{j,b}$ if $j \in \mathcal{J}_b$ and $\tilde{\mathbf{Q}}_{j,b} = \mathbf{O}$ if $j \notin \mathcal{J}_b$. This constraint can then be approximated by a reweighted ℓ_1-norm constraint, as it is often done in the compressive sensing literature [19]. The rate constraint reads then

$$\sum_{j=1}^{M} \frac{\|\tilde{\mathbf{Q}}_{j,b}\|_{\mathrm{F}}^2}{\|\tilde{\mathbf{Q}}_{j,b}\|_{\mathrm{F}}^2 + \tau} \cdot R_j \leq C_{\mathrm{B}}, \tag{8.15}$$

for some small constant regularization factor $\tau > 0$. The resulting approximated optimization problem can then be solved iteratively over a number of equivalent convex optimization problems that minimize the weighted minimum mean square errors (WMMSE) [131]. The backhaul constraints are thereby an important ingredient to enforce sparse clusters, as the more stringent the backhaul constraints are, the less MSs can be served by a BS. While this approach converges to a locally optimal precoding, it requires global CSI and high computational power, as the algorithm consists of an iterative procedure in which an entire convex optimization problem has to be solved in each iteration. In the following, we attempt to maximize the network performance in a suboptimal but more efficient way that can be distributed and extended to two-hop communication in post-cellular networks which is not directly possible with the WMMSE algorithm.

8.2.1   Leakage Based Precoding

In order to get a low-complexity algorithm for BS clustering, we apply a leakage-based beamforming scheme as in [121–123]. Although it is suboptimal in terms of achievable rate, it can be calculated in closed form, which is efficient, and is a well-proven and suitable choice for flexible BS cooperation, as it decouples the transmissions to different users. In this way, the BSs do not need to be clustered in fixed groups that can only serve a closed set of MSs, but each MS can be served by a different subset of BSs. In contrast to the block ZF approach discussed in previous chapters, this offers the possibility to apply dynamic and flexible cooperation clustering that goes beyond any cell or cooperation area boundaries.

For a given BS clustering matrix $\mathbf{C}$, let $\mathbf{H}_{j,\mathrm{L}} = \left[\{\mathbf{H}_{i,b}\}_{i \neq j, b \in \mathcal{I}_j}\right]$ be the matrix that contains all channels from the BSs $b \in \mathcal{I}_j$ that are involved in the transmission to MS j to all other MSs $i \neq j$, i.e. the channel that *leaks* interference to other users. Furthermore, we induce a per-MS power budget $\mathrm{Tr}\left\{\mathbf{Q}_j \mathbf{Q}_j^{\mathsf{H}}\right\} \leq P_{\mathrm{M}}$ instead of the per-BS transmit power constraint (8.1). This allows a scaling of the precoding matrices

$$\tilde{\mathbf{Q}}_j = \sqrt{\frac{P_{\mathrm{M}}}{\mathrm{Tr}\{\mathbf{Q}_j \mathbf{Q}_j^{\mathsf{H}}\}}} \cdot \mathbf{Q}_j \tag{8.16}$$

across the different BSs that are involved. With this, each MS has the same power budget, irrespective of how many BSs are involved in the corresponding transmit cluster. As a consequence however, the BSs might have to serve different numbers of MSs and due to the SLNR beamforming it is not a priori clear with how much power each BS transmits, or whether the BS power constraint (8.1) is violated at some BSs. How the power constraint can be incorporated is described later in Section 8.3.

The signal-to-leakage-and-noise ratio (SLNR) of the link to MS j is defined as the ratio of the receive signal power of the desired signal at MS j and the noise power plus the total leakage signal power at all other MSs. The SLNR is calculated by [123]

$$\mathrm{SLNR}_j = \frac{\mathrm{Tr}\left\{\mathbf{H}_{j,\mathcal{I}_j} \tilde{\mathbf{Q}}_j \tilde{\mathbf{Q}}_j^{\mathsf{H}} \mathbf{H}_{j,\mathcal{I}_j}^{\mathsf{H}}\right\}}{\mathrm{Tr}\left\{\mathbf{H}_{j,\mathrm{L}} \tilde{\mathbf{Q}}_j \tilde{\mathbf{Q}}_j^{\mathsf{H}} \mathbf{H}_{j,\mathrm{L}}^{\mathsf{H}}\right\} + \mathrm{Tr}\left\{\sigma_w^2 \cdot \mathbf{I}_{N_{\mathrm{M}}}\right\}} \tag{8.17}$$

$$= \frac{\mathrm{Tr}\left\{\tilde{\mathbf{Q}}_j^{\mathsf{H}} \mathbf{H}_{j,\mathcal{I}_j}^{\mathsf{H}} \mathbf{H}_{j,\mathcal{I}_j} \tilde{\mathbf{Q}}_j\right\}}{\mathrm{Tr}\left\{\tilde{\mathbf{Q}}_j^{\mathsf{H}} \mathbf{H}_{j,\mathrm{L}}^{\mathsf{H}} \mathbf{H}_{j,\mathrm{L}} \tilde{\mathbf{Q}}_j\right\} + \sigma_w^2 N_{\mathrm{M}}} \tag{8.18}$$

$$= \frac{\mathrm{Tr}\left\{\mathbf{Q}_j^{\mathsf{H}} \mathbf{H}_{j,\mathcal{I}_j}^{\mathsf{H}} \mathbf{H}_{j,\mathcal{I}_j} \mathbf{Q}_j\right\}}{\mathrm{Tr}\left\{\mathbf{Q}_j^{\mathsf{H}} \left(\mathbf{H}_{j,\mathrm{L}}^{\mathsf{H}} \mathbf{H}_{j,\mathrm{L}} + \frac{\sigma_w^2 N_{\mathrm{M}}}{P_{\mathrm{M}}} \cdot \mathbf{I}_{N_{\mathrm{B}}}\right) \mathbf{Q}_j\right\}}, \tag{8.19}$$

where the second equality follows from the cyclic shift property of the trace and the third equality from the variable substitution (8.16). The SLNR is maximized by choosing $\mathbf{Q}_j$ as the (scaled according to (8.16)) generalized eigenvector (GEV) that corresponds to the largest generalized eigenvalue of the matrix pair $\mathbf{H}_{j,\mathcal{I}_j}^{\mathsf{H}} \mathbf{H}_{j,\mathcal{I}_j}$ and $\mathbf{H}_{j,\mathrm{L}}^{\mathsf{H}} \mathbf{H}_{j,\mathrm{L}} + \frac{\sigma_n^2 N_{\mathrm{M}}}{P_{\mathrm{M}}} \cdot \mathbf{I}_{N_{\mathrm{B}}}$, in short

$$\mathbf{Q}_j^{\star} = \max \mathrm{GEV}\left\{\mathbf{H}_{j,\mathcal{I}_j}^{\mathsf{H}} \mathbf{H}_{j,\mathcal{I}_j}, \mathbf{H}_{j,\mathrm{L}}^{\mathsf{H}} \mathbf{H}_{j,\mathrm{L}} + \frac{\sigma_w^2 N_{\mathrm{M}}}{P_{\mathrm{M}}} \cdot \mathbf{I}_{N_{\mathrm{B}}}\right\}. \tag{8.20}$$

In order to support multiple spatial streams to each MS, we extend this solution and choose for $\mathbf{Q}_j$ the d_s largest generalized eigenvectors. For a further improvement of the achievable rates, the chosen generalized eigenvectors have to be scaled according to a waterfilling-like optimization. This is handled together with incorporating the power constraint in Section 8.3. Note that this form of precoding does not impose any conditions on the number of BSs, MSs, or their antennas and each MS is served by its own subset of BSs that can overlap with other subsets associated with other users.

8.2.2 Precoding for Two-Hop Networks

When the MSs are far away from the BSs, the performance can be enhanced by relays that are spread in large numbers. To this end, the relays should be of low cost and low complexity and can therefore not fulfill complicated tasks. We thus assume that no CSI is available from the second hop, i.e., the relays are not able to estimate any channels. For the BSs, however, we assume that they have *local* CSI of their first hop channels, i.e. the channels to the relays. This CSI is simple to acquire when the relays are fixed as argued in Chapter 7. In this case, the BSs only have to track quasi-static channels which simplifies channel estimation. For the second hop, we consider two different cases: First, we assume that no CSI at all is available for the relay-to-MS channels. In this case, the BSs can apply only beamforming on the first hop and the relays retransmit the amplified signals without further processing. In the second option, we assume that the BSs might have *statistical* CSI of the channels between relays and MSs. This knowledge can e.g. be obtained when the positions of the relays and MSs are known, possibly by reporting GPS information to the BSs. With this information, the pathloss (including shadowing) and thus the second order statistics of the channels are available for the calculation of the precoding. To this end, the BSs can either calculate the pathloss from the positions of the relays and MSs or these values are stored in a lookup table. In the following, we adapt the SLNR beamforming for the relaying scenario for these two cases.

Ignoring the 2nd hop

A simple and straightforward way to calculate the precoding is to treat selected relays as the receivers and to ignore the second hop. To this end, we introduce a relay clustering matrix $\mathbf{D} \in \{0,1\}^{M \times K}$ which is defined, like the BS clustering matrix $\mathbf{C}$, as the matrix whose element $\mathbf{D}[j,k]$ is 1 if MS j is served via relay k and 0 otherwise and

$\mathcal{I}_j^{(R)}$ for the corresponding index set. The (assumed) channel for the desired signal is then $\mathbf{H}_j^{(1)} = \left[\{\mathbf{H}_{k,b}^{(1)}\}_{k \in \mathcal{I}_j^{(R)}, b \in \mathcal{I}_j} \right]$, i.e. only the first hop channels of the involved links are considered, and the covariance matrix for the SLNR calculation is $\mathbf{R}_j = \mathbf{H}_j^{(1)H} \cdot \mathbf{H}_j^{(1)}$. Similarly, the leakage covariance matrix is $\mathbf{R}_{j,L} = \mathbf{H}_{j,L}^{(1)H} \cdot \mathbf{H}_{j,L}^{(1)}$, where $\mathbf{H}_{j,L}^{(1)}$ is the concatenated matrix that contains all channels from the selected BSs in $\mathcal{I}_j$ to all active relays that are not selected for transmission to MS j (inactive relays, i.e. relays with all zeros in their respective column in $\mathbf{D}$, are excluded). With this, the precoding for MS j can, as in (8.20), be found by

$$\mathbf{Q}_j^\star = \max \text{GEV} \left\{ \mathbf{R}_j, \mathbf{R}_{j,L} + \frac{\sigma_n^2 N_R |\mathcal{I}_j^{(R)}|}{P_M} \cdot \mathbf{I}_{N_B} \right\}, \qquad (8.21)$$

which maximizes the SLNR at the selected relays

$$\text{SLNR}_j = \frac{\text{Tr} \left\{ \mathbf{Q}_j^H \mathbf{R}_j \mathbf{Q}_j \right\}}{\text{Tr} \left\{ \mathbf{Q}_j^H \left(\mathbf{R}_{j,L} + \frac{\sigma_n^2 N_R |\mathcal{I}_j^{(R)}|}{P_M} \cdot \mathbf{I}_{N_B} \right) \mathbf{Q}_j \right\}}. \qquad (8.22)$$

In order to support multiple spatial streams for each user, we again use the d_s largest generalized eigenvectors and weight them with optimized power allocation as described in Section 8.3. The signals transmitted by the BSs are then received by the relays, amplified by the gain matrices $\mathbf{G}_k$ and forwarded to the MSs. For appropriately chosen relay gains and clustering matrices $\mathbf{C}$ and $\mathbf{D}$, the interference that these relays generate should be kept low. The optimization of these clustering matrices developed in Section 8.3 will take care of that.

Expectation over 2nd hop

As an alternative, the BSs can make use of the statistical CSI that is available of the second hop. Under the assumption of Rayleigh fading, the second hop channels can be written as

$$\mathbf{H}_{j,k}^{(2)} = \sqrt{\frac{1}{L_{j,k}}} \cdot \mathbf{F}_{j,k}, \qquad (8.23)$$

where $\mathbf{F}_{j,k} \sim \mathcal{CN}(\mathbf{0}, \mathbf{I})$ is the unknown small scale fading and $L_{j,k}$ is the known pathloss. With this, we can calculate the expectation with respect to $\mathbf{F}_{j,k}$ of the desired signal power at MS j

$$P_j^{(\text{sig})} = \mathsf{E}_{\mathbf{F}}\left[\mathsf{E}_{\mathbf{s}}\left[\left\|\sum_{k=1}^{K}\mathbf{H}_{j,k}^{(2)}\mathbf{G}_k\mathbf{H}_{k,\mathcal{I}_j}^{(1)}\mathbf{Q}_j\mathbf{s}_j\right\|_2^2\right]\right] \tag{8.24}$$

$$= \mathsf{E}_{\mathbf{F}}\left[\text{Tr}\left\{\sum_{k,l}\sqrt{\frac{1}{L_{j,k}L_{j,l}}}\mathbf{F}_{j,k}\mathbf{G}_k\mathbf{H}_{k,\mathcal{I}_j}^{(1)}\mathbf{Q}_j\mathbf{Q}_j^{\mathsf{H}}\mathbf{H}_{l,\mathcal{I}_j}^{(1)\mathsf{H}}\mathbf{G}_l^{\mathsf{H}}\mathbf{F}_{j,l}^{\mathsf{H}}\right\}\right] \tag{8.25}$$

$$= \sum_{k,l}\sqrt{\frac{1}{L_{j,k}L_{j,l}}}\text{Tr}\left\{\mathsf{E}_{\mathbf{F}}\left[\mathbf{F}_{j,l}^{\mathsf{H}}\mathbf{F}_{j,k}\right]\mathbf{G}_k\mathbf{H}_{k,\mathcal{I}_j}^{(1)}\mathbf{Q}_j\mathbf{Q}_j^{\mathsf{H}}\mathbf{H}_{l,\mathcal{I}_j}^{(1)\mathsf{H}}\mathbf{G}_l^{\mathsf{H}}\right\} \tag{8.26}$$

$$= \text{Tr}\left\{\sum_{k=1}^{K}\frac{1}{L_{j,k}}N_{\text{M}}\cdot\mathbf{Q}_j^{\mathsf{H}}\mathbf{H}_{k,\mathcal{I}_j}^{(1)\mathsf{H}}\mathbf{G}_k^{\mathsf{H}}\mathbf{G}_k\mathbf{H}_{k,\mathcal{I}_j}^{(1)}\mathbf{Q}_j\right\} \tag{8.27}$$

$$= \text{Tr}\left\{\mathbf{Q}_j^{\mathsf{H}}\left(N_{\text{M}}\cdot\sum_{k=1}^{K}\frac{1}{L_{j,k}}\mathbf{H}_{k,\mathcal{I}_j}^{(1)\mathsf{H}}\mathbf{G}_k^{\mathsf{H}}\mathbf{G}_k\mathbf{H}_{k,\mathcal{I}_j}^{(1)}\right)\mathbf{Q}_j\right\}, \tag{8.28}$$

which follows from the cyclic shift property of the trace and

$$\mathsf{E}_{\mathbf{F}}\left[\mathbf{F}_{j,l}^{\mathsf{H}}\mathbf{F}_{j,k}\right] = \begin{cases} N_{\text{M}}\cdot\mathbf{I}_{N_{\text{R}}}, & \text{if } k = l \\ \mathbf{0}, & \text{if } k \neq l \end{cases} \tag{8.29}$$

as the small scale fading is assumed to be independent for channels between different nodes. Likewise, the expected leakage power can be calculated by

$$P_j^{(\text{leak})} = \mathsf{E}_{\mathbf{F}}\left[\mathsf{E}_{\mathbf{s}}\left[\left\|\sum_{i\neq j}\sum_{k=1}^{K}\mathbf{H}_{i,k}^{(2)}\mathbf{G}_k\mathbf{H}_{k,\mathcal{I}_j}^{(1)}\mathbf{Q}_j\mathbf{s}_j\right\|_2^2\right]\right] \tag{8.30}$$

$$= \text{Tr}\left\{\mathbf{Q}_j^{\mathsf{H}}\left(N_{\text{M}}\cdot\sum_{i\neq j}\sum_{k=1}^{K}\frac{1}{L_{i,k}}\mathbf{H}_{k,\mathcal{I}_j}^{(1)\mathsf{H}}\mathbf{G}_k^{\mathsf{H}}\mathbf{G}_k\mathbf{H}_{k,\mathcal{I}_j}^{(1)}\right)\mathbf{Q}_j\right\} \tag{8.31}$$

and the expected noise power follows the same lines as

$$P_j^{(\text{noise})} = \mathsf{E}_{\mathbf{F}}\left[\mathsf{E}_{\mathbf{n},\mathbf{w}}\left[\left\|\sum_{k=1}^{K}\mathbf{H}_{j,k}^{(2)}\mathbf{G}_k\mathbf{n}_k + \mathbf{w}_j\right\|_2^2\right]\right] \tag{8.32}$$

$$= N_{\text{M}}\cdot\text{Tr}\left\{\sum_{k=1}^{K}\sigma_n^2\frac{1}{L_{j,k}}\mathbf{G}_k\mathbf{G}_k^{\mathsf{H}}\right\} + N_{\text{M}}\cdot\sigma_w^2. \tag{8.33}$$

The SLNR, in which N_M cancels out, is then given by

$$\text{SLNR}_j = \frac{P_j^{(\text{sig})}}{P_j^{(\text{leak})} + P_j^{(\text{noise})}} \tag{8.34}$$

$$= \frac{\text{Tr}\left\{\mathbf{Q}_j^{\mathsf{H}} \cdot \mathbf{R}_j \cdot \mathbf{Q}_j\right\}}{\text{Tr}\left\{\mathbf{Q}_j^{\mathsf{H}} \cdot \mathbf{R}_{j,\text{L+N}} \cdot \mathbf{Q}_j\right\}} \tag{8.35}$$

with

$$\mathbf{R}_j = \sum_{k=1}^{K} \frac{1}{L_{j,k}} \mathbf{H}_{k,\mathcal{I}_j}^{(1)\mathsf{H}} \mathbf{G}_k^{\mathsf{H}} \mathbf{G}_k \mathbf{H}_{k,\mathcal{I}_j}^{(1)} \tag{8.36}$$

$$\mathbf{R}_{j,\text{L+N}} = \sum_{i \neq j} \sum_{k=1}^{K} \frac{1}{L_{i,k}} \mathbf{H}_{k,\mathcal{I}_j}^{(1)\mathsf{H}} \mathbf{G}_k^{\mathsf{H}} \mathbf{G}_k \mathbf{H}_{k,\mathcal{I}_j}^{(1)} + \frac{\text{Tr}\left\{\sum_{k=1}^{K} \sigma_n^2 \frac{1}{L_{j,k}} \mathbf{G}_k \mathbf{G}_k^{\mathsf{H}}\right\} + \sigma_w^2}{P_M} \cdot \mathbf{I}_{N_B}, \tag{8.37}$$

which is again maximized by

$$\mathbf{Q}_j^{\star} = \max \text{GEV}\left\{\mathbf{R}_j, \mathbf{R}_{j,\text{L+N}}\right\}. \tag{8.38}$$

With the expectation of the second hop, no relay clustering matrix $\mathbf{D}$ is required anymore. The routing of the signal paths via the different relays is done implicitly with the beamforming. Due to the known pathloss of the second hop, the transmit signal of the BSs can be directed to the relays which provide the best expected signal power at the corresponding MS and suppresses the interference accordingly. With properly selected BS clusters, a good tradeoff between signal and interference power can thus be found and the relays that provide the best signals are selected while less useful relays are not included. With additional power control at the relays, a good performance can be achieved and an optimized relay selection can be realized, where each relay is integrated into the transmission or not depending on its relative distance to the MSs and the expected signal strengths. This optimization is developed in the following section.

8.3 Constrained Cluster Optimization

After having calculated the precoding for fixed and given clustering matrices, we now attempt to find optimal clustering matrices $\mathbf{C}$ (and $\mathbf{D}$ if required). To this end, we apply an evolutionary algorithm [83] in which we also incorporate power control to further improve the performance and to satisfy the per-BS transmit power constraint (8.1). Finding the optimal clustering matrix is a combinatorial problem that is generally hard to solve. Evolutionary algorithms are heuristic search techniques that simulate natural selection and evolution that can solve such problems efficiently. Thereby, initializations (called individuals in the jargon of evolutionary strategies) are varied by random variations as mutation or crossovers. By a selection procedure according to the resulting value of the fitness or utility function, the individuals converge toward the optimum while bad choices die out with the iterations. This procedure is quite efficient, as only evaluations of the fitness function have to be calculated and no gradients or other computations are required. Especially for optimization problems with many local optima and with large search spaces, a solution which is close to the global optimum can be found in an efficient way.

For the optimization, we consider one of the following utility or fitness functions of the clustering $\mathbf{C}$ (and thus the achievable rates):

1. the sum rate $f_\Sigma(\mathbf{C}) = \sum_{j=1}^{M} R_j$,

2. the minimum rate $f_{\min}(\mathbf{C}) = \min\{R_1, \ldots, R_M\}$, or

3. the outage probability $f_{\text{out}}(\mathbf{C}) = \Pr\{R_j < R^\star\}$ for some target rate $R^\star$.

Also other, possibly more complicated fitness functions can be applied, but we focus on these three as they provide a good overview over the different performance measures that are of importance to the communication networks.

In the optimization process, a population of N_{ind} individual clustering matrices $\mathbf{C}_n^{(t)}$, for $n = 1, \ldots, N_{\text{ind}}$ and iteration index $t = 0, 1, \ldots T_{\max}$, is generated, where $T_{\max}$ is the maximal number of iterations after which the algorithm is terminated. In each iteration step t, the evaluated fitness function of the different individuals are compared and sorted according to their value and the clusters are updated by mutation until a sufficiently good solution is found.

As initialization, N_{ind} clustering matrices $\mathbf{C}_n^{(0)}$ are randomly generated. For each of them, the SLNR precoding as well as the resulting achievable rates are calculated as described before. All individuals of the population are then sorted according to the evaluated fitness function and the N_{sur} best (surviving) clustering matrices are selected. In each iteration step t, the surviving clustering matrices are reused and N_{child} mutated children are generated from each of them. To this end, each bit of the actual clustering matrix $\mathbf{C}_n^{(t)}$ is flipped with a probability p_t that can change during the iteration steps for an accelerated convergence, i.e.

$$\mathbf{C}_n^{(t+1)} = \left(\mathbf{C}_n^{(t)} + \mathbf{C}_{\text{rand}}(p_t)\right) \quad \text{mod } 2. \tag{8.39}$$

For a good tradeoff between performance and convergence speed, we update the bit flip probabilities with $p_{t+1} = \tau \cdot p_t$, for some update factor $\tau \in (0,1)$. Additional to the survivors and the children, we also generate $N_{\text{new}} = N_{\text{ind}} - N_{\text{sur}} - N_{\text{sur}} \cdot N_{\text{child}}$ new individuals randomly in each iteration step. After T_{max} iterations, the optimization is terminated. The procedure is summarized in Algorithm 4. The option to include constraints such as the power constraint, backhaul constraint or others is already included. In the following, we discuss the extension of the scheme to such additional constraints and to incorporate power control for the BSs as well as the relays.

Algorithm 4 Evolutionary Algorithm

1: Initialization: population of N_{inv} clustering matrices $\mathbf{C}_n^{(0)}$
2: **for** $t = 0 : T_{\text{max}}$ **do**
3:     Calculate $\mathbf{Q}_k^{(t)}$ and $R_k^{(t)}$, $\forall k$ according to (8.20), (8.21), or (8.38) and (8.3)
4:     **if** Additional constraint is active **then**
5:         Set $R_k^{(t)} = 0$ if constraint is violated
6:     **end if**
7:     Choose N_{sur} best $\mathbf{C}_n^{(t)}$ according to $f(R_k^{(t)})$
8:     Mutate bits in N_{sur} surviving $\mathbf{C}_k^{(t)}$ with probability p_t
9:     Generate N_{new} new random individuals $\mathbf{C}_n^{(t)}$
10: **end for**
11: Choose $\mathbf{Q}_k = \max_n \{\mathbf{Q}_n^{(T_{\text{max}})}\}$

8.3.1 Power Control

The SLNR precoding matrices are designed based on a per-MS power budget. Due to the cooperation between multiple BSs, a per-BS transmit power constraint is not straightforward to apply, because the entire precoding matrix $\mathbf{Q}_j$ needs to be scaled with one scaling factor, otherwise the SLNR optimality would be destroyed. To this end, the per-MS power budgets $P_{\text{M},j}$ are applied which can be chosen such that the per-BS power constraint is met during the iterations of the optimization. To achieve this, the clustering matrix $\mathbf{C}$ is extended to $\tilde{\mathbf{C}} = [\mathbf{C}, \mathbf{P}]$, where row j of $\mathbf{P}$ is the binary representation of $\frac{q_j+1}{2^{N_{\text{bits}}}} \cdot P_{\text{M,max}}$ with N_{bits} bits that describes the fraction of the maximal power $P_{\text{M,max}}$ that is allocated to MS j. The extended matrix $\tilde{\mathbf{C}}$ is then optimized in the evolutionary algorithm as before, attempting not only to find the optimal BS-MS association, but also an optimal power scaling for each data stream. The per-BS power constraint can be incorporated by setting

$$
\tilde{R}_j = \begin{cases} R_j, & \text{as in (8.3) if power constraint is met} \\ 0, & \text{if power constraint is violated,} \end{cases} \tag{8.40}
$$

i.e., a rate resulting from a precoding that violates the constraint is rejected and the optimizer tries to find a valid solution.

The same procedure can also be applied to the amplification gains of the relays. Thereby, a binary matrix $\mathbf{G}$ whose j-th row is the binary representation of $\alpha_k = \frac{q_k+1}{2^{N_{\text{bits}}}} \cdot \alpha_{\max}$, i.e. the fraction of the maximal amplification, is applied. The relay power control can the be incorporated into Algorithm 4 in the same way as the BS power control.

8.3.2   Additional Constraints

In a similar way, we can incorporate additional constraints as e.g. a per-BS backhaul rate constraint

$$
R_{\Sigma,b} = \sum_{j \in \mathcal{J}_b} R_j \leq C_b, \quad \forall b, \tag{8.41}
$$

where the total delivered data rate of BS b must not exceed the capacity C_b of its backhaul connection. Note that in (8.41) only the actual user data is included in the backhaul constraint. Additional traffic (e.g. control signals) can easily be included by an additional additive or multiplicative constant. Alternatively, also other constraints are possible, such as e.g. a maximal number $M_{\max}$ of MSs a BS can serve, i.e. a constraint of the form

$$
|\mathcal{J}_b| \leq M_{\max}. \tag{8.42}
$$

For the optimization, $R_{\Sigma,b}$ is calculated in each iteration. If this value fulfills constraint (8.41), the current achievable rates are used, otherwise, they are set to zero as in (8.40). In this way, the algorithm can also deal with additional, possibly difficult, constraints.

8.3.3   Relay Networks

The performance of the relay network can also be optimized in the same way. If the SLNR scheme that ignores the second hop is applied, the relay selection matrix $\mathbf{D}$ has to be included into the optimization. It is thereby treated in the same way as the BS clustering matrix $\mathbf{C}$. For the case of the second SLNR scheme in which the statistical CSI of the second hop is considered, no such relay selection matrix is required as the relays are selected implicitly by the BS beamforming. In order to optimize the relay gain factors α_k, we can proceed as with BS power control: the applied relay gains are encoded as binary bit strings, where an optimal fraction of α_{max} is applied to each relay.

With this algorithm, the difficult problem of BS clustering, relay routing, and power allocation under practical conditions can be solved very efficiently. To this end, the achievable rates have to be evaluated for each iteration of the optimization. As the small scale fading of the second hop is assumed to be unknown, this can be realized either by a rate feedback of the MSs or by generating random virtual fading coefficients according to their distribution and to maximize (a fitness function of) the sample mean $\overline{R}_j = \frac{1}{t}\sum_{i=0}^{t} R_j^{(i)}$. As a result, the algorithm can track changes in the network topology "on the fly" without starting from the beginning, e.g. when some channels change their fading coefficients or when some MSs drop out of the network or if new nodes come in. When all MSs are served exclusively via relays, no channel coefficients to the moving users have to be known. The BSs only require covariance matrices of the desired signals, interference, and noise on the first hop channels. These can be acquired over relatively long time periods. Accordingly, the overhead for channel estimation can be kept small while all users can still profit from the advantages of BS cooperation and small relay cells.

8.4 Simulation Results & Discussion

For the evaluation of the proposed schemes in post-cellular networks, we again distinguish between direct and two-hop transmission. First, we consider networks in which M MSs are served directly by B BSs. With this, we can study the effects of BS clustering and SLNR based precoding and can compare the performance with reference schemes from the literature as well as the approaches discussed earlier in this thesis. Networks with relays are considered later. In all simulations, we apply the WINNER II based channel model with the settings of the urban environment and assume a total transmit power at each BS of $P_{\mathrm{B}} = 40\,\mathrm{W}$ if not stated otherwise and a noise variance of $\sigma_w^2 = 5 \cdot 10^{-12}\,\mathrm{W}$ at the MSs.

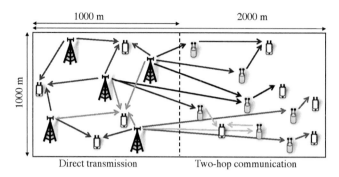

Figure 8.2.: For the simulations, direct transmission and two-hop communication via relays
are distinguished.

8.4.1 Post-Cellular Network with Direct Transmission

In order to compare the evolutionary optimization with other schemes, we first consider
a setup in which $M = 15$ MSs and $B = 5$ BSs are randomly located in a square of
$1000\,\text{m} \times 1000\,\text{m}$ according to a poisson point process [52]. The setup is schematically
depicted in the left hand side of Fig. 8.2. The BSs and MSs are equipped with $N_\text{B} = 4$
and $N_\text{M} = 2$ antennas, respectively. The evolutionary optimization as proposed in
this chapter is compared with the WMMSE approach from [28] and different static BS
clustering schemes without any optimization: In the first case, each BS serves the three
closest users (*3 closest MSs*). With this, all BS clusters have size three but different
MSs can be served by a different amount of BSs; some unlucky MSs might not be
served at all while others can receive signals from more than three BSs. In the second
case (*3 closest BSs*), each user chooses the three BSs that are closest to it. Thereby,
different BSs might have to serve a different number of MSs, but every user receives
desired signals from a BS cluster of size three. A fairer rate distribution can thus be
expected in this case. In the third static approach (*global SLNR*), all BSs form one large
virtual antenna array and jointly serve all users together. In all these static clustering
approaches, no optimization is performed. In order to satisfy the power constraint
at the BSs, the BS with the strongest transmit power is identified and all BSs are
then scaled by the same factor such that this strongest BS transmits with $P_\text{B} = 40\,\text{W}$,
while the others are scaled by the same factor. In the case of the BS clusters that
are optimized by the evolutionary algorithm, the power allocation is also optimized,
once to maximize the sum rate and once to maximize the minimum rate. Thereby, a

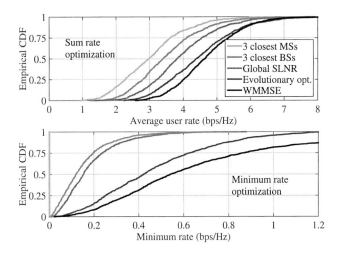

Figure 8.3.: Empirical CDFs of average user rates and minimum rates, direct transmission.

power $P_M = 40\,\mathrm{W}$ is applied first to each signal, which can then be reduced by the included power control with a resolution of $N_{bits} = 32$. The same power constraint is also applied in the WMMSE scheme. The backhaul constraint is unlimited here.

Fig. 8.3 shows the empirical CDFs of average and minimum user rates for these schemes. It can be seen that the static schemes are clearly outperformed by the optimized ones and that the suboptimal SLNR based scheme with evolutionary optimization of the BS clusters is close to the locally optimal WMMSE scheme. In the case of sum rate optimization, the evolutionary algorithm achieves almost the same performance as WMMSE, with a bit worse results in the low rate regime. The average user rates with the static global SLNR approach are however not much lower than those of the evolutionary optimization when no stringent backhaul rate constraints are applied. The BS cluster optimization usually leads to solutions in which the strongest users, which contribute most to the sum rate, are served by almost all BSs. The performance gain is to a large part due to the added power control as we will see later. The static schemes with smaller BS clusters lead to lower rates as the compound BS antenna arrays are smaller. As a result, the interference mitigation is not as efficient as with larger arrays as we have intentionally chosen a network setup with a total number of BS antennas of $B \cdot N_B = 5 \cdot 4 = 20$ which is only two thirds of the total number of MS antennas to be served $M \cdot N_M = 15 \cdot 2 = 30$. Accordingly, not all interference terms can be cancelled or sufficiently reduced and smaller antenna arrays leave more leakage in the network. The BS cluster optimization can in this setting however benefit from link selection. Applied together with power control, the interference can thus be traded with the strength of the desired signals.

With max-min optimization, the clustering optimization with included link selection and power control leads to more improvements compared to the static clustering approaches. In the 3 closest MSs approach, some MSs are not served at all. In this case, it happens that some MSs are not connected to any BS and the rate of these users is accordingly zero. When each user is served by three or all BSs, the rate distribution is fairer and all users are served, with slightly higher rates when global cooperation is applied. The improvements with the optimization schemes are here much more pronounced than for sum rate optimization. The BS cluster selection can prioritize the weak users and the power control can further improve their performance by allocating more power to them than to users that are already stronger due to their position and fading realization. The WMMSE scheme for the minimum rate maximization is here realized with a reweighted sum rate optimization as [28] cannot directly by applied with max-min optimization. In order to approximate the minimum rate, the achievable rate R_k is in the t-th iteration weighted with

$$\alpha_k = \frac{t}{\sum_{n=1}^{t} R_k^{(n)}}, \tag{8.43}$$

which gives higher weights to weaker users. Even though the evolutionary algorithm is suboptimal, it is also here close to the WMMSE algorithm which is proven to converge to a local optimum. However, the evolutionary scheme is of lower complexity and more flexible as we will see in the following.

In the next simulation, we study the influence of limited backhaul capacities. To this end, we consider the outage probability for a target rate of $R^* = 1$ bps/Hz for the different schemes shown in Fig. 8.4. The static schemes without optimization lead to very poor results when the backhaul of the BSs can only support small rates. Global cooperation with SLNR precoding but no power control is thereby the worst. When all BSs cooperate with each other, the backhaul link of each BS must support a data rate which is at least as large as the sum rate in the entire network. If this sum rate exceeds the backhaul constraint, we count that as an outage event. Improved results could be obtained with limiting the data rates to the MSs according to the backhaul constraint, i.e. each BS could lower the data rates to the MSs it serves. With an appropriate

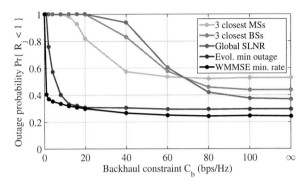

Figure 8.4.: Outage probability for different backhaul rate constraints, direct transmission.

rate scaling, the backhaul constraint could be fulfilled while some additional users can achieve their target rate. This would however involve a different type of optimization which we do not consider here. For higher backhaul capacities however, global SLNR performs better than the other static schemes as expected by the results we have seen before. When each BS serves only its three closest MSs, the backhaul rates are already limited as only the data rates of these three users needs to be supported. The outage rate behavior with small backhaul capacities is in this case lower than with global SLNR but still poor.

The optimized schemes on the other hand are quite robust with respect to the backhaul rate constraints. Even when the backhaul is restricted to a low rate, the optimization scheme manages to distribute the MSs to the different BSs and thus adapt the backhaul traffic selecting the BS clusters such that the outage probability is kept low. With BS cluster optimization, some links to MSs that achieve high rates can be excluded and the corresponding BS thus serves less users. The backhaul capacity of this BS can thereby be reduced. The performance saturates around $\frac{1}{3}$ because the number of MSs in the network is one third higher than can be served by the number of BS antennas. For comparison, the performance of the WMMSE scheme is also shown. For the latter, we again use the reweighted sum rate maximization (8.43) that approximates the minimum rate maximization. It is thus not directly comparable as the f_{out} utility is not directly applicable to the WMMSE algorithm. Again, the evolutionary scheme clearly outperforms the static ones and is close to the WMMSE algorithm. For very low backhaul capacities ($C_b \leq 4\,\text{bps/Hz}$) however, the evolutionary algorithm has

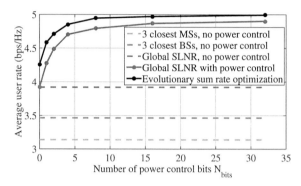

Figure 8.5.: Influence of power control on SLNR cooperation, direct transmission.

some difficulties to balance the rates appropriately as no rate scaling is performed here. The achievable rates to the MSs are evaluated and taken and are not scaled down when they exceed the target rate. With such a rate scaling, a better outage behavior can be expected but an additional optimization would be required in this case.

In the following, we look at the influence of power control on the network performance. To this end, we consider the example of sum rate maximization for different numbers of bits for the resolution of power control. The backhaul rate is not limited. The average user rates are plotted in Fig. 8.5. Without power control ($N_{\text{bits}} = 0$), the optimized clustering is only slightly better than global SLNR. As we have already argued before, the sum rate maximization without backhaul restrictions leads to solutions in which the MSs with the strongest impact on the sum rate are usually served by almost all BSs. The difference to global SLNR is therefore small in this case. When the number of bits for power control is increased, the performance improves until the performance gain saturates after about 16 bits. With 8 or 16 bits, a significant performance gain can be observed compared with the case without optimized power allocation. For comparison, we also plot the global SLNR scheme where power control with the evolutionary algorithm is also included. Also here, the average user rates increase similarly as when the BS clusters are optimized. In the case of sum rate maximization, the static clustering with full cooperation among all BSs thus achieves a similarly good performance as when link selection is applied. If the max-min case is considered, the potential for optimizing the BSs clusters is higher as weak users can be assigned to larger cooperation clusters than strong MSs. In this way, the rates can be better balanced than with a static clustering approach.

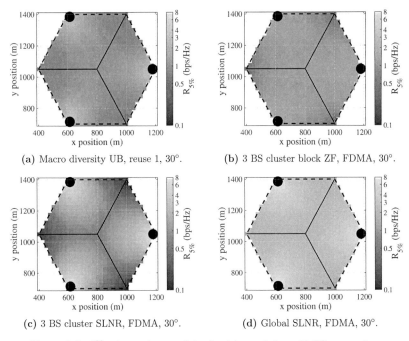

(a) Macro diversity UB, reuse 1, 30°.

(b) 3 BS cluster block ZF, FDMA, 30°.

(c) 3 BS cluster SLNR, FDMA, 30°.

(d) Global SLNR, FDMA, 30°.

Figure 8.6.: 5% outage rate area plots, direct transmission with BS cooperation.

In order to compare the SLNR precoding with the other cooperation approaches that we have introduced and studied earlier in this work, we consider the regular urban micro-cell setup with sectorized cells as described in Section 3.5. We choose the 30° orientation of the directed BS antenna arrays and separate different cooperation areas by the FDMA frequency allocation. For the simulations, we limit ourselves to static BS clustering approaches and do not apply power control. In Fig. 8.6, we show the area plots with 5% outage rates. As a reference, we use here the macro diversity UB in the reuse 1 case (as in Fig. 3.13d) as well as the 3 BS cooperation with block ZF with FDMA (Fig. 3.9d) as described in Chapter 3.

In Fig. 8.6c, we apply the SLNR based precoding and cluster the BSs in the same way as in the case of block ZF, i.e. the BSs of three adjacent sectors that point towards each other form one cooperation cluster. Always three MSs are thus served by the same set of BSs and the clusters do not overlap. For a fair comparison, the transmit power of the BSs is here set to $P_B = 80\,\mathrm{W}$ and the BSs are equipped with $N_B = 8$ antennas in all cases. We can see that the SLNR based precoding leads to 5% outage rates close to the BS that are significantly higher than with block ZF as well as with the macro diversity UB. Towards the center of the cooperation and in the corners where no BSs are located, the performance drops to low values. When this scheme would be adapted to also include power control to maximize the minimum rates, a better performance could be expected in these locations. With global SLNR however, the same high 5% outage rates close to the BSs can be maintained while the rates in all other locations are also significantly increased. Even though no power control is applied here and all BSs are scaled such that the BS with the highest transmit power fulfills the per BS power constraint with equality, the rate distribution is very homogeneous. Accordingly, a data rate of about 4 bps/Hz can be guaranteed with a high probability in the entire network. This is almost four times higher than with block ZF. With power control, the performance could further be increased. The impact of optimized BS clusters is however not expected to be large because the network geometry and the FDMA frequency allocation between the different sectors is organized for clusters that contain three BSs. The more BSs that cooperate with each other can however improve the performance significantly.

In Fig. 8.7, we compare the empirical CDFs of the global SLNR scheme with the best cooperation scenarios from Chapters 3 and 4. The SLNR based scheme achieves significantly higher data rates than all others, even higher than in the six BS super-cells

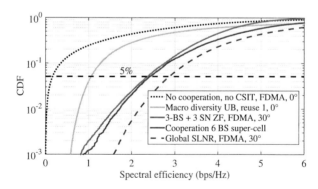

Figure 8.7.: Empirical CDFs of instantaneous user rates for different schemes in urban micro-cells.

or when six infrastructure nodes (three BSs and three SNs) jointly serve three MSs. The initial reference without cooperation and no CSIT is thereby outperformed by a factor of about 20 when the performance on the 5% outage line is considered.

This is also reflected in the KPIs shown in Table 8.1, where the global SLNR cooperation has full coverage with a target rate of 1 bps/Hz and the average 5% outage rate is 3.38 bps/Hz, which cannot be achieved with all other considered scenarios. The SLNR based scheme does thus not only achieve a significantly better performance than the other scenarios, but it is also more flexible and can dynamically be applied to changing networks and user demands. Moreover, the acquisition of the CSIT is also simpler as only covariance matrices are required instead of the full channel matrices in the case of block ZF. The global SLNR scheme however involves a large data traffic between all BSs as all have to cooperate and need to exchange all user data of the entire network. However, as we have seen before, a similar performance or even a better one can be achieved when the BS clusters are optimized and when power control is included. With limited backhaul capacities or when the size of the BS clusters is limited, the proposed clustering algorithm can still achieve very promising results.

Scheme	Configuration	Coverage	Av. $R_{5\%}$ (bps/Hz)
Reference	FDMA, 0°	15 %	0.53
Macro diversity UB	reuse, 0°	61 %	1.19
3 BS cooperation	FDMA, 30°	86 %	1.26
6 BS super-cell	reuse, super-cell	77 %	1.88
Relay selection	FDMA, 0°	25 %	0.87
3 BS + 3 SN cooperation	FDMA, 30°	99 %	2.38
3 BS SLNR cooperation	FDMA, 30°	53 %	1.51
Global SLNR cooperation	FDMA, 30°	100 %	3.38

Table 8.1.: Key performance indicators for the different schemes in dense urban micro-cells.

In the next section, we will see how this SLNR based scheme with constrained cluster optimization is applied to relay assisted networks. Thereby, we see that we can further exploit the flexibility of this scheme and can combine the dynamic BS cooperation with the advantages of the relay carpet concept.

8.4.2 Post-Cellular Relay Carpet

In the following, we consider a relay assisted post-cellular network and focus on the mobile users that are on the right hand side of the schematic depicted in Fig. 8.2. In this setup, the performance gains due to relaying are more evident. When sum rate is considered, this measure is dominated by the best MSs, which are the ones that are close to the BSs. With direct transmission, no prelog factor $\frac{1}{2}$ applies (for the results with relaying, a prelog factor of $\frac{1}{2}$ is applied here). Users that are located in the same area as the BSs can thus achieve higher rates by the direct transmission and comparing that with the relay assisted communication would therefore not reflect the advantages of the relaying schemes appropriately. MSs further away from the BSs can profit more from the relays. To this end, we consider here a setup in which $B = 5$ BSs (each with $N_B = 4$ antennas) are randomly located in the first 500 m and serve $M = 15$ MSs ($N_M = 2$ antennas) in the rest of a 2000 m long network area (the width is 1000 m as before). $K = 30$ relays with $N_R = 2$ antennas, also distributed in the same range as the MSs, assist the communication. The relay noise is set to $\sigma_n^2 = 5 \cdot 10^{-12}$ W. As all nodes are located randomly, all channels are assumed to be of NLOS condition.

For such a two-hop network, we show a typical convergence behavior of the evolutionary algorithm in Fig. 8.8. The relay noise is $\sigma_n^2 = \sigma_w^2 = 5 \cdot 10^{-12}$ W and the maximal relay amplification gain $\alpha_{k,\max}$ is chosen that the average transmit power of each relay is $P_B/10 = 4$ W. For the optimization, we use $N_{\mathrm{ind}} = 8$, $N_{\mathrm{sur}} = N_{\mathrm{child}} = 2$ and a bit flip probability update with $\tau = 0.95$. For this example scenario with a typical channel

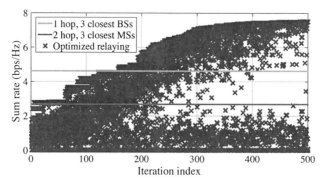

Figure 8.8.: Convergence behavior of the sum rate maximization for a typical network realization with relays.

realization, we can observe that the two-hop scheme with the relays (SLNR precoding that ignores the second hop) leads to a significantly better sum rate than if the MSs were served directly by the BSs. It can further be seen that the algorithm converges after about 500 iterations, which is quite efficient for a network size of $5 \times 15 \times 30$ nodes. Together with the power control resolution of $N_{\text{bits}} = 32$ bits, for both the BSs and the relays, this leads to an optimization space of up to $2^{M \cdot B + (M+K) \cdot N_{\text{bits}}} = 2^{1515}$ or $2^{M \cdot B + (M+K) \cdot N_{\text{bits}} + M \cdot K} = 2^{1965}$ possible solutions, depending on whether the relay routing matrix $\mathbf{D}$ is required or not. When more nodes are considered in a network or when more bits for power control are included, the required number of iterations however grows. In order to keep the simulation effort manageable, we limit ourselves to relatively small network sizes.

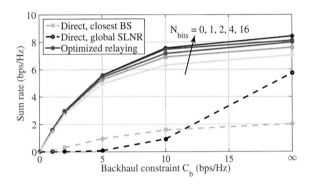

Figure 8.9.: Influence of power control on relaying for different BS backhaul constraints.

Next, we look at the influence of restricted backhaul capacities and power control and show achievable sum rates averaged over 1000 channel realizations in Fig. 8.9. For each BS we impose a backhaul rate constraint of C_b and apply power control with a resolution of a varying number of bits N_{bits}. In the case of relaying, we apply the SLNR precoding scheme that ignores the second hop channels and the same number N_{bits} is applied to power control at the BSs as well as in the relays. For comparison, we also show the results of direct transmission in the same setup without power control, once where each MS is served only by the BS that is closest to it and once where global SLNR is applied. When the backhaul constraint is violated, we count that as an outage event and set the sum rate to zero, i.e. no rate scaling is applied. In the relaying case, we can observe that the first one or two bits of the power control have the strongest impact. Beyond 16 bits, the performance gain with a higher resolution is negligible.

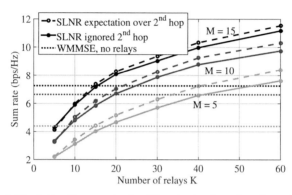

Figure 8.10.: Achievable sum rates for different network configurations with relays.

Again, we can also observe that the optimization leads to particularly good results when the backhaul rates are restricted to low values. The communication links can be balanced and the signals are routed such that a good performance is achieved while the backhaul traffic is distributed among the BSs. The power control can additionally throttle down the transmit power at the BSs as well as at the relays such that resulting data rates do not violate the backhaul constraints. Hence, no outage events occur in this case (unless when $C_b = 0$). The relay network is thus also quite robust with restrictions in the network and 16 bits for power control are sufficient. The performance gain over the direct transmission is here particularly large. However, neither the power allocation nor the BS clusters are optimized in this case. By optimizing them, the direct transmission can also be improved but relaying still leads to significantly higher rates as we will see next.

In Fig. 8.10, we show the achievable sum rates for a varying number of relays and different numbers of MSs ($M = 5, 10, 15$). The number of antennas at the $B = 5$ BSs remains fixed to $N_B = 4$. We compare the two different beamforming schemes for the two-hop communication (SLNR ignoring the second hop and SLNR with averaged second hop) and also show the results of the locally optimal WMMSE precoding for the case of direct transmission without relays. In all cases shown here, the BS clusters as well as power allocation are thus optimized. When sufficiently many relays are deployed, the two-hop schemes lead to significantly higher average sum rates than the optimized direct transmission. Thereby, exploiting the statistical CSI of the second hop leads to further gains which even grow with an increasing number of relays. In contrast

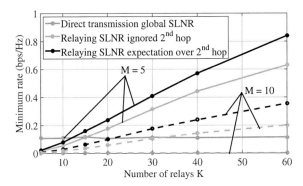

Figure 8.11.: Achievable minimum rate for different network configurations with relays.

to the scaling behavior discussed in Chapter 7, where the MSs cannot benefit from too many relays that are active simultaneously, the link selection and power control leads here to a beneficial scaling with more nodes. With the optimization, the performance continuously increases when more relays are present and due to the adapted relay amplification (which can also turn certain relays off) the network is able to balance the additional noise and interference injected by the relays. As the BSs have in total only $B \cdot N_B = 5 \cdot 4 = 20$ spatial degrees of freedom, the performance gain in sum rate starts to saturate when more than $M = 10$ users are served. The further increase in sum rate with more MSs can mostly be attributed to the increased probability that some users are closer to a transmitting node, which contributes to a higher sum rate. This form of diversity gain is more pronounced when many relays are deployed.

When we consider the minimum rate as optimization criterion, the performance gains with relaying are more pronounced. In this case, the performance is not dominated by users that can benefit from particularly good propagation conditions and are close to a transmitting node, but it is limited by remote MSs that are difficult to serve. In Fig. 8.11, we consider the same network setup as before, but apply max-min optimization. With increasing number of relays, the achievable minimum rate in the network increases almost linearly, at least up to $K = 60$ relays deployed in the network. The performance scaling is thus more beneficial as in the case of sum rate maximization and the performance gain of the SLNR approach with statistical CSI over the SLNR scheme that ignores the second hop is also larger. In order to provide a fair service in the entire area, the relays can boost also users that are far away from the wired infras-

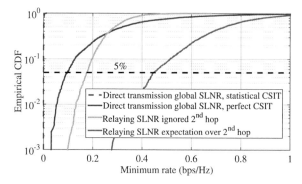

Figure 8.12.: Achievable minimum rate in a post-cellular relay carpet network with $B = 5$ BSs, $M = 15MSs$, and $K = 50$ relays, max-min optimization.

tructure provided by the BSs. When the number of MSs M gets larger, however, the probability that some users still suffer from bad propagation conditions increases. This is reflected in the decreased rates for $M = 10$ as compared to the case with $M = 5$. Nevertheless, increasing the number of relays can still compensate this performance loss and significantly higher rates can be achieved for all users compared to the direct transmission. With a large number of relays, a high QoS can thus be provided in the entire network, even when the relays are of very low complexity as assumed here.

When we consider a network without the restriction that the MSs are excluded in the area in which the BSs are located, the results look similarly good when max-min optimization is applied. In Fig. 8.12, we show empirical CDFs of minimum rates in a network where $K = 50$ relays serve $M = 15$ MSs. $B = 5$ BSs are in this case equipped with $N_B = 8$ antennas and are still located randomly in the first 500 m of the network. The MSs and relays are in this case however spread over the entire network area of 1000 m $\times$ 2000 m. This scenario resembles more the relay carpet concept with a large number of relays. Also here, we see that the two-hop transmission achieves significantly higher rates than when the users are served directly by the BSs. The 5% outage rate for the case with two-hop SLNR precoding with statistical CSI is about five times higher than with direct service from the BSs. The relaying scheme with ignored second hop is less efficient. But also in this case, the relays additionally simplify the BS signal processing as the requirements of CSIT is drastically reduced. In order to compare the curves with a direct transmission scenario where the BSs are not able to obtain accurate CSI, we also include a curve of direct transmission where the BSs have

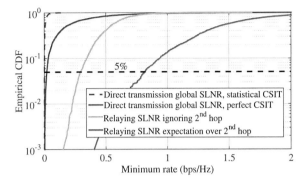

Figure 8.13.: Achievable minimum rate in a single cell with a massive MIMO BS and the help of $K = 50$ relays with max-min optimization.

only statistical CSIT. The SLNR precoding is thereby calculated in a similar way as in (8.34), but only of the direct channels, i.e. the expected SLNR with respect to $\mathbf{H}_{j,\mathcal{I}_j}$ is calculated. In this case, the comparison is fairer in the sense that channels to the possibly moving MSs cannot be estimated at any node but only the pathloss of these channels is known. The achievable rates are very poor in this case.

The advantage of simplifying the acquisition of CSIT at the BSs is particularly interesting when the BSs are equipped with very large antenna arrays. In this case, the channel estimation would impose overheads that might destroy all the gains such large antenna arrays can offer. In order to illustrate the performance of the relay carpet concept also in such a scenario, we consider a simulation where a single BS with $N_B = 50$ antennas serves an area of $1000\,\text{m} \times 1000\,\text{m}$ with $M = 15$ MSs. The BS is located in the center and the MSs are randomly placed in the entire area. Additionally, $K = 50$ relays are also spread randomly in the network area. The resulting empirical CDFs of the minimum rates are shown in Fig. 8.13. Again, we compare the relaying schemes with two direct transmission schemes: once with SLNR precoding with perfect CSIT and once where the BSs only have statistical information. In both cases, the achievable minimum rate is very poor. With relaying, the minimum rate can be increased by a factor of almost 15 when the two-hop precoding is done by ignoring the second hop and even by a factor of almost 40 when the statistical CSI of the second hop is also taken into account. The relays are thus not only an enabler of massive MIMO signal processing at sophisticated BSs, but can also provide a high QoS to large areas when the carpet of relays is sufficiently dense.

8.5 Conclusions

The post-cellular network concept with its transmission schemes combines the different approaches that were studied individually before. The evolutionary optimization framework allows to apply BS cooperation with dynamically selected cooperation clusters and power control and can also incorporate relay selection and flexible signal routing. Moreover, practical challenges such as limited backhaul connections or limited cooperation cluster sizes can also be included without increasing the complexity gravely. The proposed scheme with extended two-hop SLNR based beamforming is not only efficient but also very flexible with respect to the network topology and the user distribution and their demands. The simulation results indicate that the evolutionary algorithm is an efficient tool to optimize the BS clustering and power allocation for both classical cellular one-hop networks as well as relay assisted post-cellular networks where the joint BS selection and relay routing is a difficult problem.

Even though the SLNR based precoding that we applied here is suboptimal in terms of achievable rates, it is well suited and leads to very good results. It is more flexible than other schemes such as block ZF as no requirements on number of antennas or supported number of users need to be fulfilled and the beamforming matrices can be calculated in closed form. As only channel covariance matrices are required, the acquisition of CSIT is also simpler than with other schemes.

By the dynamic BS clustering, the network can quickly adapt to changing user distributions or rate requirements. As no fixed cooperation areas are needed, the network geometry and the deployment of infrastructure nodes does not have to be carefully designed, but every node can be used in a way that is best for the network performance. Even BSs that are far away from the users they serve can contribute to higher data rates and thus improve the overall coverage. The classical cellular network topology can thus be abandoned and BSs can be installed where this can most easily be done. By the help of the relays, high QoS can still be provided in wide areas and a high performance can be achieved even in places where no wired infrastructure is available.

With a dense deployment of relays as proposed in the relay carpet, this dynamic form of mobile communication networks can be carried further up to ad-hoc like networks in which all kinds of nodes communicate and cooperate with each other. The large number of relays allows thereby to overcome severe shadowing and the propagation conditions seen by the users can be made more stable and reliable. The beneficial combination of BS cooperation and strong densification through the small relay cells leads to ubiquitous coverage and all nodes can contribute to an efficient interference management such that all users can benefit from large bandwidths with small reuse factors. The installation of these small relay cells do not only improve the coverage and contribute to interference mitigation, but they can also be seen as a key enabler for sophisticated multi-user beamforming at the BSs with very large antenna arrays. Due to the channels between BSs and relays that are much more static than the direct links to MSs, the overheads associated with massive MIMO signal processing can be drastically reduced.

The relays considered in this chapter are very simple AF relays that can be implemented mainly in the analog domain. With this, they can be built with low costs and thus be spread in large numbers. Even though the network is turned into a two-hop network, additional delays can be avoided when FDD relays are used that can be realized as simple frequency converters. The proposed optimization framework would however also allow for more sophisticated relays. Type B relays that are able to reduce the residual interference by their filters can lead to further improvements of the performance.

Also the BS precoding that we applied here could be further improved. We have chosen the SLNR approach as it can be calculated in closed form and allows for an extension to two-hop networks where the second hop channels are only known by their statistics. The specific form of the signal powers with the properties of the trace operator allow thereby to incorporate the expectation of the channels whose fading coefficients are unknown in an easy way. Other beamforming schemes that can also handle CSIT uncertainties, as e.g. described in [27, 71, 125], could be alternatives that might lead to further improvements. Particularly the reference schemes with direct transmission considered here are thus somewhat pessimistic. Nevertheless, the SLNR based scheme does not require any iterative optimization procedures and is thus particularly attractive for application in practice. However, the evolutionary optimization can also be applied to other beamforming schemes. The link selection and power control can optimize the network performance with other approaches. So could the macro diversity scheme described in Chapter 3 also profit from improvements. Instead of choosing the serving BS for each MS individually in a binary like on-off manner (a certain BS transmits a signal to a MS with full power or not at all), a user could be served by multiple BSs with more variable and optimized signal powers. Besides this, the optimized power allocation can be extended to other forms of resource allocation to improve specific precoding schemes. Furthermore, the objective functions can be adapted or extended to include other considerations such as traffic demands for different users, resulting energy consumption, or others. To this end, a target rate can e.g. be specified for each user which should be fulfilled if it is feasible. The network performance can then be optimized with respect to increasing the rates on top of that or other objectives can be treated.

The presented solution to optimize post-cellular networks with the help of many low-complexity relays provides therefore a very dynamic and flexible approach to enhance the performance of future mobile communication networks. The different components are chosen and developed such that they can be applied to practical systems. The relays do not require any CSI and the channels to the possibly moving MSs need only to be known by their statistics. The static relays further enable massive MIMO at the BSs and the signal processing tasks are limited to involve only moderate complexity. We however assumed throughout the thesis that all nodes are perfectly synchronized. In order to bring the concept of the relay carpet to practice, imperfections at the relays such as frequency drifts, synchronization errors between cooperating BSs and others need to be taken into account. We also limited the discussion to static channels to avoid complicated models of time varying fading. We argued conceptually that the coherence time of the relay cells is much larger than the one to moving MSs. For practical implementation, however, this aspects need to be validated and the impairments due to imperfections need to be studied further. Nevertheless, we see the developed concepts as a promising way to bring mobile communication to the next generation and beyond.

9

Conclusions

In this last chapter, we recapitulate the achievements of this work and classify the implications the results have on mobile communication networks of the next generations. Besides this, we also discuss open aspects and provide directions for future work that is necessary to bring the concepts introduced in this thesis to practice.

9.1 Achievements & Insights

Cellular networks are mainly interference limited, which makes sophisticated interference management essential for a good performance. In order to provide ubiquitous access to high data rates, the available resources need to be exploited as much as possible, e.g. with coordination and cooperation. Different nodes that are of different complexities, that have diverse signal processing capabilities as well as coverage ranges have to coexist and work together to contribute to a flexible and dynamic service in large areas. Current research trends thus include aspects as accurate beamforming with massive MIMO and cooperation, heterogeneous networks with small cells, beamforming, and relaying. By the transition from cooperative cells to abandoning the cellular structure of the networks and introducing ubiquitous relaying, this thesis attempts to combine these approaches to allow future networks to achieve and fulfill their requirements.

9.1.1 Restricted Cooperation

In the first part, this thesis develops a unified framework that allows to evaluate the potential of BS cooperation in realistic setups. The focus is thereby set on block ZF as an efficient approach to realize joint DL signal processing across multiple BSs and additional supporting nodes of heterogeneous networks. Due to the nulling of the interference in clustered cooperation areas, a convex optimization problem can be formulated that allows to efficiently find a solution that maximizes the performance in the networks. Thereby, different aspects such as the sum rate, the minimal rate achievable in a cooperation area, or minimal transmit power required to achieve a certain QoS are considered. The potential performance gains that can be obtained are studied under realistic scenarios and assumptions that are inspired by the definitions of the current LTE and LTE-Advanced specifications.

The studied scenarios show that the applied cooperation schemes can lead to significant performance gains, are robust to imperfections, and are applicable to practice. Especially on cell edges between cooperating infrastructure nodes, the data rates can be considerably increased and outage probabilities reduced. The block ZF approach, however, is only a suboptimal approach that allows an efficient calculation of an approximated optimization problem. Different beamforming strategies might therefore provide better results. The leakage based precoding, as introduced later in this work, has more advantages than block ZF, also in clustered cooperation areas, as also out-of-cell interference can be considered in the beamforming and no stringent requirements on the number of antennas have to be met. Nevertheless, the framework provides important insights as how cooperative cellular networks should be planned. An appropriate orientation of the BS antennas and the placement of supporting nodes have a strong influence on the performance and the spatial distribution of the achievable data rates in the area of service. The best performance can be achieved when infrastructure nodes as BSs, RRHs, or relays are placed around the corresponding area of service and are evenly distributed. If the complexity that comes with joint beamforming and the required data exchange between cooperating nodes is feasible, as many nodes as possible should cooperate with each other. Large cooperation clusters in dense networks can solve the problem of interference limitedness and ubiquitous access to high data rates to a large extent. If the complexity is too high, the cooperation has to be locally restricted. The performance in certain areas can however be improved by installing relays or other low power nodes. One fundamental problem of conventional cellular networks remains however also in cooperative ones: With appropriate network planning, very homogenous rates can be achieved within a cooperation area. These areas, however, need to be separated from each other to reduce the inter-cluster interference, e.g. with sectorized antennas and frequency allocation, as it is done in traditional cells.

9.1.2 Distributed Cooperation with Relays

Making cellular networks heterogeneous by densifying the infrastructure nodes and installing additional nodes such as relays is a solution to increase the capacity of these networks. If more nodes with more antennas are included, more users can be served at the same time and spectral efficiency can be increased. In practical networks, however, it is often expensive, difficult, or even impossible to find appropriate sites for infrastructure nodes. For massive deployment, as it is necessary to satisfy the expected growth in the number of devices and data traffic in the future, low complexity nodes that can seamlessly be integrated with low costs are desirable. To this end, wireless relays seem to be an attractive possibility as they do not need to be connected to the wired backbone. Such relays should be as simple as possible such that large numbers of them can be installed throughout the network. DF relays that are part of cooperation clusters are however quite complex in their functionalities. AF relays might therefore be a more appropriate alternative as they can potentially be implemented with lower complexity. The second part of this thesis thus studies how such AF relays can be included into mobile communication networks.

With a distributed relay gain optimization, AF relays are able to shape the effective channel between terminal nodes in a way that interference is cancelled and the data rates increased. The developed gradient based optimization algorithm allows the source nodes (BSs) and relays to jointly optimize their precoding and combination matrices with local CSI and very limited feedback only. The overhead that comes from the distributed optimization does essentially not scale with the number of involved relays. This makes the precoding and relay gain allocation particularly suitable for channel tracking in slow fading environments and does not require explicit cooperation via the backhaul.

Additional performance gains can be achieved when AF relays form linear combinations of signals on different subcarriers. With carrier cooperative relaying, the performance gap to the (currently still unknown) capacity of relay networks that might only be achievable with complex non-linear coding can further be reduced. However, optimizing large networks over many subcarriers also increases complexity. The number of relays and their functionalities and complexity thus lead to a tradeoff. The better performance with more sophisticated relays can be balanced with a higher number of nodes of lower complexity.

9.1.3   The Relay Carpet

In order to benefit from a combination of large antenna arrays, cooperation, and relaying in large networks and to find the role of relays in which they can contribute most to high performance, the relay carpet concept is developed. Thereby, various relaying architectures and strategies are compared with each other with respect to data rates, robustness, and complexity. Simple AF relays without any CSI turn thereby out to be very effective. Implemented as FDD relays that perform a simple frequency conversion and apply a fixed-gain amplification, such devices can be implemented with very low complexity and can hence be installed in massive numbers. Especially when used with two-way relaying, high performance gains can be achieved as compared to conventional single hop networks. In this case, however, additional interference is introduced that has to be canceled by a simple form of BS cooperation.

Moreover, the relay carpet has also other advantages. The massive deployment of relays and letting the MSs be served by relays instead of the BSs directly simplifies the task of the BSs as they see the static relays as the nodes they communicate with and not the possibly moving MSs. With this, sophisticated beamforming on the first hop channel can be realized with high accuracy, even if MSs move with high velocity and the second hop channel undergoes fast fading. The static or only slowly fading first hop channels are thus an enabler of large antenna arrays and massive MIMO. Static relays can thereby also solve or at least mitigate the problem of pilot contamination. The relays in turn provide homogeneously distributed high data rates as they are in close vicinity of the MSs to be served. Shadowing and large pathloss can thus be avoided and network operators do not have to rely on random propagation channels anymore but can achieve much more predictable conditions. With appropriate amplification gains at the relays, the networks can also be operated with less power, as there are no large distances with high pathloss anymore. Ubiquitous relaying is thus an effective means to combine large antenna arrays, distributed cooperation with accurate interference management, and cell densification. The transmit power or the amplification gains of the relays need however to be adjusted appropriately, as these nodes also induce additional noise and interference into the network. Power control is therefore necessary to be able to benefit from many active relays.

9.1.4   Post-Cellular Networks

With the relay carpet, the MSs are no longer served by the BSs directly but by relays in their close vicinity, which makes cell boundaries less important. By the introduction of post-cellular networks, we go one step further and abandon such boundaries completely. If a certain area has to be served, it might be difficult or even impossible to find sufficiently many appropriate BS sites. Such areas without backhaul access points can be covered by more distant BSs and the help of many relay nodes. With a flexible and dynamic BS clustering and relay routing, each MS can then be served by its own optimized cluster of BSs and relays. In this way, the network can dynamically adapt to the network setup, the involved users, and the current traffic demand. With leakage based precoding at the BSs and simple fixed gain relays whose amplification gains are set to appropriate values, no conditions on the number of antennas have to be fulfilled and users can experience high data rates even if there is no wired backhaul access close by or the connection is limited. As the proposed algorithm can deal with arbitrary numbers of users and relays in an efficient way, more and more relays can be deployed, which further increases the network performance. With this beneficial scaling, the network can grow evolutionarily and network operators are more flexible with their network planning.

By studying cooperative communication for mobile networks, we can conclude that future networks should be highly dynamic and flexible to overcome the boundaries of current cellular networks. By dynamically adapting the available infrastructure nodes to the current network utilization, the resources available in the entire network can be allocated to the users with their different demands and under practical constraints such as different backhaul connection capacities, computational capabilities, and transmit powers. The spatial dimension thus evolves to a resource that can dynamically be allocated to the users as e.g. time or frequency blocks in current LTE/LTE-Advanced systems. For an optimization of these resources, the post-cellular network architecture with massive relaying offers an efficient possibility. The schemes developed in this thesis, however, contain in certain cases only suboptimal approaches. The considered networks are highly complex as they are large, include different types of nodes whose signals are all interdependent on each other, and have to deal with practical conditions which make rigorous mathematical treatment difficult. Nevertheless, the results obtained show promising performance gains and advantages that make the combination of large BS antenna arrays, flexible cooperation, and massive relaying a key concept to solve the challenges for mobile communication networks of the next generation and beyond.

9.2 Outlook & Future Work

In this thesis, we laid the foundation for the concepts of the relay carpet and post-cellular networks. For this network architectures, different transmission strategies are chosen and combined in a way that each node can contribute to the network performance according to its possibilities and power. These schemes are carefully chosen, adapted, and developed such that distributed cooperation between the different nodes can be realized with low complexity. Several aspects remain however suboptimal. For a practical implementation of the post-cellular relay carpet, open problems thus still have to be solved and analyzed and further investigations are necessary to gain a more fundamental understanding of the mechanisms with which the distributed nature of these networks can be further optimized. In the remainder of this chapter, we summarize the limitations of the achieved studies and outline open challenges and further work that is necessary to bring the concepts to successful implementation.

The results obtained in this work are for specific channel and signaling models and assumptions, some of which are idealized while others include certain imperfections. Further practical difficulties however have been discussed only on a conceptual level or are ignored. In order to make the relay carpet feasible for practice, more aspects need to be considered and understood. These include among others a more specific design of the BSs and relays. For the latter, we argued that when built as FDD relays, their implementation is simple. In practice, however, aspects as synchronization, frequency drifts, RF filters, etc. need to be taken into account and carefully designed and other problems might appear. Also the control link over which the BSs can control necessary settings in the relays such as the amplification gains, wake up commands, or the initiation of training phases needs to be designed in more detail. A possible implementation of that can be realized as a finite state machine that is also simple to build and of low cost. It remains however to investigate further what other practical difficulties need to be solved to be able to apply low-complexity relays and if they really are of low cost as claimed and assumed in this work.

Another step towards a successful realization of relay-carpet-like networks is to gain more fundamental insights into the tradeoff between node density, complexity (costs),

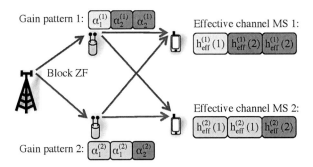

Gain pattern 1: $\alpha_1^{(1)}$ $\alpha_2^{(1)}$ $\alpha_2^{(1)}$

Block ZF

Gain pattern 2: $\alpha_1^{(2)}$ $\alpha_1^{(2)}$ $\alpha_2^{(2)}$

Effective channel MS 1: $h_{eff}^{(1)}(1)$ $h_{eff}^{(1)}(2)$ $h_{eff}^{(1)}(2)$

Effective channel MS 2: $h_{eff}^{(2)}(1)$ $h_{eff}^{(2)}(1)$ $h_{eff}^{(2)}(2)$

Figure 9.1.: Illustration of blind interference alignment with the help of relays.

and performance. The system under consideration is very complicated with many different nodes. Finding optimal solutions is therefore difficult and it remains unclear if theoretical results obtained under certain assumptions also carry over to more realistic scenarios. Nevertheless, fundamental results or bounds on the relay carpet as a whole can lead to further insights into the design of such networks and their transmission protocols. The approaches introduced and discussed in this thesis work well and are carefully chosen and developed for a beneficial interplay between the different nodes. Optimized solutions for rigorously formulated mathematical models of the entire network are however missing.

With the framework of the relay carpet and ubiquitous relaying in post-cellular networks, also other transmission strategies or ideas can be applied. As an example, the relays not only have to forward the signals to the MSs, but they can also specifically shape the effective channel between BSs and MSs into certain forms. One possibility could e.g. be to realize blind interference alignment similar as proposed in [44, 62, 96]. To this end, the relays can change their amplification gains in a specific and predefined manner such that different MSs see the links of the desired signals and interference with different fading blocks as illustrated in the minimal example in Fig. 9.1 with a BS with 2 antennas and two single antenna MSs. With this, each MS can cancel the interference intended for the other MS and each user can receive two symbols without interference in three time slots, which leads to a spatial multiplexing gain of $\frac{4}{3}$ for this small network. In contrast to [44] where a similar scheme is introduced that relies on reconfigurable antennas at the terminals, the relays can fulfill the channel shaping task and no changes in the RF chains of the terminal nodes are necessary.

Other extensions of the post-cellular relay carpet discussed in this thesis can include a combination of one- and two-hop links. Instead of serving the MSs either directly by the BSs or via relays as described in Chapter 8, each MS can receive signals from the BSs and the relays simultaneously as depicted for a simplified network in Fig. 9.2. If the MS jointly decodes the signals in both frequency bands of the BS and relays and the two links are weighted accordingly, an optimal combination of both links can be achieved which further increases the achievable rates. Moreover, the BS can also transmit additional symbols in the relay frequency band. This allows users that are close to a BS to benefit from strong single hop links and the prelog loss due to half-duplex relaying can be reduced.

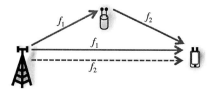

Figure 9.2.: The MSs jointly decode the combination of direct BS signals and those that are forwarded by the relays.

Furthermore, ongoing research[1] shows that also full-duplex relays can beneficially be applied to the post-cellular relay carpet. With such relays, additional loop or self interference is introduced but the performance is not affected by a prelog factor $\frac{1}{2}$ that would arise by the use of half-duplex relays. By a careful allocation of the different links, the spectral efficiency can further be increased.

Another extension is to serve users that are far way from BSs over multiple relay stages. With a multi-hop transmission, each relay forwards the received message one hop further until the message reaches its destination. As a consequence, longer distances can be overcome and strong BS clusters can serve much wider areas.

Instead of installing dedicated relays as fixed infrastructure nodes, MSs or other devices that are currently unused can also assume the functionality of forwarding signals to active users. With user cooperation, MSs that do not have relays in their close vicinity can still benefit from the advantages of two- or multi-hop communication and the network can be expanded without installing more relays. This can e.g. be particularly

---

[1]F. Marti, "Distributed Cooperation and Relaying in Post-Cellular Networks," Master Thesis, ETH Zürich, May 2015, available on request.

interesting in temporary hot spots as festivals or other events where a high density of users is present for a short time period.

Multi-hop forwarding and user cooperation can also be carried one step further to a more ad-hoc like approach. If relays and/or idle MSs start to forward messages as soon as they have received them, a much more dynamic spreading or flooding of the signals can be achieved [126]: Each node that has received a message can contribute to the cooperative transmission to increase the signal strength for the more distant devices until the message reaches the intended user. With such an approach, post-cellular networks can be made even more dynamic and the aforementioned techniques could seamlessly be incorporated into the network. Post-cellular networks with ubiquitous relaying are thus a promising approach to fulfill the requirements and demands of future networks and provide a framework that can flexibly and dynamically be extended to other emerging technologies.

Simulation Parameters

In this Appendix, we provide detailed descriptions of the parameters used for the various simulations discussed in this thesis. The parameters are sorted by the chapter in which the simulation results are presented.

Chapter 3: Locally Restricted BS Cooperation

Table A.1.: Simulation parameters for Chapter 3.

Simulation	Specific parameters		General parameters
Fig. 3.4	Tx scheme	No cooperation	Active cells/sectors: 12/36
Reference		no beamforming	Mobiles: 1 MS per sector
5% out. rate	Orientation	0°	Antennas: $N_B = 4$ per sector
	Freq. alloc.	FDMA	$N_M = 2$ per sector
Fig. 3.5	Tx scheme	No cooperation	BS antennas: directive (3.25)
Beamforming		max-min opt.	MS antennas: omni-directional
5% out. rate	Orientation	0°	Tx power: $P_B = 80\,\text{W}$
	Freq. alloc.	FDMA	Noise power: $\sigma_w^2 = 5 \cdot 10^{-12}\,\text{W}$
Fig. 3.6	Tx scheme	No cooperation	Carrier frequency: 2.6 GHz
Beamforming		max-min opt.	Bandwidth: 100 MHz
Outage prob.	Orientation	0°	Channel: urban micro-cell
Av. user rate	Freq. alloc.	FDMA	BS distance: 700 m
			OCI: spatially white + i.i.d.
			channel
Fig. 3.7	Tx scheme	Sector cooperation	BS antennas: omni-directional
Sector coop.		max-min opt.	
Av. user rate	Orientation	0°	
Outage prob.	Freq. alloc.	FDMA	

Fig. 3.8	Tx scheme	3 BS cooperation	
3 BS coop.		max-min opt.	
Outage prob.	Orientation	0° (left), 30° (right)	
5% out. rate	Freq. alloc.	FDMA	
Fig. 3.9	Tx scheme	3 BS cooperation	
3 BS coop.		max-min opt.	
5% out. rate	Orientation	0° (left), 30° (right)	
	Freq. alloc.	Reuse 1 (top),	
		FDMA (bottom)	
Fig. 3.11	Tx scheme	6 BS cooperation	
6 BS super-cells		max-min opt.	
Av. user rate	Orientation	super-cell	
5% out. rate	Freq. alloc.	Reuse 1	
Fig. 3.12	Tx scheme	Macro diversity	
Macro diversity		(max-min)	
5% out. rate	Orientation	0°	
	Freq. alloc.	FDMA	Active cells/sectors: 12/36
Fig. 3.13	Tx scheme	Macro diversity UB	Mobiles: 1 MS per sector
Macro div. UB		(privileged)	Antennas: $N_B = 4$ per sector
5% out. rate	Orientation	0°, 30°	$N_M = 2$ per sector
	Freq. alloc.	FDMA, reuse 1	BS antennas: directive (3.25)
Fig. 3.14	Tx scheme	No cooperation	MS antennas: omni-directional
No coop.		3 BS coop. (max-min)	Tx power: $P_B = 80\,\text{W}$
3 BS coop.		Macro div. UB	Noise power: $\sigma_w^2 = 5 \cdot 10^{-12}\,\text{W}$
Macro div. UB	Orientation	0°, 30°	Carrier frequency: 2.6 GHz
Empirical CDF	Freq. alloc.	FDMA, reuse 1	Bandwidth: 100 MHz
Fig. 3.15	Tx scheme	No cooperation	Channel: urban micro-cell
No coop.		3 BS coop. (max-min)	BS distance: 700 m
3 BS coop.		6 BS coop. (max-min)	OCI: spatially white + i.i.d.
Macro div. UB		Macro div. UB	channel
6 BS super-cell	Orientation	0°, 30°	
Empirical CDF	Freq. alloc.	FDMA, reuse 1	
Fig. 3.16	Tx scheme	No cooperation	
Sum rate maxi-		3 BS coop. (sum-rate)	
mization		Macro div. UB	
Empirical CDF	Orientation	0°, 30°	
	Freq. alloc.	FDMA, reuse 1	

Fig. 3.17	Tx scheme	3 BS cooperation	
3 BS coop.		max sum rate	
Sum rate	Orientation	30°	
Av. user rate	Freq. alloc.	FDMA, reuse 1	
Fig. 3.18	Tx scheme	No cooperation	
Power minimi-		Sec. coop (min power)	
zation		3 BS coop (min power)	
Empirical CDF		6 BS coop (min power)	
	Orientation	0°, 30°	
	Freq. alloc.	Reuse 1	
Fig. 3.19	Tx scheme	No ccooperation	Channel: rural macro-cell
No coop.		3 BS coop. (max-min)	BS distance: 1.5 km
3 BS coop.		6 BS coop. (max-min)	
Macro div. UB		Macro div. UB	
6 BS super-cell	Orientation	0°, 30°	
Empirical CDF	Freq. alloc.	FDMA, reuse 1	
Fig. 3.20	Tx scheme	2 BS coop. (max-min)	Channel: Rayleigh $\mathcal{CN}(0,1)$
Max-min		gradient search	Antennas: $N_B = 4$, $N_M = 2$
gradient	Num BS	2	all omni-directional
	Num MS	2	Tx power: $P_B = 1$
			Noise power: $\sigma_w^2 = 0.1$

Chapter 4: Small Cells and DF Relaying

Table A.2.: Simulation parameters for Chapter 4.

Simulation	Specific parameters		General parameters
Fig. 4.3	Tx scheme	Link selection	Active cells/sectors: 1
Relays		DF relays	Supporting nodes: 1 per sector
Varying distance		no beamforming	Mobiles: 1 per sector
5% out. rate	Network	0°, FDMA	Antennas: $N_B = 4$ (directive)
Outage prob.	BS-RS dist.	varying	$N_S = 4$ (directive)
	SN power	$P_S = 6$ W	$N_M = 2$(omni-directional)
Fig. 4.4	Tx scheme	Link selection	BS power: $P_B = 80$ W
Relays		DF relays/femto-cells	Noise power: $\sigma_w^2 = 5 \cdot 10^{-12}$ W
Femto-cells		no beamforming	Carrier frequency: 2.6 GHz
5% out. rate	Network	0°, FDMA	Bandwidth: 100 MHz

Av. user rate	BS-RS dist.	3000 m	Channel: rural macro-cell
	SN power	varying	BS distance: n.a.
Fig. 4.5	Tx scheme	Link selection	No OCI
Relays		DF relays/femto-cells	
Femto-cells		no beamforming	
Average 5%	Network	0°, FDMA	
outage rate	BS-RS dist.	varying	
	SN power	varying	
Fig. 4.6	Tx scheme	Link selection	Active cells/sectors: 12/36
Relays		DF relays/femto-cells	Supporting nodes: 2 per sector
Femto-cells		no beamforming	Mobiles: 1 per sector
Outage prob.	Network	0°, FDMA	Antennas: $N_B = 4$ (directive)
5% out. rate	SN location	cell corners	$N_S = 4$ (directive)
	SN power	$P_S = 6\,\text{W}$	$N_M = 2$(omni-directional)
Fig. 4.7	Tx scheme	Link selection	BS power: $P_B = 80\,\text{W}$
Relays		DF relays/femto-cells	Noise power: $\sigma_w^2 = 5 \cdot 10^{-12}\,\text{W}$
Femto-cells		no beamforming	Carrier frequency: 2.6 GHz
Av. user rate	Network	0°, FDMA/reuse 1	Bandwidth: 100 MHz
	SN location	cell corners	Channel: urban micro-cell
	SN power	$P_S = 6\,\text{W}$	BS distance: 700 m
Fig. 4.8	Tx scheme	Link selection	OCI: spatially white + i.i.d.
Relays		DF relays/femto-cells	channel
Femto-cells		no beamforming	
Urban micro-cell	Network	0°, FDMA/reuse 1	
Empirical CDF	SN location	cell corners	
	SN power	$P_S = 6, 80\,\text{W}$	
Fig. 4.9	Tx scheme	Link selection	Channel: rural macro-cell
Relays		DF relays/femto-cells	BS distance: 1500 m
Femto-cells		no beamforming	
Rural macro-cell	Network	0°, FDMA/reuse 1	
Empirical CDF	SN location	cell corners	
	SN power	$P_S = 6, 80\,\text{W}$	
Fig. 4.10	Tx scheme	Cooperation with	Channel: urban micro-cell
Cooperation +		assisting SNs	BS distance: 700 m
assisting SNs		(block ZF + max-min)	
Empirical CDF	Network	0°, reuse 1	
	SN location	cell corners	
	SN power	varying	

Fig. 4.11 Cooperation with/out SNs 5% out. rate	Tx scheme	Cooperation with/out assisting SNs (block ZF + max-min)	
	Network	0°/30°, FDMA/super-cell	
	SN location	cell corners	
	SN power	$P_S = 6\,\mathrm{W}$	
Fig. 4.12 Cooperation with/out SNs Empirical CDF	Tx scheme	Cooperation with/out assisting SNs (block ZF + max-min)	
	Network	0°/30°, FDMA/super-cell	
	SN location	cell corners	
	SN power	varying	
Fig. 4.13 Comparison Cooperation Empirical CDF	Tx scheme	Cooperation Relays/femto-cells	
	Network	0°/30°, FDMA/reuse1	
	SN location	cell corners	
	SN power	varying	
Fig. 4.14 Cooperation + assisting SNs Empirical CDF	Tx scheme	Cooperation with assisting SNs (block ZF + max-min)	Channel: varying CSI noise
	Network	0°, reuse 1	
	SN location	cell corners	
	SN power	varying	
Fig. 4.15 Cooperation DF relaying Av. 5% out. rate	Tx scheme	Cooperation DF relaying (max-min/link selec.)	
	Network	0°, reuse 1	
	SN location	cell corners	
	SN power	varying	
Fig. 4.16 Cooperation 5% out. rate	Tx scheme	3 BS cooperation CSI imperfections (max-min)	Channel: quantized CSI
	Network	30°, FDMA	
	SN location	cell corners	

Chapter 5: Distributed Cooperation with AF Relays

Table A.3.: Simulation parameters for Chapter 5, AF multihop relaying.

Simulation	Specific parameters		General parameters
Fig. 5.2	Tx scheme	Sum rate maximiza-	Channel: Rayleigh $\mathcal{CN}(0,1)$
Sum rate opt.		tion (gradient search)	Antennas: 2 at each node
gradient	Network	$N_0 = 2$ S-D pairs	all omni-directional
		$L = 2$ relay stages	Tx power: $P = 100$
		$N_1 = N_2 = 2$ relays	Noise power: $\sigma_w^2 = \sigma_n^2 = 1$
		per stage	No OCI
	CSI	Global, local	
Fig. 5.3	Tx scheme	Sum rate maximiza-	Tx power: varying
Sum rate opt.		tion (gradient search)	
gradient	Network	$N_0 = 2$ S-D pairs	
Empirical CDF		$L = 2$ relay stages	
		$N_1 = N_2 = 2$ relays	
		per stage	
	CSI	Local	
Fig. 5.4	Tx scheme	Sum rate maximiza-	Tx power: $P = 10$
Sum rate opt.		tion (gradient search)	
gradient		only relay/joint opt.	
Empirical CDF	Network	$N_0 = 2$ S-D pairs	
		$L = 1$ relay stage	
		$N_1 = 1, 2, 4$ relays	
	CSI	Local	
Fig. 5.5	Tx scheme	Sum rate maximiza-	Tx power: $P = 100$
Channel track-		tion (gradient search)	
ing		only relay/joint opt.	
gradient	Network	$N_0 = 2$ S-D pairs	
Empirical CDF		$L = 2$ relay stages	
		$N_1 = N_2 = 2$ relays	
		per stage	
	CSI	Varying imperfections	

Table A.4.: Simulation parameters for Chapter 5, subcarrier cooperative two-way relaying.

Simulation	Specific parameters		General parameters
Fig. 5.7 Subcarrier cooperation Sum rate single channel realization	Tx scheme	Block ZF Compare null space combinations	Channel: Rayleigh $\mathcal{CN}(0,1)$ Tx power: $P_s = P_r = 1$ Noise power: $\sigma_n^2 = \sigma_w^2 = 0.001$
	Network	$K = 2$ terminal pairs $L = 2$ relays	Antennas: omni-directional CSI: global
	Antennas	$M_k = 2$ (terminals) $N_l = 4$ (relays)	No OCI
	Subcarriers	$C = 4$	
Fig. 5.8 Subcarrier cooperation, null space selection Empirical CDF	Tx scheme	Block ZF with null space selection	Tx power: $P_s = P_r = C$ Noise power: $\sigma_n^2 = \sigma_w^2 = 0.01$
	Network	$K = 2$ terminal pairs $L = 2$ relays	
	Antennas	$M_k = 2$ (terminals) $N_l = 4$ (relays)	
	Subcarriers	$C = 4$	
Fig. 5.9 Subcarrier cooperation, null space selection Empirical CDF	Tx scheme	Block ZF with null space selection	Tx power: $P_s = P_r = 1$ Noise power: $\sigma_n^2 = \sigma_w^2 = 0.001$
	Network	$K = 2$ terminal pairs $L = 2$ relays	
	Antennas	$M_k = 2$ (terminals) $N_l = 4$ (relays)	
	Subcarriers	$C = 1, 2, 4, 6$	
Fig. 5.10 Subcarrier cooperation, null space selection Empirical CDF	Tx scheme	Block ZF with/out null space selection	Tx power: $P_s = P_r = C$ Noise power: $\sigma_n^2 = \sigma_w^2 = 0.001$
	Network	$K = 2$ terminal pairs $L = 2$ relays	
	Antennas	$M_k = 2$ (terminals) $N_l = 4$ (relays)	
	Subcarriers	$C = 4$	
Fig. 5.12 Cellular network direct, two-way relaying Empirical CDF	Tx scheme	FDMA, Block ZF with null space selection	Active sectors: 3 Mobiles: 1 MS per sector
	Network	$K = 3$ BS-MS pairs $L = 3$ relays per sector	BS antennas: directive (3.25) MS antennas: omni-directional
	Antennas	$M_k = 2$ (terminals) $N_l = 2, 3$ (relays)	Tx power: $P_B = 40\,\text{W}$ $P_M = 200\,\text{mW}$

Subcarriers $C = 1, 3$	Noise power: $\sigma_w^2 = 5 \cdot 10^{-12}\,\text{W}$
Orientation $0°$	Carrier frequency: 2.6 GHz
Freq. alloc. FDMA (reference)	Bandwidth: 100 MHz
reuse 1 (relaying)	Channel: urban micro-cell
	BS distance: 700 m

Chapter 6: Ubiquitous Relaying

Table A.5.: Simulation parameters for Chapter 6.

Simulation	Specific parameters		General parameters
Fig. 6.4	Tx scheme	Reference (block ZF)	Channel: urban micro-cell
Ubiquitous		all relay schemes	Active cells: 19
relaying	BS	Block ZF, waterfilling	Relays per cell: 6
all schemes	RS	AF/DF, 1-way/2-way	MSs per cell: 6
Av. sum rate		type A/type B	Antennas: $N_B = 24$
	MS	spatially white	$N_S = 4$
	CSI	local, perfect	$N_M = 2$
	Prelog	yes (in-band)	(all omni-directional)
Fig. 6.5	Tx scheme	Reference (block ZF)	BS power: $P_B = 40\,\text{W}$
Relaying		all relay schemes	RS power: $P_R = 6\,\text{W}$
all schemes	BS	Block ZF, waterfilling	MS power: $P_M = 200\,\text{mW}$
Empirical CDF	RS	AF/DF, 1-way/2-way	Noise power: $\sigma_n^2 = 5 \cdot 10^{-12}\,\text{W}$
Downlink		type A/type B	$\sigma_w^2 = 5 \cdot 10^{-12}\,\text{W}$
	MS	spatially white	Carrier frequency: 2.6 GHz
	CSI	local, perfect	Bandwidth: 100 MHz
	Prelog	no	Freq. alloc. reuse 1
Fig. 6.6	Tx scheme	Reference (block ZF)	BS-RS distance: 350 m
Relaying		all relay schemes	Deadzone: 233 m
all schemes	BS	Block ZF, waterfilling	OCI: same Tx scheme
Empirical CDF	RS	AF/DF, 1-way/2-way	
Uplink		type A/type B	
	MS	spatially white	
	CSI	local, perfect	
	Prelog	no	
Fig. 6.7	Tx scheme	Reference (block ZF)	BS-RS distance: varying

Relaying		Relaying schemes	Deadzone: varying
BS-RS distance	BS	Block ZF, waterfilling	
Mean user rate	RS	AF, 1/2-way, type B	
Downlink		DF, 1/2-way, type B	
	MS	spatially white	
	CSI	local, perfect	
	Prelog	no	
Fig. 6.8	Tx scheme	Reference (block ZF)	
Relaying		Relaying schemes	
BS-RS distance	BS	Block ZF, waterfilling	
Mean user rate	RS	AF, 1/2-way, type B	
Uplink		DF, 1/2-way, type B	
	MS	spatially white	
	CSI	local, perfect	
	Prelog	no	
Fig. 6.9	Tx scheme	Reference (block ZF)	BS-RS distance: $350\,\mathrm{m}$
Relaying		Relaying schemes	Deadzone: $233\,\mathrm{m}$
Tx power	BS	Block ZF, waterfilling	Tx power: $P_\mathrm{B} = P_\mathrm{R} = P_\mathrm{M}$
Mean user rate	RS	AF, 1/2-way, type A/B	varying
Downlink		DF, 1/2-way, type B	
	MS	spatially white	
	CSI	local, perfect	
	Prelog	no	
Fig. 6.10	Tx scheme	Reference (block ZF)	
Relaying		Relaying schemes	
Tx power	BS	Block ZF, waterfilling	
Mean user rate	RS	AF, 1/2-way, type A/B	
Uplink		DF, 1/2-way, type B	
	MS	spatially white	
	CSI	local, perfect	
	Prelog	no	
Fig. 6.11	Tx scheme	Reference (block ZF)	BS power: $P_\mathrm{B} = 40\,\mathrm{W}$
Relaying		Relaying link selection	RS power: $P_\mathrm{R} = 6\,\mathrm{W}$
Av. user rate	BS	Block ZF, waterfilling	
area plots	RS	AF type A/B	
Downlink		Link selection	
	MS	n.a.	
	CSI	local, perfect	

	Prelog	no	
Fig. 6.12 Relaying all schemes Empirical CDF Downlink	Tx scheme	Reference (block ZF) all relay schemes	For comparison:
	BS	Block ZF, waterfilling	• Reference, FDMA, 0°
	RS	AF/DF, 1-way/2-way type A/type B	• 3 BS cooperation: 12 cells, FDMA, 30°
	MS	spatially white	• 6 BS super cells
	CSI	local, perfect	
	Prelog	no	
Fig. 6.13 Relaying Increasing den- sity Mean user rates Up-/downlink	Tx scheme	Reference (block ZF) Relaying schemes	Relays per cell: K varying
			MSs per cell: $M = K$ varying
	BS	Block ZF, waterfilling	Node location: random
	RS	AF, 1/2-way, type A/B DF, 1/2-way, type B	BS antennas: varying
			$(N_{\mathrm{B}} = M \cdot N_{\mathrm{R}})$
	MS	spatially white	
	CSI	local, perfect	
	Prelog	no	
Fig. 6.15 Relaying all schemes Mean user rates Imperfect CSI Up-/Downlink	Tx scheme	Reference (block ZF) all relay schemes	Relays per cell: 6
			MSs per cell: 6
	BS	Block ZF, waterfilling	Bs Antennas: $N_{\mathrm{B}} = 24$
	RS	AF/DF, 1-way/2-way type A/type B	Channel: CSI imperfections
	MS	spatially white	
	CSI	imperfections	
	Prelog	no	
Fig. 6.16 Relaying Increasing den- sity Mean user rates Imperfect CSI Up + downlink	Tx scheme	Reference (block ZF) Relaying schemes	Relays per cell: K varying
			MSs per cell: $M = K$ varying
	BS	Block ZF, waterfilling	Node location: random
	RS	AF, 1/2-way, type A/B DF, 1/2-way, type B	BS antennas: varying
			$(N_{\mathrm{B}} = M \cdot N_{\mathrm{R}})$
	MS	spatially white	Channel: CSI imperfections
	CSI	imperfections	
	Prelog	no	

Chapter 7: The Cellular Relay Carpet

Table A.6.: Simulation parameters for Chapter 7.

Simulation	Specific parameters		General parameters
Fig. 7.6	Tx scheme	Reference (block ZF)	Channel: urban micro-cell
Relaying		Relaying	MSs per cell: 6
AF 1-way type A	BS	Block ZF, waterfilling	Antennas: $N_B = 24$
Av. sum rate	RS	AF 1-way type A	$N_S = 2$
Down-/uplink	MS	spatially white	$N_M = 2$
	CSI	local, perfect	(all omni-directional)
	Prelog	yes (in-band)	BS power: $P_B = 40\,\text{W}$
	Active cells	7	RS gain: $\alpha_{c,k} = L_{c,k} \cdot \gamma_{c,k}$
	RS per MS	1, 2, 3	MS power: $P_M = 200\,\text{mW}$
Fig. 7.7	Tx scheme	Reference (block ZF)	Noise power: $\sigma_n^2 = 5 \cdot 10^{-12}\,\text{W}$
Relaying		Relaying	$\sigma_w^2 = 5 \cdot 10^{-12}\,\text{W}$
AF 1-way type A	BS	Block ZF, waterfilling	Carrier frequency: 2.6 GHz
Av. sum rate	Antennas	$N_B = K_m M N_R$	Bandwidth: 100 MHz
Downlink	RS	AF 1-way type A	Freq. alloc. reuse 1
		Selection/combination	BS-RS distance: 350 m
	MS	spatially white	Deadzone: 233 m
	CSI	local, perfect	OCI: same Tx scheme
	Prelog	yes	
	Active cells	1	
	RS per MS	$K_m = 1, \ldots, 10$	
Fig. 7.8	Tx scheme	Reference (block ZF)	
Relaying		Relaying	
AF 1-way type A	BS	Block ZF, waterfilling	
Av. sum rate	Antennas	$N_B = K_m M N_R$	
Downlink	RS	AF 1-way type A	
		Selection/combination	
	MS	spatially white	
	CSI	local, perfect	
	Prelog	yes	
	Active cells	7	
	RS per MS	$K_m = 1, \ldots, 10$	
Fig. 7.9	Tx scheme	Reference (block ZF)	Channel: urban micro-cell

Relaying		Relaying	Active cells: 7
min. power	BS	Block ZF, waterfilling	Relays per cell: 6
Empirical CDF	RS	AF, 1-way type A/B	MSs per cell: 6
	Control	min. power	Antennas: $N_B = 24$
	CSI	local, perfect	$N_S = 4$
	Prelog	no	$N_M = 2$
Fig. 7.10	Tx scheme	Reference (block ZF)	(all omni-directional)
Relaying		Relaying	BS power: $P_B = 40\,\mathrm{W}$
Max-min	BS	Block ZF, waterfilling	RS power: $P_R = 6\,\mathrm{W}$
Empirical CDF	RS	AF 1-way, type A/B	MS power: $P_M = 200\,\mathrm{mW}$
	Control	Max-min rate	Noise power: $\sigma_n^2 = 5 \cdot 10^{-12}\,\mathrm{W}$
	CSI	local, perfect	$\sigma_w^2 = 5 \cdot 10^{-12}\,\mathrm{W}$
	Prelog	no	Carrier frequency: 2.6 GHz
Fig. 7.11	Tx scheme	Reference (block ZF)	Bandwidth: 100 MHz
Relaying		Relaying	Freq. alloc. reuse 1
Min. outage	BS	Block ZF, waterfilling	BS-RS distance: 350 m
Empirical CDF	RS	AF 1-way, type A/B	Deadzone: 233 m
	Control	min. outage	OCI: same Tx scheme
	CSI	local, perfect	
	Prelog	no	

Chapter 8: Post-Cellular Networks

Table A.7.: Simulation parameters for Chapter 8.

Simulation	Specific parameters		General parameters
Fig. 8.3	Tx scheme	BS cooperation	Channel: urban post-cellular
Dynamic		Static clustering	Area: $1000\,\mathrm{m} \times 1000\,\mathrm{m}$
BS cooperation		Evolutionary opt.	Num BS: 5
Empirical CDF		• max sum rate	Num MS: 15
Downlink		• max min rate	No relays
		WMMSE	Node location: random
	BS	SLNR, power control	Antennas: $N_B = 4$
	Control	$N_{\mathrm{bits}} = 32$	$N_M = 2$
	Backhaul	$C_B = \infty$	N_R : n.a.
	RS	no	(all omni-directional)
	CSI	local, perfect	Max BS power: $P_B = 40\,\mathrm{W}$
	Prelog	1 (direct transmission)	Max RS power: n.a.

Fig. 8.4	Tx scheme	BS cooperation	Noise power: $\sigma_w^2 = 5 \cdot 10^{-12}\,\mathrm{W}$
Dynamic		Static clustering	Carrier frequency: 2.6 GHz
BS cooperation		Evolutionary opt.	Bandwidth: 100 MHz
Outage prob.		• min outage	Freq. alloc. reuse 1
Downlink		WMMSE	No OCI
	BS	SLNR, power control	All nodes same Tx scheme
	Control	$N_{\mathrm{bits}} = 32$	
	Backhaul	C_{B} varying	
	RS	no	
	CSI	local, perfect	
	Prelog	1 (direct transmission)	
Fig. 8.5	Tx scheme	BS cooperation	
Dynamic		Static clustering	
BS cooperation		Evolutionary opt.	
Av. user rate		• max sum rate	
Downlink	BS	SLNR, power control	
	Control	N_{bits} varying	
	Backhaul	$C_{\mathrm{B}} = \infty$	
	RS	no	
	CSI	local, perfect	
	Prelog	1 (direct transmission)	
Fig. 8.6	Tx scheme	Macro div. UB	Active cells/sectors: 12/36
BS cooperation		3 BS coop (Block ZF	Mobiles: 1 MS per sector
5% out. rate		+ max min)	Antennas: $N_{\mathrm{B}} = 4$ per sector
Downlink		3 BS SLNR (static)	$N_{\mathrm{M}} = 2$ per sector
		Global SLNR (static)	BS antennas: directive (3.25)
	Orientation	30°	MS antennas: omni-directional
	Freq. alloc.	FDMA, reuse 1	Tx power: $P_{\mathrm{B}} = 80\,\mathrm{W}$
Fig. 8.7	Tx scheme	No cooperation	Noise power: $\sigma_w^2 = 5 \cdot 10^{-12}\,\mathrm{W}$
BS cooperation		Macro div. UB	Carrier frequency: 2.6 GHz
Empirical CDF		3 BS coop (Block ZF)	Bandwidth: 100 MHz
Downlink		3 BS + SN coop.	Channel: urban micro-cell
		6 BS super-cell	BS distance: 700 m
		Global SLNR (static)	OCI: spatially white + i.i.d.
	Orientation	0°, 30°	channel
	Freq. alloc.	FDMA, reuse 1	
Fig. 8.8	Tx scheme	BS cooperation	Channel: urban post-cellular

Convergence with relays Sum rate Downlink		with relaying Evolutionary opt: • max sum rate	Area: $2000\,\text{m} \times 1000\,\text{m}$ Num BS: 5 Num MS: 15
	BS	SLNR, power control	Num RS: 30
	RS	Spatially white, gain control	Node location: random Antennas: $N_\text{B} = 4$
	Control	$N_\text{bits} = 32$	$N_\text{M} = 2$
	Backhaul	$C_\text{B} = \infty$	$N_\text{R} = 2$
	CSI BS	Local, perfect	(all omni-directional)
	CSI RS	No	Max BS power: $P_\text{B} = 40\,\text{W}$
	Prelog	$\frac{1}{2}$ (2-hop)	Max RS power: $P_\text{R} = 4\,\text{W}$
Fig. 8.9 Dynamic coop. with relays Sum rate Downlink	Tx scheme	BS cooperation with relaying Evolutionary opt. • max sum rate	Noise power: $\sigma_n^2 = 5 \cdot 10^{-12}\,\text{W}$ $\sigma_w^2 = 5 \cdot 10^{-12}\,\text{W}$ Carrier frequency: 2.6 GHz Bandwidth: 100 MHz
	BS	SLNR, power control	Freq. alloc. reuse 1
	RS	Spatially white, gain control	No OCI All nodes same Tx scheme
	Control	N_bits : varying	
	Backhaul	C_B : varying	
	CSI BS	Local, perfect	
	CSI RS	No	
	Prelog	$\frac{1}{2}$ (2-hop)	
Fig. 8.10 Dynamic coop. with relays Sum rate Downlink	Tx scheme	BS cooperation with relaying Evolutionary opt. • max sum rate	Num RS: 5, ..., 60 Num MS: 5, 10, 15
	BS	SLNR, power control	
	RS	Spatially white, gain control	
	Control	$N_\text{bits} = 32$	
	Backhaul	$C_\text{B} = \infty$	
	CSI BS	Local, perfect	
	CSI RS	No	
	Prelog	$\frac{1}{2}$ (2-hop)	
Fig. 8.11 Dynamic coop. with relays	Tx scheme	BS cooperation with relaying Evolutionary opt.	

Min rate		• max min rate	
Downlink	BS	SLNR, power control	
	RS	Spatially white,	
		gain control	
	Control	$N_{\text{bits}} = 32$	
	Backhaul	$C_B = \infty$	
	CSI BS	Local, perfect	
	CSI RS	No	
	Prelog	$\frac{1}{2}$ (2-hop)	
Fig. 8.12	Tx scheme	BS cooperation	Area: $1000\,\text{m} \times 1000\,\text{m}$
Dynamic coop.		with relaying	Num BS: 5, $N_B = 8$
Relay carpet		Evolutionary opt.	Num RS: 50
Empirical CDF		• max min rate	Num MS: 15
Minimum rate	BS	SLNR, power control	
	RS	Spatially white,	
		gain control	
	Control	$N_{\text{bits}} = 32$	
	Backhaul	$C_B = \infty$	
	CSI BS	Local, perfect	
	CSI RS	No	
	Prelog	$\frac{1}{2}$ (2-hop)	
Fig. 8.13	Tx scheme	BS cooperation	Area: $1000\,\text{m} \times 1000\,\text{m}$
Massive MIMO		with relaying	Num BS: 1, $N_B = 50$ (center)
with relays		Evolutionary opt.	Num RS: 50
Empirical CDF		• max min rate	Num MS: 15
Minimum rate	BS	SLNR, power control	
	RS	Spatially white,	
		gain control	
	Control	$N_{\text{bits}} = 32$	
	Backhaul	$C_B = \infty$	
	CSI BS	Local, perfect	
	CSI RS	No	
	Prelog	$\frac{1}{2}$ (2-hop)	

List of Figures

Acronyms

2G	second generation.
3G	third generation.
4G	fourth generation.
5G	fifth generation.
AF	amplify-and-forward.
AWGN	additive white Gaussian noise.
BC	broadcast channel.
BD	block diagonalization.
bps	bit per second.
bps/Hz	bit per second per Hertz.
BS	base station.
CDF	cumulative distribution function.
CDMA	code division multiple access.
CF	compress-and-forward.
CG	conjugate gradient.
CoMP	coordinated multi-point.
CSCG	circularly symmetric complex Gaussian.
CSI	channel state information.
CSIR	channel state information at receiver.
CSIT	channel state information at transmitter.
DAS	distributed antenna system.
DF	decode-and-forward.

DL   downlink.
DMT   diversity multiplexing tradeoff.
DPC   dirty paper coding.
DSL   digital subscriber line.

EDGE   Enhanced Data Rates for Global Evolution.
eNodeB   evolved node B – base station.

FDD   frequency division duplex.
FDMA   frequency division multiple access.

GEV   generalized eigen-value.
GPRS   General Packet Radio Service.
GPS   Global Positioning System.
GSM   Global System for Mobile Communications.

HNB   home nodeB – femto-cell BS.
HSDPA   High Speed Downlink Packet Access.
HSPA   High Speed Packet Access.

i.i.d.   independent and identically distributed.
IC   interference channel.
ICI   intra-cluster interference.
IO   input-output.
IPv6   internet protocol version 6.
ISI   intersymbol interference.
ITU   International Telecommunication Union.

KKT   Karush-Kuhn-Tucker.

LOS   line-of-sight.
LTE   Long Term Evolution.

LTE-A	Long Term Evolution Advanced.
M2M	machine-to-machine.
MAC	multiple access channel.
MF	matched filter.
MIMO	multiple-input multiple-output.
MMSE	minimum mean square error.
MS	mobile station.
MVDR	minimum variance distortionless response.
n.a.	not available.
NLOS	non-line-of-sight.
OCI	out-of-cluster interference.
OFDM	orthogonal frequency division multiplexing.
OFDMA	orthogonal frequency division multiple access.
PDF	probability density function.
PFR	partial frequency reuse.
PHY	physical layer.
PMF	probability mass function.
QF	quantize-and-forward.
QoS	quality of service.
RRH	remote radio head.
RS	relay station.
s.t.	subject to.
SDMA	space division multiple access.
SIC	successive interference cancellation.
SINR	signal-to-interference-plus-noise ratio.

SIR	signal-to-interference ratio.
SISO	single-input single-output.
SLNR	signal-to-leakage-plus-noise ratio.
SN	supporting node.
SNR	signal-to-noise ratio.
SVD	singular value decomposition.
TDD	time division duplex.
TDMA	time division multiple access.
UB	upper bound.
UL	uplink.
UMTS	Universal Mobile Telecommunications System.
VLSI	very large scale integration.
WLAN	wireless local area network.
WMMSE	weighted minimum mean square error.
XOR	exclusive or operation.
ZF	zero-forcing.

Notation

c	speed of light, $c = 299\,792\,458\,\text{m/s}$.
a, A	scalars a and A.
$\mathbf{a}$	vector $\mathbf{a}$.
$\mathbf{A}$	matrix $\mathbf{A}$.
$\mathbf{A}^{\mathsf{T}}$	transpose of matrix $\mathbf{A}$.
$\mathbf{A}^{\mathsf{H}}$	Hermitian (complex conjugate) transpose of matrix $\mathbf{A}$.
$\mathbf{A}^*$	complex conjugate of matrix $\mathbf{A}$.
$\mathbf{A}^{-1}$	inverse of matrix $\mathbf{A}$.
$\mathbf{A}[i, j]$	the element on the ith row and j column of matrix $\mathbf{A}$.
$\mathbf{I}_n$	$n \times n$ identity matrix.
$\mathbf{1}_{n \times m}$	$n \times m$ all one matrix.
$\mathbf{O}_{n \times m}$	$n \times m$ all zero matrix.
$\mathbf{E}_{i,j}$	matrix with all entries 0 except $\mathbf{E}[i, j] = 1$.
$\|\mathbf{a}\|_p$	p-norm of vector $\mathbf{a}$.
$\|\mathbf{A}\|_{\mathrm{F}}$	Frobenius-norm of matrix $\mathbf{A}$.
$\text{vec}\{\mathbf{A}\}$	all entries of matrix $\mathbf{A}$ stacked into a vector.
$\mathcal{A}$	set $\mathcal{A}$.
$\overline{\mathcal{A}}$	set of elements that are not in $\mathcal{A}$.
$\|\mathcal{A}\|$	cardinality of set $\mathcal{A}$.
$\mathcal{A} \cup \mathcal{B}$	union of set $\mathcal{A}$ and set $\mathcal{B}$.
$\mathbb{N}$	set of natural numbers.
$\mathbb{Z}$	set of integers.
$\mathbb{R}$	set of real numbers.
$\mathbb{C}$	set of complex numbers.
$\mathbb{R}^{n \times m}$	$n \times m$ matrix of real numbers.

$\mathbb{C}^{n\times m}$	$n \times m$ matrix of complex numbers.
$\min\{\mathcal{A}\}$	minimum of the elements in $\mathcal{A}$.
$\max\{\mathcal{A}\}$	maximum of the elements in $\mathcal{A}$.
$a \mod b$	a modulo b.
$(\cdot)^+$	$\max\{0,\cdot\}$.
$\mathrm{diag}(\mathbf{A})$	diagonal elements of $\mathbf{A}$.
$\mathrm{diag}(\{x_1,\ldots,x_n\})$	diagonal matrix with entries $x_1,\ldots,x_n$ on the diagonal.
$\mathrm{blkdiag}(\mathbf{A}_1,\ldots,\mathbf{A}_n)$	block diagonal matrix with matrices $\mathbf{A}_1,\ldots,\mathbf{A}_n$.
$\det(\mathbf{A})$	determinant of matrix $\mathbf{A}$.
$\mathrm{Tr}\{\mathbf{A}\}$	trace of matrix $\mathbf{A}$.
$\mathrm{null}\{\mathbf{A}\}$	null space of matrix $\mathbf{A}$.
$\mathrm{GEV}\{\mathbf{A},\mathbf{B}\}$	generalized eigenvector of matrix pair $\mathbf{A}$ and $\mathbf{B}$.
$\log_a(x)$	logarithm with base a of x.
$\ln(x)$	natural logarithm of x, $\ln(x)=\log_e(x)$.
$\partial f/\partial x$	partial derivative of f with respect to x.
$\nabla_{\mathbf{x}}f$	gradient of f with respect to $\mathbf{x}$.
$\odot$	element wise product.
$\otimes$	Kronecker product.
$\lim_{x\to a}$	limit when x approaches a.
$o(f)$	grows slower than f.
$M!$	factorial $M!=1\cdot 2\cdots M$.
$\Pr\{\cdot\}$	probability of an event.
$\mathrm{E}[\cdot]$	expectation.
$\mathcal{N}(\mu,\sigma^2)$	normal distribution with mean μ and variance σ^2.
$\mathcal{CN}(\mu,\sigma^2)$	circularly symmetric complex Gaussian distribution with mean μ and variance σ^2.

Bibliography

[1] 3rd Generation Partnership Project 3GPP, "Further advancements for E-UTRA physical layer aspects (release 9)," *3GPP TR 36.814, V9.0.0*, March 2010.

[2] ——, "Coordinated multi-point operation for LTE physical layer aspects (release 11)," *3GPP TR 36.819, V11.1.0*, December 2011.

[3] ——, "Overview of 3GPP release 11 v0.2.0," *3GPP Release 11, V0.2.0*, September 2014.

[4] ——, "Spatial channel model for multiple input multiple output (MIMO) simulations," *3GPP TR 25.996, V12.0.0*, September 2014.

[5] J. Andrews, S. Buzzi, W. Choi, S. Hanly, A. Lozano, A. Soong, and J. Zhang, "What will 5G be?" *IEEE Journal on Selected Areas in Communications*, vol. 32, no. 6, pp. 1065–1082, June 2014.

[6] A. G. Armada, M. Sánchez-Fernández, and R. Corvaja, "Constrained power allocation schemes for coordinated base station transmission using block diagonalization," *EURASIP Journal on Wireless Communications and Networking*, vol. 2011, October 2011.

[7] E. Auger, "Wideband multi-user cooperative networks: Theory and measurements," Ph.D. dissertation, ETH Zürich, 2012.

[8] T. Bai and R. Heath, "Asymptotic coverage and rate in massive MIMO networks," in *IEEE Global Conference on Signal and Information Processing (GlobalSIP)*, December 2014, pp. 602–606.

[9] S. Berger, T. Unger, M. Kuhn, A. Klein, and A. Wittneben, "Recent advances in amplify-and-forward two-hop relaying," in *IEEE Communications Magazine*, vol. 47, no. 7, July 2009, pp. 50–56.

[10] S. Berger and A. Wittneben, "Cooperative distributed multiuser MMSE relaying in wireless ad-hoc networks," in *Asilomar Conference on Signals, Systems and Computers*, Pacific Grove, CA, USA, October 2005, pp. 1072–1076.

[11] P. Bhat *et al.*, "LTE-Advanced: an operator perspective," *IEEE Communications Magazine*, vol. 50, no. 2, pp. 104–114, February 2012.

[12] E. Bjornson, M. Bengtsson, G. Zheng, and B. Ottersten, "Computational framework for optimal robust beamforming in coordinated multicell systems," in *IEEE International Workshop on Computational Advances in Multi-Sensor Adaptive Processing (CAMSAP)*, December 2011, pp. 245–248.

[13] E. Bjornson, M. Kountouris, and M. Debbah, "Massive MIMO and small cells: Improving energy efficiency by optimal soft-cell coordination," in *International Conference on Telecommunications (ICT)*, May 2013, pp. 1–5.

[14] H. Bölcskei, R. Nabar, O. Oyman, and A. Paulraj, "Capacity scaling laws in MIMO relay networks," *IEEE Transactions on Wireless Communications*, vol. 5, no. 6, pp. 1433–1444, June 2006.

[15] S. Boyd and L. Vandenberghe, *Convex Optimization*.   New York, NY, USA: Cambridge University Press, 2004.

[16] D. Brandwood, "A complex gradient operator and its application in adaptive array theory," *IEE Proceedings F: Communications, Radar and Signal Processing*, vol. 130, no. 1, pp. 11–16, February 1983.

[17] V. Cadambe and S. Jafar, "Interference alignment and degrees of freedom of the K-user interference channel," *IEEE Transactions on Information Theory*, vol. 54, no. 8, pp. 3425–3441, August 2008.

[18] G. Caire and S. Shamai, "On the achievable throughput of a multiantenna Gaussian broadcast channel," *IEEE Transactions on Information Theory*, vol. 49, no. 7, pp. 1691–1706, July 2003.

[19] E. Candes, M. Wakin, and S. Boyd, "Enhancing sparsity by reweighted ℓ_1 minimization," *Journal of Fourier Analysis and Applications*, vol. 14, no. 5, pp. 877–905, 2008.

[20] M. Cao, X. Wang, S.-J. Kim, and M. Madihian, "Multi-hop wireless backhaul networks: A cross-layer design paradigm," *IEEE Journal on Selected Areas in Communications*, vol. 25, no. 4, pp. 738 –748, May 2007.

[21] H. Chen, S. Shahbazpanahi, and A. Gershman, "Filter-and-forward distributed beamforming for two-way relay networks with frequency selective channels," *IEEE Transactions on Signal Processing*, vol. 60, no. 4, pp. 1927–1941, April 2012.

[22] M. Chiang *et al.*, "Power control in wireless cellular networks," *Foundations and Trends in Networking*, vol. 2, no. 4, pp. 381–533, July 2008.

[23] W. Choi and J. Andrews, "Downlink performance and capacity of distributed antenna systems in a multicell environment," *IEEE Transactions on Wireless Communications*, vol. 6, no. 1, pp. 69 –73, January 2007.

[24] Cisco, "Cisco visual networking index: Global mobile data traffic forecast update, 2013-2018," *Cisco white paper*, February 2014. [Online]. Available: http://www.cisco.com/c/en/us/solutions/collateral/service-provider/visual-networking-index-vni/white_paper_c11-520862.html

[25] R. Corvaja, J. Garcia Fernandez, and A. Garcia Armada, "Mean achievable rates in clustered coordinated base station transmission with block diagonalization," *IEEE Transactions on Communications*, vol. 61, no. 8, pp. 3483–3493, August 2013.

[26] M. Costa, "Writing on dirty paper (corresp.)," *IEEE Transactions on Information Theory*, vol. 29, no. 3, pp. 439–441, May 1983.

[27] H. Cox, R. M. Zeskind, and M. M. Owen, "Robust adaptive beamforming," *IEEE Transactions on Acoustics, Speech, and Signal Processing*, vol. 35, no. 10, pp. 1365–1376, October 1987.

[28] B. Dai and W. Yu, "Sparse beamforming design for network MIMO system with per-base-station backhaul constraints," in *IEEE Workshop on Signal Processing Advances for Wireless Communications (SPAWC)*, June 2014.

[29] C. Eşli, S. Berger, and A. Wittneben, "Optimizing zero-forcing based gain allocation for wireless multiuser networks," in *IEEE International Conference on Communications (ICC)*, June 2007, pp. 5825–5830.

[30] C. Eşli, "Design and optimization of distributed multiuser cooperative wireless networks," Ph.D. dissertation, ETH Zürich, 2010.

[31] C. Eşli, J. Wagner, and A. Wittneben, "Distributed gradient based gain allocation for coherent multiuser AF relaying networks," in *IEEE International Conference on Communications (ICC)*, June 2009.

[32] O. El Ayach, S. Peters, and J. Heath, R.W., "The practical challenges of interference alignment," *IEEE Wireless Communications*, vol. 20, no. 1, pp. 35–42, February 2013.

[33] S.-E. Elayoubi, O. Ben Haddada, and B. Fourestié, "Performance evaluation of frequency planning schemes in OFDMA-based networks," *IEEE Transactions on Wireless Communications*, vol. 7, no. 5, pp. 1623–1633, May 2008.

[34] A. Fehske, G. Fettweis, J. Malmodin, and G. Biczok, "The global footprint of mobile communications: The ecological and economic perspective," *IEEE Communications Magazine*, vol. 49, no. 8, pp. 55–62, August 2011.

[35] G. Foschini, K. Karakayali, and R. Valenzuela, "Coordinating multiple antenna cellular networks to achieve enormous spectral efficiency," *IEE Proceedings - Communications*, vol. 153, no. 4, pp. 548–555, August 2006.

[36] R. Ganesan, T. Weber, and A. Klein, "Interference alignment in multi-user two way relay networks," in *IEEE Vehicular Technology Conference (VTC Spring)*, May 2011.

[37] I. Garcia, N. Kusashima, K. Sakaguchi, and K. Araki, "Dynamic cooperation set clustering on base station cooperation cellular networks," in *IEEE International Symposium on Personal, Indoor and Mobile Radio Communications (PIMRC)*, September 2010, pp. 2127 –2132.

[38] D. Gesbert, M. Kountouris, R. Heath, C.-B. Chae, and T. Salzer, "Shifting the MIMO paradigm," *IEEE Signal Processing Magazine*, vol. 24, no. 5, pp. 36–46, Sept 2007.

[39] D. Gesbert *et al.*, "Multi-cell MIMO cooperative networks: a new look at interference," *IEEE Journal on Selected Areas in Communications*, vol. 28, No. 9, pp. 1380–1408, December 2010.

[40] F. Ghavimi and H.-H. Chen, "M2M communications in 3GPP LTE/LTE-A networks: Architectures, service requirements, challenges and applications," *IEEE Communications Surveys & Tutorials*, vol. 17, no. 2, pp. 525–549, October 2014.

[41] A. Ghosh *et al.*, "Heterogeneous cellular networks: From theory to practice," *IEEE Communications Magazine*, vol. 50, no. 6, pp. 54 –64, June 2012.

[42] K. Gomadam, V. Cadambe, and S. Jafar, "Approaching the capacity of wireless networks through distributed interference alignment," in *IEEE Global Communications Conference (GLOBECOM)*, 2008, pp. 1–6.

[43] S. Goode and A. Annin, *Differential Equations and Linear Algebra*, 3rd ed. Pearson Prentice Hall, 2007.

[44] T. Gou, C. Wang, and S. Jafar, "Aiming perfectly in the dark-blind interference alignment through staggered antenna switching," *IEEE Transactions on Signal Processing*, vol. 59, no. 6, pp. 2734–2744, June 2011.

[45] A. Gupta and R. Jha, "A survey of 5G network: Architecture and emerging technologies," *IEEE Access*, vol. PP, no. 99, 2015.

[46] R. Habendorf, I. Riedel, and G. Fettweis, "Reduced complexity vector precoding the multiuser downlink," in *IEEE Global Telecommunications Conference (GLOBECOM)*, November 2006.

[47] H. Halbauer, S. Saur, J. Koppenborg, and C. Hoek, "3D beamforming: Performance improvement for cellular networks," *Bell Labs Technical Journal*, vol. 18, no. 2, pp. 37–56, September 2013.

[48] I. Hammerström and A. Wittneben, "On the optimal power allocation for nonregenerative OFDM relay links," in *IEEE International Conference on Communications (ICC)*, June 2006.

[49] T. Han and N. Ansari, "On greening cellular networks via multicell cooperation," *IEEE Wireless Communications*, vol. 20, no. 1, pp. 82–89, February 2013.

[50] H. Harashima and H. Miyakawa, "Matched-transmission technique for channels with intersymbol interference," *IEEE Transactions on Communications*, vol. 20, no. 4, pp. 774–780, August 1972.

[51] S. Haykin, *Adaptive Filter Theory*, 3rd ed. Prentice-Hall, Inc., 1996.

[52] R. Heath and M. Kountouris, "Modeling heterogeneous network interference," in *Information Theory and Applications Workshop (ITA)*, February 2012, pp. 17–22.

[53] C. Hellings and W. Utschick, "Carrier-cooperative transmission in parallel MIMO broadcast channels: Potential gains and algorithms," in *International Symposium on Wireless Communication Systems (ISWCS)*, November 2011.

[54] ——, "On the inseparability of parallel MIMO broadcast channels with linear transceivers," *IEEE Transactions on Signal Processing*, vol. 59, no. 12, pp. 6273–6278, December 2011.

[55] B. Hochwald, C. Peel, and A. Swindlehurst, "A vector-perturbation technique for near-capacity multiantenna multiuser communication-part ii: perturbation," *IEEE Transactions on Communications*, vol. 53, no. 3, pp. 537–544, March 2005.

[56] K. Hosseini *et al.*, "Massive MIMO and small cells: How to densify heterogeneous networks," in *IEEE International Conference on Communications (ICC)*, June 2013.

[57] K. Huang and J. Andrews, "A stochastic-geometry approach to coverage in cellular networks with multi-cell cooperation," in *IEEE Global Telecommunications Conference (GLOBECOM 2011)*, December 2011, pp. 1–5.

[58] S. Hur, T. Kim, D. Love, J. Krogmeier, T. Thomas, and A. Ghosh, "Millimeter wave beamforming for wireless backhaul and access in small cell networks," *IEEE Transactions on Communications*, vol. 61, no. 10, pp. 4391–4403, October 2013.

[59] IEEE 802.16 Broadband Wireless Access Group, "Multihop-hop relay systems evaluation methodology (channel model and performance metric)," IEEE, Tech. Rep. 802.16j-06/013r3, February 2007.

[60] International Telecommunication Union – IMT, "Future technology trends of terrestrial IMT systems," *Report ITU-R M.2320-0*, November 2014.

[61] International Telecommunication Union, Radiocommunication sector – ITU-R, "Framework and overall objectives of the future development of IMT-2000 and systems beyond IMT-2000," *Recommendation ITU-R M.1645*, June 2003.

[62] S. A. Jafar, "Exploiting channel correlations – simple interference alignment schemes with no CSIT," in *IEEE Global Communications Conference (GLOBECOM)*, 2009.

[63] M. Jankiraman, *Space-Time Codes and MIMO Systems*. Artech House, Inc. Norwood, MA, USA, 2004.

[64] S.-W. Jeon, S.-Y. Chung, and S. A. Jafar, "Degrees of freedom of multi-source relay networks," in *Allerton Conference on Communication, Control, and Computing*, September 2009.

[65] N. Jindal, S. Vishwanath, and A. Goldsmith, "On the duality of Gaussian multiple-access and broadcast channels," *IEEE Transactions on Information Theory*, vol. 50, no. 5, pp. 768–783, May 2004.

[66] M. Joham, W. Utschick, and J. Nossek, "Linear transmit processing in MIMO communications systems," *IEEE Transactions on Signal Processing*, vol. 53, no. 8, pp. 2700–2712, August 2005.

[67] J. Jose, A. Ashikhmin, T. Marzetta, and S. Vishwanath, "Pilot contamination and precoding in multi-cell TDD systems," *IEEE Transactions on Wireless Communications*, vol. 10, no. 8, pp. 2640–2651, 2011.

[68] M. K. Karakayali, G. J. Foschini, R. A. Valenzuela, and R. Yates, "On the maximum common rate achievable in a coordinated network," in *IEEE International Conference on Communications (ICC)*, June 2006.

[69] M. K. Karakayali, G. J. Foschini, and R. A. Valenzuela, "Network coordination for spectrally efficient communications in cellular systems," *IEEE Wireless Communications*, August 2006.

[70] M. Khandaker and Y. Rong, "Interference MIMO relay channel: Joint power control and transceiver-relay beamforming," *IEEE Transactions on Signal Processing*, vol. 60, no. 12, pp. 6509–6518, 2012.

[71] S.-J. Kim, A. Magnani, A. Mutapcic, S. P. Boyd, and Z.-Q. Luo, "Robust beamforming via worst-case SINR maximization," *IEEE Transactions on Signal Processing*, vol. 56, no. 4, pp. 1539–1547, April 2008.

[72] G. Kramer, M. Gastpar, and P. Gupta, "Cooperative strategies and capacity theorems for relay networks," *IEEE Transactions on Information Theory*, vol. 51, no. 9, pp. 3037–3063, 2005.

[73] B. Krasniqi, M. Wrulich, and C. Mecklenbräuker, "Network-load dependent partial frequency reuse for LTE," in *International Symposium on Communications and Information Technology (ISCIT)*, September 2009, pp. 672–676.

[74] M. Kuhn, R. Rolny, and M. Kuhn, "Impact of relays and supporting nodes on locally restricted cooperation in future cellular networks," in *International Symposium on Wireless Communication Systems (ISWCS)*, November 2011, pp. 202–206.

[75] M. Kuhn, J. Wagner, and A. Wittneben, "Cooperative processing for the WLAN uplink," in *IEEE Wireless Communications and Networking Conference (WCNC)*, March 2008, pp. 1294–1299.

[76] P. Kyösti *et al.*, "WINNER II channel models," WINNER, Tech. Rep. IST-4-027756 WINNER II D1.1.2 V1.2 Part I Channel Models, September 2007.

[77] A. Lapidoth, *A Foundation in Digital Communication.* Cambridge University Press, 2009.

[78] L. Le and E. Hossain, "Multihop cellular networks: Potential gains, research challenges, and a resource allocation framework," *IEEE Communications Magazine*, vol. 45, no. 9, pp. 66 –73, September 2007.

[79] D. Lee, H. Seo, B. Clerckx, E. Hardouin, D. Mazzarese, S. Nagata, and K. Sayana, "Coordinated multipoint transmission and reception in LTE-advanced: deployment scenarios and operational challenges," *IEEE Communications Magazine*, vol. 50, no. 2, pp. 148–155, February 2012.

[80] J. Lee *et al.*, "Coordinated multipoint transmission and reception in LTE-advanced systems," *IEEE Communications Magazine*, vol. 50, no. 11, pp. 44–50, November 2012.

[81] J. Lee and N. Jindal, "Dirty paper coding vs. linear precoding for MIMO broadcast channels," in *Asilomar Conference on Signals, Systems and Computers*, Pacific Grove, CA, USA, October 2006, pp. 779–783.

[82] ——, "High SNR analysis for MIMO broadcast channels: Dirty paper coding vs. linear precoding," *IEEE Transactions of Information Theory*, vol. 53, no. 12, pp. 4787–4792, December 2007.

[83] K. Y. Lee and M. A. El-Sharkawi, *Modern Heuristic Optimization Techniques*, 1st ed. J. Wiley and Sons, Inc., 2008.

[84] C. Leow, Z. Ding, K. Leung, and D. Goeckel, "On the study of analogue network coding for multi-pair, bidirectional relay channels," *IEEE Transactions on Wireless Communications*, vol. 10, no. 2, pp. 670–681, 2011.

[85] X. Lin, R. Heath, and J. Andrews, "The interplay between massive MIMO and underlaid D2D networking," *IEEE Transactions on Wireless Communications*, vol. 14, no. 6, pp. 3337–3351, June 2015.

[86] L. Litwin and M. Pugel, "The principles of OFDM," *Mobile Dev & Design*, pp. 1–4, January 2001.

[87] K. J. R. Liu, A. K. Sadek, W. Su, and A. Kwasinski, *Cooperative Communications and Networking*. Cambridge University Press, 2009.

[88] J. Löfberg, "YALMIP : A toolbox for modeling and optimization in MATLAB," in *Proceedings of the CACSD Conference*, Taipei, Taiwan, 2004. [Online]. Available: http://users.isy.liu.se/johanl/yalmip

[89] A. Lozano, J. Andrews, and R. Heath, "Spectral efficiency limits in pilot-assisted cooperative communications," in *IEEE International Symposium on Information Theory (ISIT)*, July 2012, pp. 1132–1136.

[90] A. Lozano, R. Heath, and J. Andrews, "Fundamental limits of cooperation," *IEEE Transactions on Information Theory*, vol. 59, no. 9, pp. 5213–5226, 2013.

[91] R. Machado and A. Wyglinski, "Software-defined radio: Bridging the analog-digital divide," *Proceedings of the IEEE*, vol. 103, no. 3, pp. 409–423, March 2015.

[92] E. A. Marland, *Early Electrical Communication.* Abelard-Schumann Ltd, London, UK, 1964.

[93] P. Marsch and G. P. Fettweis, *Coordinated Multi-Point Mobile Communications: From Theory to Practice.* Cambridge University Press, 2011.

[94] T. Marzetta, "Noncooperative cellular wireless with unlimited numbers of base station antennas," *IEEE Transactions on Wireless Communications*, vol. 9, no. 11, pp. 3590–3600, 2010.

[95] X. Mingbo, N. Shroff, and E. Chong, "A utility based power-control scheme in wireless cellular systems," *IEEE/ACM Transactions on Networking*, vol. 11, no. 2, pp. 210–221, April 2003.

[96] M. Morales-Cespedes, J. Plata-Chaves, D. Toumpakaris, S. Jafar, and A. Garcia Armada, "Blind interference alignment for cellular networks," *IEEE Transactions on Signal Processing*, vol. 63, no. 1, pp. 41–56, January 2015.

[97] O. Munoz, J. Vidal, and A. Agustin, "Non-regenerative MIMO relaying with channel state information," in *IEEE International Conference on Acoustics, Speech, and Signal Processing (ICASSP)*, March 2005, pp. 361–364.

[98] T. Nakamura *et al.*, "Trends in small cell enhancements in LTE-Advanced," *IEEE Communications Magazine*, vol. 51, no. 2, pp. 98–105, February 2013.

[99] A. Osseiran *et al.*, "Scenarios for 5G mobile and wireless communications: The vision of the METIS project," *IEEE Communications Magazine*, vol. 52, no. 5, pp. 26–35, May 2014.

[100] A. Ozgur, O. Leveque, and D. Tse, "Hierarchical cooperation achieves optimal capacity scaling in ad hoc networks," *IEEE Transactions on Information Theory*, vol. 53, no. 10, pp. 3549–3572, October 2007.

[101] A. Y. Panah, K. T. Truong, S. W. Peters, and J. Heath, R. W., "Interference management schemes for the shared relay concept," *EURASIP Journal on Advances in Signal Processing*, vol. 2010, September 2010.

[102] A. Paulraj, R. Nabar, and D. Gore, *Introduction to Space-Time Wireless Communications.* Cambridge University Press, 2003.

[103] C. Peel, B. Hochwald, and A. Swindlehurst, "A vector-perturbation technique for near-capacity multiantenna multiuser communication-part i: channel inversion and regularization," *IEEE Transactions on Communications*, vol. 53, no. 1, pp. 195–202, January 2005.

[104] S. Peters, A. Panah, K. Truong, and R. Heath, "Relay architectures for 3GPP LTE-Advanced," *EURASIP Journal on Wireless Communications and Networking*, vol. 2009, no. 1, pp. 1:1–1:14, March 2009.

[105] K. B. Petersen and M. S. Pedersen, "The matrix cookbook," November 2012. [Online]. Available: http://www2.imm.dtu.dk/pubdb/p.php?3274

[106] J. G. Proakis, *Digital Communication*. McGraw-Hill Higher Education, 2001.

[107] M. Qiana, Y. Wanga, Y. Zhoua, L. Tiana, and J. Shia, "A super base station based centralized network architecture for 5G mobile communication systems," *Digital Communications and Networks*, vol. 1, no. 2, pp. 152 – 159, April 2015.

[108] Z. Qingling and J. Li, "Rain attenuation in millimeter wave ranges," in *International Symposium on Antennas, Propagation EM Theory (ISAPE)*, October 2006, pp. 1–4.

[109] S. A. Ramprashad, G. Caire, and H. C. Papadopoulos, "Cellular and network MIMO architectures: MU-MIMO spectral efficiency and costs of channel state information," in *Asilomar Conference on Signals, Systems, and Computers*, November 2009.

[110] B. Rankov and A. Wittneben, "Achievable rate regions for the two-way relay channel," in *IEEE International Symposium on Information Theory (ISIT)*, July 2006, pp. 1668–1672.

[111] ——, "On the capacity of relay-assisted wireless MIMO channels," in *IEEE Workshop on Signal Processing Advances for Wireless Communications (SPAWC)*, July 2004.

[112] ——, "Spectral efficient signaling for half-duplex relay channels," in *Asilomar Conference on Signals, Systems, and Computers*, Pacific Grove, CA, USA, November 2005.

[113] T. S. Rappaport, *Wireless Communications: Principles and Practice*, 2nd ed. Upper Saddle River, NJ, USA: Prentice Hall PTR, 2001.

[114] T. Rappaport *et al.*, "Millimeter wave mobile communications for 5G cellular: It will work!" *IEEE Access*, vol. 1, pp. 335–349, 2013.

[115] T. Rappaport, J. Murdock, and F. Gutierrez, "State of the art in 60-GHz integrated circuits and systems for wireless communications," *Proceedings of the IEEE*, vol. 99, no. 8, pp. 1390–1436, August 2011.

[116] I. Riedel, P. Rost, P. Marsch, and G. Fettweis, "Creating desirable interference by optimized sectorization in cellular systems," in *IEEE Global Telecommunications Conference (GLOBECOM)*, December 2010, pp. 1–5.

[117] R. Rolny, *Distributed Multiuser Multihop Networks: Performance Limits and Decentralized Algorithm Design.* Master Thesis, ETH Zürich, September 2009, available on request.

[118] R. Rolny, J. Wagner, C. Eşli, and A. Wittneben, "Distributed gain matrix optimization in non-regenerative MIMO relay networks," in *Asilomar Conference on Signals, Systems, and Computers*, Pacific Grove, CA, November 2009.

[119] Y. Rong and Y. Hua, "Optimality of diagonalization of multi-hop MIMO relays," *IEEE Transactions on Wireless Commununications*, vol. 8, no. 12, pp. 6068 – 6077, December 2009.

[120] F. Rusek *et al.*, "Scaling up MIMO: Opportunities and challenges with very large arrays," *IEEE Signal Processing Magazine*, vol. 30, no. 1, pp. 40–60, January 2013.

[121] T. Rüegg, A. U. T. Amah, and A. Wittneben, "On the trade-off between transmit and leakage power for rate optimal MIMO precoding," in *IEEE Workshop on Signal Processing Advances for Wireless Communications (SPAWC)*, June 2014.

[122] T. Rüegg, M. Kuhn, and A. Wittneben, "Decentralized target rate optimization for MU-MIMO leakage based precoding," in *Asilomar Conference on Signals, Systems, and Computers*, November 2014.

[123] M. Sadek and S. Aissa, "Leakage based precoding for multi-user MIMO-OFDM systems," *IEEE Transactions on Wireless Communications*, vol. 10, no. 8, pp. 2428–2433, August 2011.

[124] A. Sanderovich, S. Shamai, Y. Steinberg, and G. Kramer, "Communication via decentralized processing," *IEEE Transactions on Information Theory*, vol. 54, no. 7, pp. 3008–3023, July 2008.

[125] M. Sanjabi, M. Hong, M. Razaviyayn, and Z.-Q. Luo, "Joint base station clustering and beamformer design for partial coordinated transmission using statistical channel state information," in *IEEE Workshop on Signal Processing Advances for Wireless Communications (SPAWC)*, June 2014.

[126] A. Scaglione and Y.-W. Hong, "Opportunistic large arrays: cooperative transmission in wireless multihop ad hoc networks to reach far distances," *IEEE Transactions on Signal Processing*, vol. 51, no. 8, pp. 2082–2092, August 2003.

[127] D. Schmidt, C. Shi, R. Berry, M. Honig, and W. Utschick, "Minimum mean squared error interference alignment," in *Asilomar Conference on Signals, Systems and Computers*, Pacific Grove, CA, USA, November 2009, pp. 1106–1110.

[128] S. Schwarz *et al.*, "Pushing the limits of LTE: A survey on research enhancing the standard," *IEEE Access*, vol. 1, pp. 51–62, May 2013.

[129] S. Schwarz and M. Rupp, "Exploring coordinated multipoint beamforming strategies for 5G cellular," *Access, IEEE*, vol. 2, pp. 930–946, September 2014.

[130] J. R. Shewchuk, "An introduction to the conjugate gradient method without the agonizing pain," Carnegie Mellon University, Pittsburgh, PA, USA, Tech. Rep., 1994.

[131] Q. Shi, M. Razaviyayn, Z.-Q. Luo, and C. He, "An iteratively weighted MMSE approach to distributed sum-utility maximization for a MIMO interfering broadcast channel," *IEEE Transactions on Signal Processing*, vol. 59, no. 9, pp. 4331–4340, September 2011.

[132] S. Shim, J. S. Kwak, R. Heath, and J. Andrews, "Block diagonalization for multiuser MIMO with other-cell interference," *IEEE Transactions on Wireless Communications*, vol. 7, no. 7, pp. 2671–2681, July 2008.

[133] Q. Spencer, A. Swindlehurst, and M. Haardt, "Zero-forcing methods for downlink spatial multiplexing in multiuser MIMO channels," *IEEE Transactions on Signal Processing*, vol. 52, no. 2, pp. 461 – 471, February 2004.

[134] V. Stankovic and M. Haardt, "Generalized design of multi-user MIMO precoding matrices," *IEEE Transactions on Wireless Communications*, vol. 7, no. 3, pp. 953–961, March 2008.

[135] S. Stotas and A. Nallanathan, "On the throughput and spectrum sensing enhancement of opportunistic spectrum access cognitive radio networks," *IEEE Transactions on Wireless Communications*, vol. 11, no. 1, pp. 97–107, January 2012.

[136] S. Sun, T. Rappaport, R. Heath, A. Nix, and S. Rangan, "MIMO for millimeter-wave wireless communications: beamforming, spatial multiplexing, or both?" *IEEE Communications Magazine*, vol. 52, no. 12, pp. 110–121, December 2014.

[137] S. Sun, Y. Ju, and Y. Yamao, "Overlay cognitive radio OFDM system for 4G cellular networks," *IEEE Wireless Communications*, vol. 20, no. 2, pp. 68–73, 2013.

[138] M. Taranetz and M. Rupp, "Performance of femtocell access point deployments in user hot-spot scenarios," in *Australasian Telecommunication Networks and Applications Conference ATNAC*, Brisbane, Australia, November 2012.

[139] E. Telatar, "Capacity of multi-antenna Gaussian channels," *European Transactions on Telecommunications*, vol. 10, pp. 585–595, June 1999.

[140] A. Tenenbaum and R. Adve, "Linear processing and sum throughput in the multiuser MIMO downlink," *IEEE Transactions on Wireless Communications*, vol. 8, no. 5, pp. 2652–2661, May 2009.

[141] K. C. Toh, M. J. Todd, and R. H. Tütüncü, "SDPT3 – a MATLAB software for semidefinite-quadratic-linear programming," 2006.

[142] M. Tomlinson, "New automatic equaliser employing modulo arithmetic," *Electronics Letters*, vol. 7, no. 5, pp. 138–139, March 1971.

[143] D. Tse and P. Viswanath, *Fundamentals of Wireless Communication*. Cambridge University Press, 2005.

[144] S. Verdú, *Multiuser Detection*. Cambridge University Press, 1998.

[145] P. Viswanath and D. Tse, "Sum capacity of the vector Gaussian broadcast channel and uplink-downlink duality," *IEEE Transactions on Information Theory*, vol. 49, no. 8, pp. 1912–1921, August 2003.

[146] H. Viswanathan and S. Mukherjee, "Performance of cellular networks with relays and centralized scheduling," *IEEE Transactions on Wireless Communications*, vol. 4, no. 5, pp. 2318 – 2328, September 2005.

[147] J. Wagner, "Distributed forwarding in multiuser multihop wireless networks," Ph.D. dissertation, ETH Zürich, 2012.

[148] J. Wagner and A. Wittneben, "On the diversity-multiplexing tradeoff of multiuser ampfliy & forward multihop networks," in *Asilomar Conference on Signals, Systems, and Computers*, Nov. 2009.

[149] C.-X. Wang, X. Hong, X. Ge, X. Cheng, G. Zhang, and J. Thompson, "Cooperative MIMO channel models: A survey," *IEEE Communications Magazine*, vol. 48, no. 2, pp. 80 –87, February 2010.

[150] T. Weber, A. Sklavos, and M. Meurer, "Imperfect channel-state information in MIMO transmission," *IEEE Transactions on Communications*, vol. 54, no. 3, pp. 543–552, 2006.

[151] A. Wittneben and B. Rankov, "Impact of cooperative relays on the capacity of rank-deficient MIMO channels," in *Proceedings of the IST Summit on Mobile and Wireless Communications*, June 2003, pp. 421–425.

[152] G. Zeitler, G. Bauch, and J. Widmer, "Quantize-and-forward schemes for the orthogonal multiple-access relay channel," *IEEE Transactions on Communications*, vol. 60, no. 4, pp. 1148–1158, 2012.

[153] Y. Zeng, E. Gunawan, and Y. L. Guan, "Modified block diagonalization precoding in multicell cooperative networks," *IEEE Transactions on Vehicular Technology*, vol. 61, no. 8, pp. 3819–3824, October 2012.

[154] H. Zhang, H. Dai, and Q. Zhou, "Base station for multiuser MIMO: Joint transmission and BS selection," *Conference on Information Sciences and Systems*, March 2005.

[155] H. Zhang and H. Dai, "Cochannel interference mitigation and cooperative processing in downlink multicell multiuser MIMO networks," *EURASIP Journal on Wireless Communication Networks*, vol. 2004, no. 2, pp. 222–235, December 2004.

[156] J. Zhang *et al.*, "Networked MIMO with clustered linear precoding," *IEEE Transactions on Wireless Communications*, vol. 8, no. 4, April 2009.

[157] R. Zhang, "Cooperative multi-cell block diagonalization with per-base-station power constraints," *IEEE Journal on Selected Areas in Communications*, vol. 28, no. 9, pp. 1435–1445, December 2010.

[158] J. Zhao, M. Kuhn, A. Wittneben, and G. Bauch, "Self-interference aided channel estimation in two-way relaying systems," in *IEEE Global Communications Conference (GLOBECOM)*, November 2008.

[159] L. Zheng and D. Tse, "Diversity and multiplexing: a fundamental tradeoff in multiple-antenna channels," *IEEE Transactions on Information Theory*, vol. 49, no. 5, pp. 1073–1096, May 2003.

[160] J. Zhu, G. Liu, Y. Wang, and P. Zhang, "A hybrid inter-cell interference mitigation scheme for OFDMA based E-UTRA downlink," in *Asia-Pacific Conference on Communications (APCC)*, August 2006, pp. 1–5.

Curriculum Vitae

Name:	**Raphael Thomas Livius Rolny**
Date of Birth:	September 13, 1982
Citizen:	Regensdorf ZH, Switzerland
Nationality:	Swiss

Education

11/2009-01/2016   **ETH Zürich, Switzerland**
PhD studies at the Communication Technology Laboratory, Department of Information Technology and Electrical Engineering.

10/2003-10/2009   **ETH Zürich, Switzerland**
Bachelor and Master studies in Information Technology and Electrical Engineering. Degree: Master of Science (MSc ETH).

08/1998-09/2002   **Kantonsschule Wattwil, Switzerland**
Matura with focus on applied mathematics and physics.

Experience

11/2009-01/2016   **ETH Zürich, Switzerland**
Research assistant at the Communication Technology Laboratory headed by Prof. Dr. Armin Wittneben.

- Research in wireless communications and signal processing.
- Teaching and presentation experience.
- Organization of the annual "International Seminar on Mobile Communications" in cooperation with TU Vienna and TU Munich.
- Project work in industry projects with Swisscom (Schweiz) AG.

- Supervision of Master Theses and student projects.
- Reviewer of international conference and journal submissions.

09/2008-01/2009 **Phonak AG, Stäfa, Switzerland**
Internship in the department "Advanced Products." Research work on signal classification and signal processing for hearing aids.

02/2003-10/2003 **Military Service**
Sergeant in the Swiss Air Force with leadership experience (group leader of 5-12 people).

11/2002-02/2003 **Intrum Justitia, Schwerzenbach, Switzerland**
Student job in solvency and credit assessment.

Award

09/2013 **Best Paper Award at PIMRC 2013, London, UK.**
For the paper entitled "The Relay Carpet: Ubiquitous Two-Way Relaying in Cooperative Cellular Networks".

Publications

- **Journal Papers**

  - **The Cellular Relay Carpet: Distributed Cooperation with Ubiquitous Relaying**
    R. Rolny, T. Rüegg, M. Kuhn, and A. Wittneben, *Springer International Journal of Wireless Information Networks*, June 2014.

- **Conference, Symposium, and Workshop Papers**

  - **Constrained Base Station Clustering for Cooperative Post-Cellular Relay Networks**
    R. Rolny, M Kuhn, and A. Wittneben, *IEEE Wireless Communications and Networking Conference (WCNC)*, New Orleans, LA, USA, March 2015.

  - **Power Control for Cellular Networks with Large Antenna Arrays and Ubiquitous Relaying**
    R. Rolny, C. Dünner, and A. Wittneben, *IEEE Workshop on Signal Processing Advances for Wireless Communications (SPAWC)*, Toronto, Canada, June 2014.

– MIMO Relaying with Compact Antenna Arrays: Coupling, Noise
Correlation and Superdirectivity
Y. Hassan, R. Rolny, and A. Wittneben, *IEEE International Symposium
on Personal, Indoor and Mobile Radio Communications (PIMRC)*, London,
UK, September 2013.

– The Relay Carpet: Ubiquitous Two-Way Relaying in Cooperative
Cellular Networks
R. Rolny, M. Kuhn, and A. Wittneben, *IEEE International Symposium
on Personal, Indoor and Mobile Radio Communications (PIMRC)*, London,
UK, September 2013 (best paper award).

– Multi-Cell Cooperation Using Subcarrier-Cooperative Two-Way
Amplify-and-Forward Relaying
R. Rolny, M. Kuhn, A. U. T. Amah, and A. Wittneben, *International Sym-
posium on Wireless Communication Systems (ISWCS)*, Ilmenau, Germany,
August 2013, (invited paper).

– The Impact of Combined Physical Layer Cooperation and Scheduling for the Downlink of LTE-Advanced Networks
M. Kuhn, T. Sarmiento , M. Kuhn, and R. Rolny, *IEEE Vehicular Technology Conference (VTC) Spring*, Dresden, Germany, June 2013.

– Relaying and Base Station Cooperation: a Comparative Survey
for Future Cellular Networks
R. Rolny, M. Kuhn, A. Wittneben, and T. Zasowski, *Asilomar Conference
on Signals, Systems, and Computers*, Pacific Grove, CA, USA, November
2012.

– Universal Computation with Low-Complexity Wireless Relay Net-
works
E. Slottke, R. Rolny, and A. Wittneben, *Asilomar Conference on Signals,
Systems, and Computers*, Pacific Grove, CA, USA, November 2012.

– Impact of Relays and Supporting Nodes on Locally Restricted
Cooperation in Future Cellular Networks
M. Kuhn, R. Rolny, and M. Kuhn, *IEEE International Symposium on Wireless Communication Systems (ISWCS)* 2011, Aachen, Germany, November
2011.

– The Potential of Restricted PHY Cooperation for the Downlink
of LTE-Advanced
M. Kuhn, R. Rolny, A. Wittneben, M. Kuhn, and T. Zasowski, *IEEE Vehicular Technology Conference (VTC) Fall*, San Francisco, September 2011.

– Distributed Gain Allocation in Non-Regenerative Multiuser Multihop MIMO Networks
R. Rolny, J. Wagner, and A. Wittneben, *Asilomar Conference on Signals,
Systems, and Computers*, Pacific Grove, CA, USA, November 2010, (invited
paper).

- **Distributed Gain Matrix Optimization in Non-Regenerative MIMO Relay Networks**
  R. Rolny, J. Wagner, C. Eşli, and A. Wittneben, *Asilomar Conference on Signals, Systems, and Computers*, Pacific Grove, CA, USA, November 2009.

- **Patents**

  - **Method for Deploying a Cellular Communication Network**
    T. Zasowski, M. Kuhn, and R. Rolny, *Schweizer Patentanmeldung; European Patent Office*, Application/ Patent No 11405379.6 - 1525, December 2011.

Future Mobile Communication: From Cooperative Cells to the Post-Cellular Relay Carpet

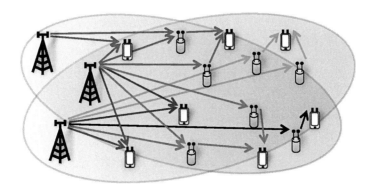

Raphael Rolny

λογος

Series in Wireless Communications
edited by:
Prof. Dr. Armin Wittneben
Eidgenössische Technische Hochschule
Institut für Kommunikationstechnik
Sternwartstr. 7
CH-8092 Zürich

E-Mail: wittneben@nari.ee.ethz.ch
Url: http://www.nari.ee.ethz.ch/

Bibliographic information published by the Deutsche Nationalbibliothek

The Deutsche Nationalbibliothek lists this publication in the
Deutsche Nationalbibliografie; detailed bibliographic data are
available in the Internet at http://dnb.d-nb.de .

ISBN 978-3-8325-3332-8
ISSN 1611-2970

Logos Verlag Berlin GmbH
Comeniushof, Gubener Str. 47,
10243 Berlin
Tel.: +49 030 42 85 10 90
Fax: +49 030 42 85 10 92
INTERNET: http://www.logos-verlag.de

Diss. ETH No. 23243

Future Mobile Communication: From Cooperative Cells to the Post-Cellular Relay Carpet

A thesis submitted to attain the degree of

DOCTOR OF SCIENCES of ETH ZURICH

(Dr. sc. ETH Zurich)

presented by

RAPHAEL THOMAS LIVIUS ROLNY

Master of Science (MSc), ETH Zurich

born on September 13, 1982

citizen of Regensdorf (ZH), Switzerland

accepted on the recommendation of

Prof. Dr. Armin Wittneben, examiner

Prof. Dr. Gerhard Bauch, co-examiner

2016

Day of Doctoral Examination: January 18, 2016

You see, wire telegraph is a kind of a very, very long cat. You pull his tail in New York and his head is meowing in Los Angeles. Do you understand this? And radio operates exactly the same way: you send signals here, they receive them there. The only difference is that there is no cat.

- Attributed to Albert Einstein when asked to describe radio.

Abstract

The increasing demand for ubiquitous data service sets high expectations on future cellular networks. They should not only provide data rates that are higher by orders of magnitude than today's systems, but also have to guarantee high coverage and reliability. Thereby, sophisticated interference management is inevitable. The focus of this work is to develop cooperative transmission schemes that can be applied to cellular networks of the next generation and beyond. For this, conventional network architectures and communication protocols have to be challenged and new concepts need to be developed. Starting from cellular networks with base station (BS) cooperation, this thesis investigates how classical network architectures can evolve to future networks in which the mobile stations (MSs) are no longer served by BSs in their close vicinity, but by a dynamic and flexible heterogeneity of different nodes.

Based on recent information theoretic results, we develop a practical and robust linear BS cooperation scheme based on block zero-forcing in a limited area (BS cluster) and an approximation of the residual interference with which a convex optimization problem can be formulated and efficiently be solved. The transmission scheme is then extended to heterogeneous networks that can also include remote radio heads and/or decode-and-forward (DF) relays. While such relays can improve coverage range as well as the data rates in interference limited areas, they need to be involved in the cooperation process and have to exchange signals with their cooperation partners (e.g. BSs), what makes the relays rather complex.

Amplify-and-forward (AF) relays of low complexity that cooperate in a distributed fashion are therefore an attractive alternative. With properly selected amplification gains, such AF relays can cancel interference and assist the communication between BSs and MSs. In order to find appropriate amplification factors, we develop a distributed gradient-based optimization algorithm that allows each node to calculate its factors with local channel state information (CSI) only, even when applied to multi-hop setups. We show that the overhead of the scheme does not scale with the number of involved nodes and that it can be further improved by subcarrier cooperation, i.e. when signals are combined over multiple subcarriers.

In order to increase the capacity of cellular systems by the required factors, we combine large antenna arrays (massive MIMO) at few BS locations with massively deployed small relay cells. In this "relay carpet" concept, we can benefit from the advantages of both approaches and can simplify channel estimation at the BSs, which would limit the performance gains in conventional networks. The relays, that are of very low complexity and low cost, turn the network into a two-hop network where the BSs as well as the MSs see the relays as the nodes they communicate with. This enables sophisticated multi-user MIMO beamforming at the BSs and the performance of such a network scales beneficially with the number of involved nodes. Especially very simple AF relays without the requirement of any CSI can thus lead to large rates when the network is dense. These relays are also beneficial for coverage and power savings.

In a further evolution, we abandon the cellular network layout completely and let backhaul access points operate in places where they can most easily be installed. For such a "post-cellular" network architecture, we apply dynamic cooperation clusters and many distributed low-complexity relays. By optimizing the BS clusters as well as the relay routing under practical conditions and power control, we show that coverage and high performance can be achieved with aggressive spatial multiplexing and cooperative leakage-based precoding.

With a transition from classical cell-based networks to relay enabled post-cellular networks, we trade off node complexity with density. Aggressive spatial multiplexing can thereby deliver high data rates to large areas in a very efficient way, even when the backhaul capacity is limited or when in certain areas no backhaul access is available at all. The beneficial performance scaling shows that such post-cellular networks can offer a flexible and dynamic solution for mobile communication of future generations.

Kurzfassung

Die immer grösser werdende Nachfrage an mobile Datendienste stellt hohe Anforderungen an zukünftige Mobilfunknetze. Diese sollen nicht nur Datenraten liefern, die um Grössenordnungen höher liegen als in gegenwärtigen Systemen, sondern auch hohe Verfügbarkeit und Zuverlässigkeit garantieren. Dabei ist insbesondere ein effektiver Umgang mit Interferenz von hoher Bedeutung. In dieser Arbeit entwickeln wir deshalb effiziente kooperative Übertragungstechniken, die im Mobilfunk der nächsten Generation und darüber hinaus angewendet werden können. Dafür müssen konventionelle Netzwerkarchitekturen und Kommunikationsprotokolle überdacht und neue Konzepte erarbeitet werden. Ausgehend von Netzwerken mit kooperierenden Basisstationen untersucht diese Arbeit, wie klassische zelluläre Netzwerkarchitekturen zu zukünftigen Netzwerken weiterentwickelt werden können, in denen die mobilen Nutzer nicht mehr von nahe gelegenen Basisstationen bedient werden, sondern durch eine dynamische und flexible Vielzahl von verschiedenen Infrastrukturknoten.

Basierend auf aktuellen informationstheoretischen Forschungsresultaten entwickeln wir praktische und robuste lineare Kooperationsverfahren. Dabei wird die Interferenz in einem beschränkten Gebiet durch eine Gruppe von mehreren Basisstationen komplett ausgelöscht. Durch eine Approximation der restlichen Interferenz von ausserhalb des Kooperationsgebietes kann ein konvexes Optimierungsproblem formuliert werden, das effizient gelöst werden kann. Das Übertragungsverfahren wird dann so erweitert, dass es auch in heterogenen Netzwerken angewendet werden kann, in denen zusätzliche Decode-and-Forward (DF) Relais, Femto-Zellen-Basisstationen oder andere Hilfsknoten die Basisstationen unterstützen. Damit die Hilfsknoten nicht nur die Netzabdeckung vergrössern, sondern auch die Datenraten in interferenzlimitierten Umgebungen verbessern können, müssen diese in den Kooperationsprozess der Basisstationen eingebunden werden und mit diesen Signale sowie Kanalinformation austauschen, was zu einer relativ komplexen Implementierung führt.

Amplify-and-Forward (AF) Relais, die durch verteilte Verfahren miteinander kooperieren, sind deshalb eine attraktive Alternative, da diese als einfache Umsetzer mit geringer Komplexität realisiert werden können. Mit geeignet gewählten Verstärkungsfaktoren können solche AF Relais Interferenz abschwächen oder auslöschen und somit die Kommunikation zwischen Basisstationen und Mobilgeräten unterstützen. Um die Verstärkungsfaktoren zu optimieren, entwickeln wir einen verteilten gradientenbasierten Algorithmus, der es jedem Knoten erlaubt, seine Faktoren mit lokalem Wissen zu berechnen. In Übertragungen über zwei oder mehr Stufen skaliert der Overhead dieses Verfahrens nicht mit der Anzahl Relais pro Stufe. In Breitbandsystemen kann darüber hinaus eine weitere Verbesserung der erzielbaren Datenraten durch optimierte Linearkombinationen über mehrere Unterträger erzielt werden.

Um die Kapazität der Systeme um die benötigten Grössenordnungen zu erhöhen, kombinieren wir grosse Antennenarrays an Basisstationen mit kleinen Relaiszellen, die flächendeckend verteilt sind. In diesem "Relaisteppich" kann man von den Vorteilen ausgeklügelter Mehrfachnutzerübertragung an den Basisstationen und der verteilten Signalverarbeitung der Relais profitieren. Die vielen Relais verwandeln das Netzwerk in ein Zwei-Hop-System, was grosse Antennenarrays an den Basisstationen ohne den sonst damit verbundenen hohen Overhead ermöglicht. Mit einfachen AF Relais, die keinerlei Kanalwissen benötigen, kann eine äussert günstige Skalierung der Leistungsfähigkeit des Netzwerkes erreicht werden, besonders wenn viele Knoten dicht beieinander liegen. Neben deutlich erhöhten Datenraten können die Relais auch benötigte Sendeleistung einsparen, was die Systeme auch noch energieeffizienter macht.

In einer weiteren Evolution verzichten wir schliesslich ganz auf die zelluläre Struktur und lassen die mobilen Nutzer nur durch spärlich platzierte Basisstationen bedienen, die auch weit weg von den Nutzern liegen können. In solchen "post-zellulären" Netzwerken müssen die Basisstationen nicht mehr überall verteilt sein, sondern können da platziert werden, wo es möglichst (kosten-) günstig möglich ist. Durch eine optimierte Zuteilung der kooperierenden Basisstationsgruppen und Relais unter praktischen Bedingungen und Leistungsregulierung können sehr hohe Netzabdeckungen und Datenraten erreicht werden.

Mit dem Übergang von klassischen zellbasierten Netzwerken zur post-zellulären Netzwerkarchitektur können wir die Komplexität der Knoten mit deren Anzahl aufwiegen. Aggressives Multiplexing in der räumlichen Dimension kann hohe Datenraten in grossen Gebieten erzielen, auch wenn die kabelgebundene Infrastruktur limitiert oder in Teilen des Netzwerkes gar nicht vorhanden ist. Die vorteilhafte Skalierung mit der Anzahl der Knoten zeigt deshalb, dass post-zelluläre Netzwerke eine äusserst dynamische und flexible Lösung für die Anforderungen zukünftiger Mobilfunknetze bieten.

Acknowledgements

First and foremost, I would like to thank my advisor Prof. Dr. Armin Wittneben for guiding me through this thesis. Your continuous support, your motivation and inspiration helped me in all the time of research and writing of this thesis. I could not have imagined a better research environment as the one provided by the Communication Technology Laboratory at ETH Zurich. Thank you for allowing me to grow as an engineer and as a scientist.

Besides my advisor, I would also like to thank Prof. Dr. Gerhard Bauch for being the co-referee of this thesis and for serving as committee member at my defense. I also want to thank you for the exam questions (even at hardship), your comments and suggestions. Thank you for letting my defense be an enjoyable moment.

A sincere thank you also goes to Dr. Marc Kuhn, who always provided me with valuable suggestions, comments, and assistance. The discussions with you, whether at work or during lunch, were always fruitful, interesting, and fun. I thank my fellow labmates Zemene, Eric, Yehia, Tim, Gregor, Christoph, Heinrich, and all the other current and former members of the Communication Technology Lab for the stimulating discussions, the vivid talks during or after work, the exciting conference trips, ski weekends, summer events, and for all the fun we have had in the last years. You were more than just colleagues. Special thanks also go to Jutta, Priska, Lara, Barbara, and Claudia. Your active support in all administrative things contributed greatly to an enjoyable working atmosphere.

Also I thank my friends Doris, Harri, Sandro, David, and Beni. The weekly Signalöl always provided very exhilarating discussions (of course only about serious scientific topics). Furthermore, I also thank Noppa, Kevin, Markus, Lea, and Alessia for being my bandmates of Walter Calls Ambulance and the patience when I was stressed during the writing of my thesis. The rehearsals, concerts, and all the time we have spent together were always a good balance for the time at work. You guys rock!

I want to give special thanks to my family: my parents and to my sister for the unconditional support throughout the writing of this thesis and my whole life. I would not have completed this thesis if it was not for you.

Last but not the least, I would also like to thank you, dear reader, for putting this book into your hands and anyone else who I have forgotten to mention here. Thank you very much! I try to consider you in the next book ;-)

Zurich, February 2016

Contents

Contents